Introduction to Linear Programming

LEONID NISON VASERSTEIN

Penn State University

in collaboration with

CHRISTOPHER CATTELIER BYRNE

Penn State University

Pearson Education, Inc.

Upper Saddle River, New Jersey 07458

Library of Congress Cataloging-in-Publication Data

Vaserstein, Leonid, N.
 Introduction to linear programming/Leonid Nison Vaserstein in collaboration
with Christopher Cattelier Byrne.
              p. cm.
 Includes bibliographical references and index
 ISBN 0-13-035917-3
 1. Linear programming. I. Byrne, Christopher Cattelier. II. Title

T57.74 . V365 2003   2002070164
519.7'2 –dc21

Acquisitions Editor: *George Lobell*
Editor in Chief: *Sally Yagan*
Production Editor: *Lynn Savino Wendel*
Vice President/Director of Production and Manufacturing: *David W. Riccardi*
Senior Managing Editor: *Linda Mihatov Behrens*
Assistant Managing Editor: *Bayani Mendoza DeLeon*
Executive Managing Editor: *Kathleen Schiaparelli*
Manufacturing Buyer: *Michael Bell*
Manufacturing Manager: *Trudy Pisciotti*
Editorial Assistant: *Jennifer Brady*
Marketing Manager: *Halee Dinsey*
Marketing Assistant: *Rachel Beckman*
Art Director: *Jayne Conte*
Cover Designer: *Kiwi Designs*
Cover Photo Credits: *Vasily Kandinsky* "Points" 1935, oil and lacquer on canvas,
32 in. × 39 3/8 in. (81.28 cm × 100.01 cm), Verso in black paint: K/No. 621/
1935; Recto lower left in grey 79-6.2. Long Beach Museum of Art , The Milton
Wichner Collection. © 2002 Artists Rights Society (ARS), New York/ADAGP,
Paris.

Printed in the United States of America

10 9 8 7 6 5 4 3 2 1

ISBN 0-13-035917-3

Pearson Education, Ltd., London
Pearson Education Australia Pty. Limited, Sydney
Pearson Education Singapore, Pte., Ltd
Pearson Education North Asia Ltd, Hong Kong
Pearson Education Canada, Ltd., Toronto
Pearson Educacion de Mexico, S.A. de C.V.
Pearson Education – Japan, Tokyo
Pearson Education Malaysia, Pte. Ltd

For my students

Contents

Chapter 6. Transportation Problems

Chapter 7. Matrix Games

Chapter 8. Linear Approximation

Appendix. Guide to Mathematical Programming

Preface

Why This Book Was Written

This text has evolved from an advanced undergraduate math course serving students with different mathematical backgrounds and various majors, including mathematics, computer science, statistics, engineering, secondary education, actuarial science, computer engineering, science, and business. Some students are in the integrated five-year science/business program leading to a Master's in Business Administration, and some are Ph.D. students.

Since this course does not require deep mathematical theories such as calculus, differential equations, abstract algebra, topology or number theory, it offers an opportunity for students with modest mathematical backgrounds to learn some useful and important mathematics. With this in mind, I tried to avoid, whenever possible, deep or complicated mathematical concepts such as vector spaces, determinants, echelon forms, limits, and derivatives.

Many students enroll because linear programming is widely used in business and other areas. They need to learn how to formulate real-life problems, how to adapt the formulation for specific computer software, and how to interpret and apply the results of computations back to real-life problems. In the case when a computer does not produce any result or produces nonsense, they should be able to adjust the problem or choose an appropriate software.

There are many excellent textbooks on linear programming, but most of them require a strong mathematical background and are accessible only to mathematics majors or written for advanced students and contain much more than can be covered in a one-semester course.

It is a real challenge to keep both advanced students and beginners in one class! Although linear algebra is a prerequisite for linear programming at Penn State, some students in class have difficulty solving systems of linear equations. On the other hand, some students in class are strong in mathematics or computer science.

So I tried to avoid texts that are a bit like the porridge that Goldilocks rejected—they either presented material that was too "cold" in that it was trivialized or they presented material that was too "hot" in that it was predicated on a rigorous mathematical background. In the former case, many students were bored; in the latter case, many students found the material to be inaccessible.

The text starts from the beginning, assuming very little mathematical background. Thus I offer the reader the opportunity to learn linear programming integrally by first considering relevant tools from linear algebra and logic before seeing the simplex method. The section on logic is an important, albeit often ignored, component of linear programming. Throughout the book, I present a rich palette of examples and applications and I ask students to attempt exercises of differing degrees of difficulty. Students like this approach to linear programming, as evidenced by the enrollment and remarks they have made on their end-of-semester evaluation forms.

The widespread availability of computers has not eliminated the need for computational skills, but it has increased the relative importance of logical skills. It is a curiosity rather than an important asset, nowadays, if you can compute 100 digits of π by hand. Today computers can compute the first 10^{10} digits. But is it logically possible to compute the 10^{100}-th digit of π?

How to Use This Book

The text is written at three levels. Most of it is accessible to students who even do not know linear algebra or calculus. Remarks and some exercises are addressed to more sophisticated students. The appendix at the end of the book gives a general idea of further developments in linear and mathematical programming. It is intended as a guide for further studies. Also it gives details on topics mentioned in Chapters 1 through 8 requiring a stronger mathematical background and a level of experience and sophistication on the part of students that is beyond what is typically achieved by undergraduate students in the United States.

Many examples with solutions are given in the text, so I feel that it is not necessary to give students solutions for numerous exercises. However, at the end of book, answers are given to some exercises including those that could be tricky. Exercises are of different difficulties, but all of them can be solved by hand. Computers are allowed but not required in class. I do not give problems where computer skills give a big advantage. The exercises in the first section of Chapter 1, besides checking understanding of definitions, also test the mathematical background of students.

Acknowledgments and References

My class notes evolved over several years, and many students and graders contributed to their improvement by pointing out misprints and errors and asking questions. Reviewers and editors of Prentice Hall made numerous corrections and improvements.

I deliberately chose not to tie this book to a particular software package, because I believe that if students learn the material in this text, they can intelligently apply this knowledge when using any one of the excellent software packages now available. Another reason is that any particular package will be soon outdated with the appearance of new packages and advances in computers and operating systems.

However, the students in class are allowed to use any hardware and software they like, even during tests. Software that can do linear programming includes *Mathematica*, *Maple*, and *Excel*.

More software for linear programming is available via the Internet either to be downloaded for free or to be used online. There is a lot of useful information about linear programming on the Internet. I list some URLs, but keep in mind that the Web changes rapidly:

- http://carbon.cudenver.edu/ hgreenbe/glossary/
  (Mathematical Programming Glossary)
- http://www.mathprog.org/
  (Mathematical Programming Society)
- http://iris.gmu.edu/ asofer/siagopt.html
  (SIAM Activity Group on Optimization)
- http://solon.cma.univie.ac.at/ neum/glopt.html
  (Global Optimization, Wien)
- http://www.informs.org/Resources/
  (INFORMS)

Searching the Web with the keyword "linear programming" will yield many other Web sites. There are many books on linear programming. The Web site

$$http://www.amazon.com$$

listed 771 matches for "linear programming" on August 16, 2002.

There are also many journals that publish papers on linear and nonlinear programming. The Web site

$$http://www.informs.org/Resources/$$

lists 36 hardcopy journals and 14 online journals on operations research (as of August 16, 2002). It lists also 35 societies related to operations research.

Leon Vaserstein
vstein@math.psu.edu

Chapter 1

Introduction

§1. What Is Linear Programming?

Perhaps the earliest examples of mathematical models for analyzing and optimizing the economy were provided almost 250 years ago by a French economist. In his *Tableau Économique,* written in 1758, François Quesnay (1694–1774) explained the interrelation of the roles of the landlord, peasant, and artisan in eighteenth-century France by considering several factors separately. For example, there are "The Economical Tableau considered relative to National Cash," and "The Economical Tableau considered in the Estimation of the Produce and Capital Stock of Every Kind of Riches."

The nineteenth-century French mathematician Jean-Baptiste-Joseph Fourier (1768–1830) had some knowledge of the subject of linear programming, as evidenced by his work in linear inequalities as early as 1826 (see §A.10 in the Appendix). He also suggested the simplex method for solving linear programs arising from linear approximation (see Chapter 8). In the late 1800s, the writings of the French economist L. Walras (1834–1910) demonstrated his use of linear programming. However, with a few other notable exceptions, such as Kantorovich's 1939 monograph *Mathematical Methods for Organization and Planning of Production,* there was comparatively little attention paid to linear programming preceding World War II.

The fortuitous synchronization of the advent of the computer and George B. Dantzig's reinvention of the simplex algorithm in 1947 contributed to the dizzyingly explosive development of linear programming with applications to economics, business, industrial engineering, actuarial sciences, operations research, and game theory. Progress in linear programming is noteworthy enough to be reported in the *New York Times.* In 1970 P. Samuelson (b. 1915) was awarded the Nobel Prize in Economics, and in 1975 L. Kantorovich (1912–1986) and T. C. Koopmans (1910–1985) received the

1

Nobel Prize in Economics for their work in linear programming. The subject of linear programming even made its way into Len Deighton's suspense spy story, *The Billion Dollar Brain,* published in 1966:

"I don't want to bore you," Harvey said, "but you should understand that these heaps of wire can practically think—linear programming—which means that instead of going through all alternatives they have a hunch which is the right one."

Optimization problems come in two flavors: maximization problems and minimization problems. In a maximization problem, we want to maximize a function over a set, and in a minimization problem, we want to minimize a function over a set,

In both cases, the function is real valued and it is called the *objective function.* The set is called the *feasible region* or the set of *feasible solutions.* To solve an optimization (maximization or minimization) problem means usually to find both the *optimal value* (maximal or minimal value, respectively) over the feasible region and an *optimal solution* or *optimizer* [i.e., how (where) to reach the optimal value, if it is possible]. It is not required unless otherwise instructed to find all optimal solutions. This is different from solving a system of linear equations, where a complete answer describes all solutions.

The optimal value is also known as *optimum* or *extremum.* Depending on the flavor, the terms *maximum* (max for short) and *minimum* (min for short) are also used. The set of all optimal solutions (maximizers or minimizers) is called the *optimality region.*

Now we consider a simple example.

Imagine that you are asked to solve the following optimization problem:

$$\begin{cases} \text{Maximize} & x \\ \text{subject to} & 2 \le x \le 3. \end{cases}$$

Clearly the goal is to find the largest value for x, given that this variable is limited as to the values it can assume. Since these limitations are explicitly stated as functions of the variable under consideration, called the *objective variable,* there is no difficulty in solving the problem; just take the maximum value. Thus, you can correctly conclude that the maximum value for x is 3, attained at $x = 3$.

However, it is more often the case that the range of values for the objective variable is given implicitly by placing limitations on another variable or other variables related with the objective variable. These variables are called *decision* or *control variables*. These variables are under our control: We are free to decide their values subject to given constraints. They are different from data that form an input for our optimization problem. The objective function is always a function of decision variables. Sometimes it has a name called the *objective variable*.

For instance, in a problem such as finding which rectangles of fixed perimeter encompass the largest area, the objective variable is "area," and the decision variables are $l = $ length of the rectangle and $w = $ width of the rectangle. In general, when the objective variable is given as a function of decision variables, we use the term *objective function* to describe the function we want to optimize. These limitations on the decision variables, however they might be described, are called the *constraints* or *restraints* of the problem.

Thus, a *mathematical program* is an optimization problem where the objective function is a function of real variables (decision variables) and the feasible region is given by conditions (constraints) on the variables. So a feasible solution is a set of values for all the decision variables satisfying all the constraints in the problem. Mathematical programs are addressed in *mathematical programming*.

What is *linear programming* then? Linear programming is the part of mathematical programming that studies optimization (extremal) problems having objective functions and constraints of particularly simple form. Mathematically, a *linear program* is an optimization problem of the following form: Maximize (or, sometimes, minimize) an *affine function* subject to a finite set of *linear constraints*. Contrary to modern perception, the word *programming* here does not refer to computer programming. In our context, which goes back to military planning, *programming* means something like "detailed planning."

Now we define the terms *affine function* and *linear constraint*. In this book, unless indicated otherwise, a *number* means a "real number" and a function means a "real-valued function."

Definition 1.1. A function f of variables $x_1, \ldots, x_n$ is called a *linear form* if it can be written as $c_1 x_1 + \cdots + c_n x_n$, where the coefficients c_i are given real numbers (constants). A function f is called *affine* if it is the sum of a linear form and a constant. ∎

Of course, it is not necessary to denote these variables as x_i, the coefficients as c_i, or the function as f. For example, $g(x,y) = 2x - a^2 y$, where a is some fixed real number, is a linear form in two variables, which are denoted x and y instead of x_1 and x_2. Note that if a were a variable and y were a fixed nonzero number, then $f(x, a) = 2x - a^2 y$ is not a linear form in x and a (see Problem 1.2). Here are three affine functions of two variables, $x, y : x - 4y - 3$, $y + 2$, $x + y$.

Problem 1.2. Show that the function $g(x, a) = 2x - a^2 y$, where y is a fixed *nonzero* number, is not a linear form in x and a.

Solution. Suppose, to the contrary, that $g(x, a) = 2x - a^2 y$ is a linear form in x and a; that is, $g(x, a) = 2x - a^2 y = c_1 x + c_2 a$ with coefficients c_1, c_2 independent of x and a. Then $g(1, 0) = 2 = c_1$ and $g(0, 1) = -y = c_2$. Thus, $g(x, a) = 2x - a^2 y = 2x - ya$, hence $a^2 y = ya$ for all a. Taking $a = 2$, we see that $y = 0$. But, since y cannot equal zero by hypothesis, we have arrived at our hoped-for contradiction. ∎

The term *linear function* means "linear form" in some textbooks and "affine function" in others. The term *linear functional* in linear programming means "linear form."

Linear constraints come in three flavors, of type $=$, $\geq$, or $\leq$. The linear constraints of type $=$ are familiar linear equations, that is, the equalities of the form

$$\text{an affine function} = \text{an affine function.}$$

Most often, they come in the standard form

$$\text{a linear form} = \text{a constant.}$$

For illustration, $x = 2$, $x - y = 0$, $5y = 7$ are three linear equations for two variables x, y written in standard form, while $2 = x$, $x = y$, $3y + x + 3 = x - 2y - 4$ are the same equations written differently.

Two other types of linear constraints are inequalities of the form

$$\text{an affine function } (\leq \text{ or } \geq) \text{ an affine function.}$$

Often they are written as

$$\text{a linear form } (\leq \text{ or } \geq) \text{ a constant.}$$

Thus, a linear constraint consists of two affine functions (the left-hand side and the right-hand side) connected by one of three symbols: $=$, $\leq$, $\geq$. Strict linear inequalities such as $x > 0$ are not considered to be linear constraints.

Example 1.3

(i) $y = \sin 5$ is a linear constraint on the variable y.

(ii) $x \geq 0$ is a linear constraint on the variable x.

(iii) $2x + 3y \leq 7$ is a linear constraint on the variables x and y. Note, however, that

(iv) $y + \sin x = 1$

is not a linear constraint on the variables x and y, since $\sin x$ is *not* a linear form in x.

Definition 1.4. A *linear program* (LP for short), or *linear programming problem,* is any optimization problem where we are required to maximize (or minimize) an affine function subject to a *finite* set of linear constraints. ∎

For example, the following is a linear program:

$$\begin{cases} \text{minimize} & f(x_1, \ldots, x_n) = c_1 x_1 + \cdots + c_n x_n + d, \\ \text{subject to} & \displaystyle\sum_{i=1}^{n} a_{ji} x_i \leq b_j \quad \text{for } j = 1, \ldots, m \\ & x_i \geq 0 \quad \text{for } i = 1, \ldots, n \end{cases} \quad (1.5)$$

where m, n are given natural numbers, $d, c_i, , b_j, a_{j,i}$ are constants, and x_i are decision (control) variables (unknowns). We call (1.5) a linear program in *canonical* form.

The finite set of constraints in Definition 1.4 can be empty. In other words, the number of constraints is allowed to be zero. If there are no constraints in an optimization problem, we talk about *unconstrained* optimization. Note that, unless otherwise instructed, we cannot ignore any of the given constraints in an optimization problem.

Recall that constant terms are not allowed in linear forms, but we allow constant terms in the objective functions of linear programs. Thus, according to our definitions, the function $x - 2y + 3$ of two variables x and y is not a linear form but it is an affine function, and it can be the objective function of a linear program. Some textbooks make different choices in definitions.

It is possible to have an optimization problem or even a linear program for which there are no feasible solutions (see Example 1.6). Such a problem is called *infeasible* or *inconsistent*. It is also possible for an optimization problem to have feasible solutions but no optimal solutions. For example, maximize x subject to $x < 1$. This explains why we do not allow these kind of constraints in linear programming.

An optimization problem is called *unbounded* if the objective function takes arbitrary large values in the case of the maximization problem and arbitrary small values in the case of the minimization problem (Example 1.7). We will see in Chapter 4 that any feasible linear program either has an optimal solution or is unbounded.

Note that there may be more than one optimal solution (or none at all, as in Example 1.6) among the feasible solutions (Example 1.8). However, the optimal (maximal or minimal) value of an optimization problem is unique (if it exists). Had we found two different values, one would be better, so the other would not be optimal.

Example 1.6. *An Infeasible LP*

$$\begin{cases} \text{Maximize} & 4x + 5y \\ \text{subject to} & 2x + y \leq 4 \\ & -2x - y \leq -5 \\ & x \geq 0, \quad y \geq 0. \end{cases}$$

Note that if x and y satisfy the constraint $2x + y \leq 4$, then, by multiplying by -1, we obtain $-2x - y \geq -4$. However, the second constraint demands that $-2x - y \leq -5$. Obviously, the two given constraints are mutually exclusive and therefore there are no feasible solutions. This linear program is *infeasible*. ∎

Example 1.7. *An Unbounded LP*

$$\begin{cases} \text{Maximize} & x - 2y \\ \text{subject to} & -3x + 2y \leq -2 \\ & -6x - 5y \leq -1 \\ & x \geq 0, \quad y \geq 0. \end{cases}$$

This linear program does have feasible solutions (for example $x = 2/3$, $y = 0$), but none of them is optimal. For any real number M, there is a feasible solution x, y such that $x - 2y > M$. An example of such a feasible solution is $x = 2/3 + M$ and $y = 0$. In a sense, there are so many feasible solutions that none of them even gets close to being optimal. This linear program is *unbounded*. ∎

Example 1.8 *A LP with Many Optimal Solutions*

$$\begin{cases} \text{Minimize} & x + y \\ \text{subject to} & x, y, z \geq 0. \end{cases}$$

In this problem with three variables x, y, z the optimal solutions are $x = y = 0$, $z \geq 0$ arbitrary nonnegative number. The optimal value is 0. ∎

Example 1.9. *A LP with One Optimal Solution*

$$\begin{cases} \text{Minimize} & x + y + z \\ \text{subject to} & x \geq -1, y \geq 2, z \geq 0. \end{cases}$$

In this linear program with three variables x, y, z the optimal solution is $x = -1, y = 2, z = 0$. The optimal value is 1. Note that a solution should contain values for all variables involved. ∎

Example 1.10. *A Nonlinear Problem*

$$\begin{cases} \text{Minimize} & x^2 + y^3 + z^4 \\ \text{subject to} & |x| \geq 1, |y| \leq 2. \end{cases}$$

This is a mathematical program with three variables and two constraints that is not linear because the objective function is not affine and the constraints are not linear. (However, the second constraint can be replaced by two linear constraints, and the feasible region is the disjoint union of two parts; each can be given by three linear constraints.) Nevertheless, using common sense, it is clear that the problem splits into three separate optimization problems with one variable each. So there are exactly two optimal solutions, $x = \pm 1, y = -3, z = 0$ and min $= -26$. ∎

All numbers in linear programming are real numbers. In fact, it is hard to imagine a linear program arising out of business and industrial concerns, with numbers not being actually rational numbers. Why? You might ask yourself if the price of a product could be stated as an irrational number, for example, $\sqrt{2}$. We will see later that to solve a linear programming problem with rational data we do not need irrational numbers. However, this is not the case with nonlinear problems, as you can see when you solve the (nonlinear) equation $x^2 = 2$.

To develop your own appreciation of optimization problems, try to solve the following two problems. Are they linear programs?

Problem 1.11

$$\begin{cases} \text{Maximize} & x \\ \text{subject to} & 2 \leq x \leq 3. \end{cases}$$

Solution. As we noted earlier, the maximal value is 3 (max $= 3$ for short) and it is reached at $x = 3$. ∎

One of the main applications of the first derivative of a function, which you study in calculus, is to find the maxima or minima of a function by looking at the critical points. Yet you can see from Problem 1.11 that first-year calculus is not sufficient to solve linear programs. Suppose you are trying to find the maximum and minimum of the linear form $f(x) = x$ on the interval $2 \leq x \leq 3$ by determining where the first derivative equals zero. You observe that the first derivative, 1, never equals zero. Yet the objective function reaches its maximum, 3, and minimum, 2, on this interval.

Problem 1.12

$$\begin{cases} \text{Maximize} & x + 2y + z \\ \text{subject to} & x + y = 1, \\ & z \geq 0. \end{cases}$$

Solution. The objective function takes arbitrarily large values as y goes to $+\infty$, $x = 1 - y$, $z = 0$. Informally, we can write max $= \infty$. This is an *unbounded* linear program.   ∎

Problems 1.11 and 1.12 are both linear programs because the objective functions and all the constraints are linear. Note that the optimal (maximal or minimal) value of an optimization problem is unique. (Had we found two different values, one would be better, so the other would not be optimal.) The optimal value always exists if we add the symbols $-\infty$, $+\infty$ to the set of real numbers as possible values for the optimal value. But if it is reached at all, there could be more than one way to reach it. That is, we may have many optimal solutions for the same optimal value.

Now you have encountered the following terms: *linear form, affine function, linear constraint, linear program, objective function, optimal value, optimal solution, feasible solution.* Try to explain the meaning of each of these terms.

Remark. We have already mentioned that linear programming is a part of mathematical programming. In its turn, mathematical programming is a tool in *operations research* (or operational research), which is an application of scientific methods to the management and administration of organized military, governmental, commercial, and industrial processes. Historically, the terms *programming* and *operations* came from planning military operations.

The terms *systems engineering* and *management science* mean almost the same as operations research with less or more stress on the human factor. As a part of operations research, linear programming is concerned not only with solving of linear programs but also with

- acquiring and processing data required to make decisions
- problem formulation and model construction
- testing the models and interpreting solutions
- implementing solutions into decisions
- controlling the decisions
- organizing and interconnecting different aspects of the process

In this book we stress mathematical aspects of linear programming, but we are also concerned with translating word problems into mathematical language, transforming linear programs into differents forms, and making connections with game theory and statistics.

How is linear programming connected with *linear algebra*? The main concern in linear algebra is solving systems of linear equations. We will see in Chapter 5 that solving linear programs is equivalent to finding feasible solutions for systems of linear constraints. Thus, from a mathematical point of view, linear programming is about more general and difficult problems.

Remark. Besides mathematical programming, there are other areas of mathematics and computer science where optimization plays a prominent role. For example, both *control theory* and *calculus of variations* are concerned with optimization problems that cannot be described easily with a finite set of variables. The feasible solutions could be functions satisfying certain conditions. We may ask what is the shortest curve connecting two given points in plane. Or we can ask about the most efficient way to sort data of any size. Sometimes mathematical programming can help to solve those problems.

Historic Remark. The mathematicians mentioned in this book are well known, and their bios can be found in encyclopedias, biographies, history books, and on the Web.

Joseph Fourier, a French mathematician well known also as an Egyptologist and administrator, is famous for his Fourier series, which are very important in mathematical physics and engineering. His work on linear approximation and linear programming is not so well known. A son of a tailor, he had 11 siblings and 3 half-siblings. His mother died when he was nine years old, and his father died the following year. He received military and religious education and was involved in politics. His life was in danger a few times.

Leonid Kantorovich, a Soviet mathematician with very important contributions to economics, was almost unknown in the United States until the simplex method was successfully implemented for computers and widely used. He got his Ph.D. in mathematics at age 18. The author had the pleasure of meeting Kantorovich several times at mathematical talks and at business meetings involving optimization of advanced planning in the former U.S.S.R. One of many things he did in mathematics was introducing the notion of a distance between probability distributions, which was rediscoved later in different forms by other mathematicians, including the author (the Vaserstein distance). This distance is the optimal value for a problem similar to the transportation problem (see Example 2.4 and Chapter 6).

Exercises

1–13. State whether the following are true or false. Explain your reasoning.

1. $1 \leq 2$

2. $-10 \leq -1$

3. $3 \leq 3$

4. $-5/12 \geq -3/7$

5. $x^2 + |y| \geq 0$ for all numbers x, y

6. $3x \geq x$ for all numbers x

7. $3x^3 \geq 2x^2$ for all numbers x

8. Every linear program should have at least one linear constraint

9. Every linear program has an optimal solution

10. Each variable in a linear program should be nonnegative

11. Any linear program has a unique optimal solution

12. The total number of constraints in a linear program is always larger than the number of variables

13. The constraint $2x + 5 = 6x - 3$ is equivalent to a linear equation for x ■

14–17. Determine whether the following functions of x and y are linear forms.

14. $2x$

15. $x + y + 1$

16. $(\sin 1)\, x + e^z\, y$

17. $x \sin a + y\,z$ ■

18–23. Is this a linear constraint for x?

18. $x > 2$

19. $|x| \leq 1$

20. $0 = 1$

21. $0 \geq 1$

22. $xy^2 = 3$

23. $ax = b$. ■

24–26. Do you agree with the following statements? Why or why not?

24. $|x| \leq 1$ is equivalent to a system of two linear constraints

25. $|x| \geq 1$ is equivalent to a system of two linear constraints

26. The equation $(x - 1)^2 = 0$ is equivalent to a linear constraint for x ■

27. Solve the equation $ax = b$ for x, where a and b are given numbers.

28–30. Solve the following three linear systems of equations for x and y.

28.
$$\begin{cases} x & + & 2y & = & 3 \\ 5x & + & 9y & = & 4 \end{cases}$$

29.
$$\begin{cases} x & + & 2y & = & 3 \\ 5x & + & 10y & = & 15 \end{cases}$$

30.
$$\begin{cases} x & + & 2y & = & 3 \\ 3x & + & 6y & = & 0 \end{cases}$$

■

31. Minimize (over x, y, z) $(x + y)^2 + (z + 1)^2$.
32. Minimize $|x + 2| + |x + 3|$ subject to $|x| \leq 2.5$.
33. Maximize $1/(1 + x^2)$.
34. Maximize $(x + y)^2 + (z + 1)^2$.
35. Minimize $|x + y| + (z + 1)^6 + (x - y + z)^2$.
36–42 Is this a linear form of two variables, x and y? (Answer Yes or No.)

36. $2x + 3y$ **37.** $2x + 3y = 1$
38. $x + y^2$ **39.** xy
40. y **41.** 0
42. $(x + 1)^2 + 2y - x^2 - 1$ **43.** x/y ■

44–49. Is this constraint for two variables, x and y linear?

44. $xy = 0$ **45.** $x = 0$
46. $x < 0$ **47.** $x + y = 0$
48. x is an integer   **49.** $x \geq 1$ or $y = 0$ ■

50–57. Is this a linear constraint for x and y? (Answer Yes or No.)

50. $x + 2y$ **51.** $x \geq 1$
52. $0 = 0$ **53.** $0 = 1$
54. $x + y \leq 0$ **55.** $x^2 = 2$
56. $x \geq 0$ **57.** $xy = 0$ ■

57. Show that any linear form $f(x, y)$ of two variables x, y has the following two properties:
• (proportionality) $f(ax, ay) = af(x, y)$ for every number a
• (additivity) $f(x_1 + x_2, y_1 + y_2) = f(x_1, y_1) + f(x_2, y_2)$
58. Conversely, show that every function $f(x, y)$ of two variables x, y with these two properties is linear.

59. Minimize $|x| + (x - 2y)^2 + \sin z + 2^u + \log(v + 101)$ subject to $|x|, |y|, |z|, |u|, |v| \leq 100$. How many optimal solutions are there? *Hint*: Try to minimize every term in the objective function separately.

§2. Examples of Linear Programs

As you recall, we mentioned in §1 that linear programs arise in business and industry. Our goal for the next two sections is to present some cases of real-life situations that can be described as linear programs.

Example 2.1. *A Diet Problem*

In this age of health consciousness, many people are analyzing the nutritive content of the food they eat. Let us see how this can be set up as a linear program.

The general idea is to select a mix of different foods for a person's diet in such a way that basic nutritional requirements are satisfied at minimum cost.

Of course, a realistic problem of this type would be quite complicated. We would have to rely on nutritionists to learn what the basic nutritional requirements are (and these would vary with the individual). Additionally, in order to have variety and avoid nutritional boredom, we would have to consider a long list of possible foods. Our example is drastically simplified.

According to the recommendations of a nutritionist, a person's daily requirements for protein, vitamin A, and calcium are as follows: 50 grams of protein, 4000 IUs (international units) of vitamin A, 1000 milligrams of calcium. For illustrative purposes, let us consider a diet consisting only of apples (raw, with skin), bananas (raw), carrots (raw), dates (domestic, natural, pitted, chopped), and eggs (whole, raw, fresh) and let us, if we can, determine the amount of each food to be consumed in order to meet the Recommended Dietary Allowances (RDA) at minimal cost.

Food	Unit	Protein	Vit. A	Calcium
		(g)	(IU)	(mg)
apple	1 medium (138 g)	0.3	73	9.6
banana	1 medium (118 g)	1.2	96	7
carrot	1 medium (72 g)	0.7	20253	19
dates	1 cup (178 g)	3.5	890	57
egg	1 medium (44 g)	5.5	279	22

Since our goal is to meet the RDA with minimal cost, we also need to compile the costs of these foods:

Food	Cost (in cents)
1 apple	10
1 banana	15
1 carrot	5
1 cup of dates	60
1 egg	8

Using these data, we can now set up a linear program. Let a, b, c, d, e be variables representing the quantities of the five foods we are going to use in the diet. The objective function to be minimized is the total cost function (in cents),

$$C = 10a + 15b + 5c + 60d + 8e,$$

where the coefficients represent cost per unit of the five items under consideration.

What are the constraints ? Obviously,

$$a, b, c, d, e \geq 0. \qquad (i)$$

These constraints are called *nonnegativity constraints*.

Then, to ensure that the minimum daily requirements of protein, vitamin A, and calcium are satisfied, it is necessary that

$$\begin{cases} 0.3a & + & 1.2b & + & 0.7c & + & 3.5d & + & 5.5e \geq 50 \\ 73a & + & 96b & + & 20253c & + & 890d & + & 279e \geq 4000 \\ 9.6a & + & 7b & + & 19c & + & 57d & + & 22e \geq 1000, \end{cases} \qquad (ii)$$

where, for example, in the first constraint, the term $0.3\,a$ expresses the number of grams of protein in each apple multiplied by the quantity of apples needed in the diet, the second term $1.2\,b$ expresses the number of grams of protein in each banana multiplied by the quantity of bananas needed in the diet, and so forth.

Notice that the terms of the first constraint are written in grams, all terms in the second constraint are written in IUs, and all terms in the third constraint are written in milligrams.

Recall that solving this linear program requires finding an optimal value and an optimal solution. Let us attempt to find the solution of this problem by trial and error. Since carrots are cheap, let us consider first the all-carrot diet. This means that a, b, d and e equal zero. The three constraints in (ii) reduce to

$$\begin{cases} 0.7c & \geq & 50 & \text{(g)} \\ 20253c & \geq & 4000 & \text{(IU)} \\ 19c & \geq & 1000 & \text{(mg)} \end{cases} \qquad (iii)$$

Thus, with $c = 500/7$ (approximately 71 carrots) the protein requirement in the diet is exactly satisfied, whereas the requirements for vitamin A and calcium are grossly exceeded. Since carrots cost 5 cents each, we find that the cost of this diet is $\$25/7$. Can we do better? Can we find another diet consisting of something other than carrots and that costs less than $\$25/7 \approx \$3.57$ a day? (The person being asked to eat 71 carrots per day hopes so!)

Note that since eggs are an excellent source of protein, the co-efficient of e is comparatively large in the constraint describing the protein requirement. This observation suggests that we could meet the nutritional requirements we have established for ourselves while avoiding monotony if we incorporate some eggs into our daily menu. We will try to reduce the amount of carrots, c, in our diet by increasing the number of eggs, e. Since we keep a, b and d equal zero, the three constraints are now

$$\begin{cases} 0.7c+ & 5.5e & \geq 50 \ \text{(grams of protein)} \\ 20253c+ & 279e & \geq 4000 \ \text{(IUs of vitamin A)} \\ 19c+ & 22e & \geq 1000 \ \text{(milligrams of calcium)}. \end{cases} \qquad (iv)$$

It is easy to see that we satisfy these constraints with $c = 50, e = 3$. Since carrots cost 5 cents each and eggs cost 8 cents each, the cost of this new diet is $\$2.74$. Voila! Our new diet of 50 carrots and 3 eggs per day offers a welcome respite from the 71-carrot diet as well as being substantially cheaper. Shall we try to do even better?

Although the trial-and-error method has helped us to do some analysis, it gives us no guidelines as to whether we can lower the cost further by considering other combinations of the five foods in our daily diet. We will return to this example later when we study the simplex method (Chapter 4).

Remark. Here is some more food for thought about the diet problem.

(a) Each ingredient can be measured in its own units. Always include the name of the unit to avoid confusion. Indicate also the unit for the cost objective function (cost can be expressed in cents, dollars, thousands of dollars, etc.).

(b) Could we have allowed the number of eggs in the answer to be expressed as a fraction, say 1/2? In a formal solution to a linear program, yes, we could have (the divisibility assumption of linear programming). Does it make sense? It depends on the situation discussed. Sometimes we have to require that certain variables are integers. Adding these nonlinear constraints turns a linear program into an integer linear program.

(c) How do we solve the diet problem? We will answer this question when we discuss the simplex method in Chapter 4. For now, we have tried a few iterations of the trial-and-error method, and, although we were able to come up with two feasible solutions, we still do not know what the optimal solution is.

(d) Should we measure apples in weight units rather than in pieces, since apples could be of different size? It depends. The unit you choose depends on real-life situations. If you buy apples in a convenience store, they are about the same size and you pay for each piece of fruit. So *piece* is appropriate in this situation. On the other hand, if you buy your apples by the bushel at a farm, then *bushel* is an appropriate unit. Just switching to a different weight unit is not a complete cure for variations in ingredients although sometimes it helps. The numbers in real life are often not exact (if they are known at all). Can you plan your diet for the next months if you are not even sure about the prices tomorrow? Other people do it, and sometimes you must do it, too. ∎

Let us try to set up another problem where the goal is to minimize cost.

Example 2.2. *A Blending Problem*

Many coins in different countries are made from cupronickel (75% copper, 25% nickel). Suppose that the four available alloys (scrap metals), A, B, C, D, to be utilized to produce the coin contain the percentages of copper and nickel shown in the following table:

Alloy	A	B	C	D
% copper	90	80	70	60
% nickel	10	20	30	40
$/lb		1.2	1.4	1.7	1.9

The cost in dollars per pound of each alloy is given as the last row in the same table.

Notice that none of the four alloys contains the desired percentages of copper and nickel. Our goal is to combine these alloys into a new blend containing the desired percentages of copper and nickel for cupronickel while minimizing the cost. This lends itself to a linear program.

Let a, b, c, d be the amounts of alloys A, B, C, D in pounds to make a pound of the new blend. Thus,

$$a, \ b, \ c, \ d \geq 0. \tag{i}$$

Since the new blend will be composed exclusively from the four alloys, we have

$$a + b + c + d = 1. \tag{ii}$$

The conditions on the composition of the new blend give

$$\begin{cases} .9a & + & .8b & + & .7c & + & .6d & = & .75 \\ .1a & + & .2b & + & .3c & + & .4d & = & .25. \end{cases} \tag{iii}$$

For example, the first equality states that 90% of the amount of alloy A, plus 80% of the amount of alloy B, plus 70% of the amount of alloy C, plus 60% of the amount of alloy D will give the desired 75% of copper in a pound of the new blend. Likewise, the second equality gives the desired amount of nickel in the new blend.

Taking the preceding constraints into account, we minimize the cost function

$$C = 1.2a + 1.4b + 1.7c + 1.9d$$

In this problem all the constraints, except (i), are *equalities*. In fact, there are three linear equations and four unknowns. However, the three equations are not independent. For example, the sum of the equations in (iii) gives (ii). Thus (ii) is redundant.

In general, a constraint is said to be *redundant* if it follows from the other constraints of our system. Since it contributes no new information regarding the solutions of the linear program, it can be dropped from consideration without changing the feasible set.

Example 2.3. *A Manufacturing Problem*

We are now going to state a program in which the objective function, a profit function, is to be maximized. A factory produces three products: P1, P2, and P3. The unit of measure for each product is the standard-sized boxes into which the product is placed. The profit per box of P1, P2, and P3 is $2, $3 and $7, respectively. Denote by x_1, x_2, x_3 the number of boxes of P1, P2, and P3, respectively. So the profit function we want to maximize is

$$P = 2x_1 + 3x_2 + 7x_3$$

The five resources used are raw materials R1 and R2, labor, working area, and time on a machine. There are 1200 lbs of R1 available, 300 lbs of R2, 40 employee-hours of labor, 8000 m² of working area, and 8 machine-hours on the machine.

The amount of each resource needed for a box of each of the products is given in the following table (which also includes the aforementioned data):

Resource	Unit	P1	P2	P3	|| Available
R1	lb	40	20	60	||	1200
R2	lb	4	1	6	||	300
Labor	hour	.2	.7	2	||	40
Aarea	m^2	100	100	800	||	8000
Machine	hour	.1	.3	.6	||	8
Profit	$	2	3	7	||	→ max

As we see from this table, to produce a box of P1 we need 40 pounds of R1, 4 pounds of R2, 0.2 hours of labor, 100 m² of working area, and 0.1 hours on the machine. Also, the amount of resources needed to produce a box of P2 and P3 can be deduced from the table.

The constraints are

$$x_1, x_2, x_3 \geq 0, \qquad\qquad (i)$$

and

$$\begin{cases} 40x_1 & + & 20x_2 & + & 60x_3 & \leq & 1200 & \text{(pounds of R1)} \\ 4x_1 & + & x_2 & + & 6x_3 & \leq & 300 & \text{(pounds of R2)} \\ .2x_1 & + & .7x_2 & + & 2x_3 & \leq & 40 & \text{(labor)} \\ 100x_1 & + & 100x_2 & + & 800x_3 & \leq & 8000 & \text{(area in m}^2) \\ .1x_1 & + & .3x_2 & + & .8x_3 & \leq & 8 & \text{(machine).} \end{cases} \qquad (ii)$$

Note that a naive first approximation of the optimal solution is to produce only boxes of P3, since the profit from each one of them is bigger than the profit from boxes of P1 and P2. This means setting up $x_1 = 0$ and $x_2 = 0$. The constraints (ii) now become

$$\begin{cases} 60x_3 & \leq & 1200 \\ 6x_3 & \leq & 300 \\ 2x_3 & \leq & 40 \\ 800x_3 & \leq & 8000 \\ .8x_3 & \leq & 8. \end{cases} \qquad (iii)$$

One solution to these inequalities is $x_3 = 10$ with profit $P = 70$. However, when we figure out how much of the resources were used to produce these 10 boxes of P3, we see that 600 lbs of R1 remain unused, more than half of R2 was not used, and half of the labor was wasted. Please keep in mind that the maximal profit involves an optimal use of the resources in order to get the best return on our investment. Although this naive first approximation is feasible, we are guessing that it is not an optimal solution to this problem. We will see later that the optimal solution is $x_1 = 20, x_2 = 20, x_3 = 0$ with max $= 100$.

Example 2.4. *A Transportation Problem*
Another concern that manufacturers face daily is *transportation costs* for their products. Let us look at the following hypothetical situation and try to set it up as a linear program.

A manufacturer of widgets has warehouses in Atlanta, Baltimore, and Chicago. The warehouse in Atlanta has 50 widgets in stock, the warehouse in Baltimore has 30 widgets in stock, and the warehouse in Chicago has 50 widgets in stock. There are retail stores in Detroit, Eugene, Fairview, Grove City, and Houston. The retail stores in Detroit, Eugene, Fairview, Grove City, and Houston need at least 25, 10, 20, 30, 15 widgets, respectively. Obviously, the manufacturer needs to ship widgets to all five stores from the three warehouses and he wants to do this in the cheapest possible way.

This presents a perfect backdrop for a linear program, to minimize shipping cost. To start, we need to know the cost of shipping one widget from each warehouse to each retail store. This is given by a shipping cost table

	1.D	2.E	3.F	4.G	5.H
1. Atlanta	55	30	40	50	40
2. Baltimore	35	30	100	45	60
3. Chicago	40	60	95	35	30

Thus, it costs $30 to ship one unit of the product from Baltimore to Eugene (E), $95 from Chicago to Fairview (F), and so on.

In order to set this up as a linear program, we introduce variables that represent the number of units of product shipped from each warehouse to each store. We have numbered the warehouses according to their alphabetical order and we have enumerated the stores similarly. Let x_{ij}, for all $1 \leq i \leq 3$, $1 \leq j \leq 5$, represent the number of widgets shipped from warehouse $\#i$ to store $\#j$. This gives us 15 unknowns. The objective function (the quantity to be minimized) is the shipping cost given by

$$C = 55x_{11} + 30x_{12} + 40x_{13} + 50x_{14} + 40x_{15}$$
$$+ 35x_{21} + 30x_{22} + 100x_{23} + 45x_{24} + 60x_{25}$$
$$+ 40x_{31} + 60x_{32} + 95x_{33} + 35x_{34} + 30x_{35}$$

where $55x_{11}$ represents the cost of shipping one widget from the warehouse in Atlanta to the retail store in Detroit (D) multiplied by the number of widgets that will be shipped, and so forth.

What are the constraints? First, our 15 variables satisfy the condition that

$$x_{ij} \geq 0, \text{ for all } 1 \leq i \leq 3, \ 1 \leq j \leq 5 \tag{i}$$

since shipping a negative amount of widgets makes no sense. Second, since the warehouse $\#i$ cannot ship more widgets than it has in stock, we get

$$\begin{cases} x_{11} + x_{12} + x_{13} + x_{14} + x_{15} \leq 50 \\ x_{21} + x_{22} + x_{23} + x_{24} + x_{25} \leq 30 \\ x_{31} + x_{32} + x_{33} + x_{34} + x_{35} \leq 50 \end{cases} \tag{ii}$$

Next, working with the amount of widgets that each retail store needs, we obtain the following five constraints:

$$\begin{cases} x_{11} & + & x_{21} & + & x_{31} & \geq & 25 \\ x_{12} & + & x_{22} & + & x_{32} & \geq & 10 \\ x_{13} & + & x_{23} & + & x_{33} & \geq & 20 \\ x_{14} & + & x_{24} & + & x_{34} & \geq & 30 \\ x_{15} & + & x_{25} & + & x_{35} & \geq & 15 \end{cases} \qquad (iii)$$

The problem is now set up. We will study an efficient method for solving such problems later. For now, we are still limping along with the trial-and-error method.

Geographically and financially speaking, it seems reasonable for the warehouse in Atlanta to ship widgets to the retail stores in Eugene and Fairview, while Grove City and Houston are attractive markets for Chicago. Thus,

$$\begin{cases} x_{12} & = & 10, & x_{22} & = & x_{32} & = & 0 \\ x_{13} & = & 20, & x_{23} & = & x_{33} & = & 0 \\ x_{34} & = & 30, & x_{35} & = & 15. \end{cases}$$

We have now decided on the values for 8 of the 15 variables. Continuing, since the retail stores in Eugene and Fairview will get the widgets they need from the warehouse in Atlanta and the retail stores in Grove City and Houston will receive widgets from the warehouse in Chicago, it makes no sense to ship additional widgets to these stores from another warehouse since that would increase the shipping cost. Thus we set

$$x_{14} = x_{15} = x_{24} = x_{25} = 0.$$

This leaves only the three variables x_{11}, x_{21}, x_{31} to be determined. By using the constraints, we determine that

$$x_{21} = 25, x_{11} = x_{31} = 0$$

Some Remarks about This Feasible Solution

(a) Notice that in our feasible solution we require that the total number of units shipped to Detroit is 25, the number of units shipped to Eugene is 10, and so on. Shipping more units than what each store needs would only increase the cost. On the other hand, we did not use the whole supply of widgets since the supply is greater than the demand.

(b)   In real life, the total shipping cost is not always expressible by a linear form. For example, discounts could be available for bulk shipping. Nonlinear problems are usually more difficult to solve than the linear ones.

Example 2.5. *Job Assignment Problem*

Suppose that a production manager must assign n workers to do n jobs. If every worker could perform each job at the same level of skill and efficiency, the job assignments could be issued arbitrarily. However, as we know, this is seldom the case. Thus, each of the n workers is evaluated according to the time he or she takes to perform each job. The time, given in hours, is expressed as a number greater than or equal to zero. Obviously, the goal is to assign workers to jobs in such a way that the total time is as small as possible. In order to set up the notation, we let c_{ij} be the time it takes to worker $\#i$ to perform job $\#j$. Then the times could naturally be written in a table. For example, take $n = 3$ and let the times be given as in the following table:

	a	b	c
A	10	70	40
B	20	60	10
C	10	20	90

So if worker **A** gets assigned to job **a**, worker **B** fills job **b** and worker **C** does job **c**, then the total time is $10 + 60 + 90 = 160$. This is not a minimum since, if **A** does **b**, **B** does **c** and **C** does **a**, then the total time equals $70 + 10 + 10 = 90$. For $n = 3$, the total number of possible ways of assigning jobs is $3 \times 2 = 6$. This can be seen from the information contained in the following table:

Assignment			Total Time
Aa	Bb	Cc	160
Aa	Bc	Cb	40
Ab	Ba	Cc	180
Ab	Bc	Ca	90
Ac	Ba	Cb	80
Ac	Bb	Ca	110

From the table we can see that the minimum value of the total time is 40; we conclude that the production manager would be wise to assign worker **A** to job **a**, worker **B** to job **c** and worker **C** to job **b**.

In general, this method of selection is not good. The total number of possible ways of assigning jobs is $n! = n \times (n-1) \times (n-2) \times \cdots \times 2 \times 1$. This is an enormous number even for moderate n. For $n = 70$,

$$n! = 11978571669969891796072783721689098736458938142546425857555362864628000958278984531968000000000000000.$$

It has been estimated that if a Sun Workstation computer had started solving this problem at the time of the Big Bang, by looking at all possible job assignments, then by now it would not have finished yet its task.

Although is not obvious, the job assignment problem can be expressed as a linear program. As such, it can be solved by the simplex method. When $n = 70$ it takes seconds. We will return to this and related problems in Chapter 7 after we discuss the simplex method.

We conclude this section with a quotation. In 1980 Eugene Lawler wrote that

[Linear programming] is used to allocate resources, plan production, schedule workers, plan investment portfolios and formulate marketing (and military) strategies. The versatility and economic impact of linear programming in today's industrial world is truly awesome.

Exercises

1. Suppose that for your balanced diet you need only protein and vitamins A, B_1, C, B_6, B_{12} and you are allowed to eat only cereals. Go to a grocery store and choose 10 different boxes of cereals. Write down the percentages of U.S. recommended daily allowances (RDAs) for the aforementioned ingredients and the price of each cereal brand. Then write down your diet problem, indicating the units, date, and store. If necessary, you may choose your gender, age, and calorie intake to determine your RDA. If you cannot find your protein RDA on boxes and elsewhere, take it to be 50 g. Write down the name of the store and the date. You may take data from a store on the Internet.

2. Solve the blending problem in Example 2.2. *Hint*: Note that in *(iii)* of Example 2.2 we have a system of two linear equations in four unknowns. Thus, two variables can be eliminated and the resulting problem in two variables can be solved by graphical methods (see §1 of Chapter 2).

3. Solve the linear program in Example 2.3 with the additional condition $x_3 = 0$. *Hint*: Use the graphical method (see §1 of Chapter 2).

4. You have 100 quarters and 90 dimes and no other money. You have to pay a given amount C. No change is given to you. You do not want to overpay too much. State this word optimization problem in mathematical form (i.e., in terms of decision variables, objective function, and constraints). Is this problem linear? Solve it for $C = 15$ cents, for $C = \$1.02$, and for $C = \$100$.

5. You have a string loop of length 100. You want to make a rectangle of maximal area. State this problem mathematically and solve it.

6. Example 2.6 has a nice variation. Suppose that the production manager devises a system whereby each employee is given a numerical rating depending on how well he or she performs a particular job. Then the manager would want to assign people to jobs in a way that the sum of their ratings is as large as possible, so that the level of efficiency of the company is also as high as possible. Your company is small and you have just rated four employees as to how well they can do four jobs which need to be assigned. Solve this maximization problem, using the following matrix of ratings:

	a	b	c	d
A	10	70	40	55
B	20	60	10	67
C	10	20	90	43
D	15	37	89	23

7. Another interpretation of the same mathematical problem is the *matching problem*. You want to match n boys with n girls into n couples with objective to produce the maximal bliss (or minimal grief). Here is your data, where the numbers are dollars that the couples are expected to pay you for happy marriage (negative numbers mean they would try to get some money from you for an unhappy marriage) and where $n = 4$:

	a	b	c	d
A	1	-2	2	0
B	2	0	1	1
C	3	1	0	-1
D	-1	0	1	2

Solve this problem. *Hint:* Check all $4! = 24$ matches unless you see a faster way.

8–10. Solve the following matching problems where the numbers in tables are the expected numbers of happy years together for each couple and you want to maximize the total. If you cannot solve the problem in a reasonable timeframe, try to find as good a feasible solution as possible. Data are not taken from real life.

8.	a	b	c	d	e
A	8	2	9	0	0
B	2	9	1	1	3
C	3	1	7	1	1
D	1	6	1	2	9
E	8	8	1	9	1

9.	a	b	c	d	e	f	g
A	8	2	9	0	3	8	7
B	2	0	1	1	3	7	9
C	3	1	1	1	1	6	9
D	1	0	1	2	9	5	8
E	8	8	1	1	1	5	7
F	1	6	1	9	9	5	8
G	6	6	6	5	1	5	4

10.	a	b	c	d	e	f	g
A	8	2	5	0	0	1	7
B	2	9	1	1	3	7	5
C	3	1	7	1	1	6	9
D	1	6	1	2	1	5	8
E	8	8	1	0	1	5	7
F	0	6	1	9	1	5	8
G	6	6	1	5	1	5	4

11. We want to find a maximal number among given numbers $c_1, \ldots, c_n$. State this as a linear program.

12. Given three distinct numbers, a, b, c, we want to find the *median* (i.e., the number x that is one of the given numbers but is not maximal or minimal). State this as a linear program. *Hint:* This is a difficult problem, but it will be solved in Chapter 8.

§3. Graphical Method

Example 3.1. Consider the following simple linear program:

$$\begin{cases} \text{Maximize} & y = 3x \\ \text{subject to} & 0 \leq x \leq 5. \end{cases}$$

The value of x that optimizes this function becomes obvious when we look at the following figure:

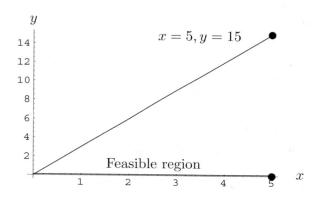

In general, when the number of decision variables is no more than 2, we can use some pictures in the plane to solve our optimization problem. Unfortunately, it is much harder to make and use pictures in a higher dimension.

When we have a linear program with one decision variable x, the graph of the objective function y in the (x, y)-plane is a straight line and the feasible region is a set in the x-axis. In general, any finite system of linear constraints for one variable x is equivalent to either a single linear constraint or to a system of two linear constraints. To see this, replace every constraint by a constraint of one of the following five types: $0 = 0, 0 = 1, x \geq b, x \leq b, x = b$. Therefore, the feasible region for a linear program with one variable x has one of the following six shapes:

- the whole line
- the ray $x \leq c$ going from $-\infty$ up to a number c
- the ray $x \geq c$ going from a number c up to $+\infty$
- the closed interval $a \leq x \leq b$ with endpoints $a\,b$ $(a < b)$
- a point $x = c$
- the empty set (that is, there are no feasible solutions)

We leave it as an exercise for the reader to verify the following facts:

Fact 3.2. If the objective function is a constant function, $y = c$, then the optimal value is $y = c$ and it is attained at any point x in the feasible region.

Fact 3.3. Now assume that the objective function is of the form $y = \alpha x$, where α is a *nonzero* real number.

1.   If the feasible region is the whole line, no linear program has an optimal solution.

2.   If the feasible region is a ray $x \le c$ (respectively, $x \ge c$), a linear program has an optimal solution if and only if it is a maximization problem (respectively, a minimization problem). The optimal value is $y = \alpha c$ attained at the point $x = c$.

3.   If the feasible region is a closed interval, $a \le x \le b$, $a \le b$, the optimal value is attained at one of the endpoints a or b.

4.   If the feasible region is the empty set, there are no optimal solutions.

When the number of decision variables is 2, it is possible to draw a picture for our linear program in the Cartesian plane in order to see all feasible and optimal solutions. This is called the *graphing method* for solving linear programs. By doing this, we gain geometrical insight that can be applied to problems with any number of variables.

Sometimes, problems with a large number of variables can be reduced to problems with a smaller number of variables. For instance, in Example 2.2, we can reduce the number of variables (not counting the objective variable C) from 4 to 2.

Here is an example when the number of variables is 1 from the beginning.

Example 3.4.

$$\begin{cases} \text{Minimize} & y & = & 0.3x \\ \text{subject to} & 2x & \le & 50, \\ & 3x & \le & 120, \\ & -5x & \ge & -250, \\ & 0.5x & \ge & -3. \end{cases}$$

We can picture this linear program as follows. Each inequality gives a ray on the x-axis: The inequality $2x \le 50$ represents the ray $(-\infty, 25]$, the constraint $3x \le 120$ represents the ray $(-\infty, 40]$ and the constraint $-5x \geq -250$ represents the ray $(-\infty, 50]$. What ray is obtained from the fourth constraint in Example 3.4? Since the contribution made by each and every constraint must be taken into account, the feasible region must be the common part or intersection of these rays. This gives the interval $-6 \leq x \leq 25$ as the feasible region; that is, the set of all feasible solutions.

In Figure 3.5, we plot the objective function on the feasible region.

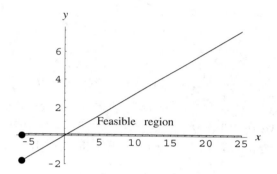

Figure 3.5

Now it is clear that the minimal value for the objective function in our problem is -1.8. It is reached at $x = -6$, which is the unique optimal solution. *Answer*: min $= -1.8$ at $x = -6$. ∎

The set of optimal solutions also has one of those six forms. And it is a subset of the feasible region. When the feasible set is an interval, one endpoint or the other is an optimal solution; we have exactly one optimal solution (which must be an endpoint) if and only if the objective function is not constant.

We now consider a linear program with two variables.

Example 3.6.

$$\begin{cases} \text{Maximize} & f = x + 9y \\ \text{subject to} & x \geq 0, \ y \geq 0, \\ & x - y \leq 3, \\ & x - 3y \geq -5, \\ & 5x + 7y \leq 35 \end{cases}$$

First we draw the feasible region, F, in the (x, y)-plane, where F consists of all points (x, y) satisfying the linear constraints. Since each constraint is a linear inequality, the set of all points satisfying this inequality describes a half-plane. Thus the feasible region F is the intersection of all these half-planes (Figure 3.7).

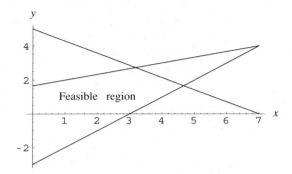

Figure 3.7. The feasible region F.

This region F is a convex (see §3 of Chapter 4) polygon with five vertices (a pentagon). The vertices with y-coordinate equal to zero are $(0,0)$ and $(3,0)$. The vertex that lies on the y-axis is $(0, \frac{5}{3})$. The fourth vertex is $(\frac{14}{3}, \frac{5}{3})$ which corresponds to the intersection of the straight lines $x - y = 3$ and $5x + 7y = 35$ and the highest vertex is the intersection of the lines $x - 3y = -5$ and $5x + 7y = 35$ and has coordinates $(x, y) = (\frac{35}{11}, \frac{30}{11})$.

In Figure 3.8, we draw the graph of the function $y = -x/9$ in the (x, y)-plane. This corresponds to the value $f = 0$ of the objective function.

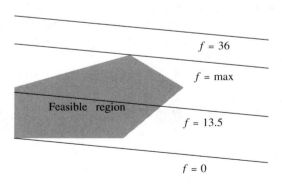

Figure 3.8. The feasible region F and levels for f.

The value of f is constant along parallel straight lines passing through the feasible region; that is, $x + 9y = f$ with $f \geq 0$. Now it is clear that the minimal value of f is 0, the value $f = 36$ is infeasible, and the maximal value of f in F is reached at the vertex $(\frac{35}{11}, \frac{30}{11})$. This optimal value is $f = \frac{35}{11} + 9(\frac{30}{11}) = \frac{305}{11}$. The objective function f takes values between 0 and $305/11$ on F. *Answer:* $\max = \frac{305}{11} \approx 27.7$ at $x = \frac{35}{11}, y = \frac{30}{11}$. ∎

In general, the feasible region for a linear program with two variables x, y is a polygonal region of one of the following shapes: the empty set; a point; an interval; a ray; a straight line, a bounded convex polygon, a half-plane, the whole plane; a strip, an angle, an unbounded polygonal region with $s \geq 3$ sides. The set of optimal solutions is a subset of one of the listed shapes.

When the feasible region is a (bounded nonempty) polygon, it is clear that we have at least one optimal solution for any linear objective function. Moreover, a vertex (depending on the objective function) is an optimal solution (corner principle).

Of course, the corner principle does not work for nonlinear problems. Here is an example.

Example 3.9. Minimize $g = x^2 - 3x + 2y^2 - 4y$ subject to the same constraints as in Example 3.6.

We plot the levels of our nonlinear objective function in the feasible region (Figure 3.10).

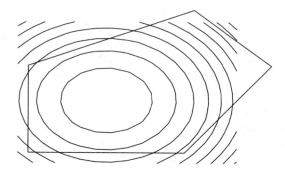

Figure 3.10. The feasible region F and levels for g in Example 3.9

The figure shows clearly that the optimal solution is inside, not at the boundary of F. So we can ignore the constraints. To minimize g without constraints, we can use derivatives or do a simple algebraic transformation: $g = (x - 3/2)^2 + 2(y - 1)^2 - 9/4 - 2$, hence min $=$ -4.25 at $x = 1.5, y = 1$. It will not hurt if we double check directly that this solution is feasible. ∎

Another way to obtain a nonlinear problem from Example 3.6 is to add nonlinear constraints.

Example 3.11. *An integer program*
Solve the linear problem in Example 3.6 with the additional condition that both x and y are integers.

Again we can draw a figure (Figure 3.12):

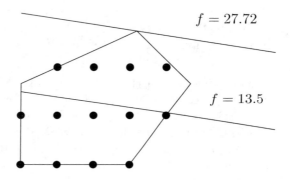

Figure 3.12. The feasible points and levels for f in Example 3.11

It is clear that the feasible region (the integer points in F) consists of 13 points with possible exceptions of two points, $(x, y) = (1, 2), (4, 1)$, which apparently are on the boundary. Checking the constraint $x - y \leq 3$ for $x = 4, y = 1$ and $x - 3y \geq -5$ for $x = 1, y = 2$ confirms the figure. We can see from Figure 3.12 that the optimal solution is $x = 4, y = 2$, so min $= 4 + 9 \cdot 2 = 22$. The additional condition reduced the maximum from $305/11 \approx 27.7$ to 22.

In general, *integer programming* is concerned with linear programs where some or all variables are required to be integers. This additional condition usually makes solving the problem much more difficult. However, this is not the case with this example. ∎

Next we consider another modification of Example 3.6 with the objective function changed and a nonlinear constraint added.

Example 3.13.

$$\begin{cases} \text{Maximize } h = x^2 - y^2 \\ \text{subject to } x \geq 0, \ y \geq 0, \ y \text{ an integer} \\ x - y \leq 3, x - 3y \geq -5, 5x + 7y \leq 35. \end{cases}$$

Now the feasibly region consists of three horizontal line segments and levels of the nonlinear objective function h are disconnected (Figure 3.14).

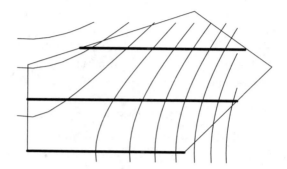

Figure 3.14. The feasible lines and levels for h in Example 3.13.

The figure shows that the optimal solutions are restricted to the following two points: the right ends of the upper and middle intervals. These two points are $x = 4.25, y = 2$ and $x = 4, y = 1$. The values of h at these points are 14.0625 and 15. So the max= 15 at $x = 4, y = 1$. ∎

In the next example we are required to solve a family of linear programs depending on two parameters.

Example 3.15.
Maximize $x + y$ subject to $0 \le x \le a, 0 \le y \le b$, where a, b are given numbers.

It is clear that the feasible region in this problem is
a rectangle when $a, b > 0$,
a line segment when $a = 0$ and $b > 0$ or $b = 0$ and $a > 0$,
a point when $a = b = 0$,
empty when $a < 0$ or $b < 0$.
So the answer is
max $= a + b$ at $x = a, y =$ b when $a, b \ge 0$,
the program is infeasible otherwise.
The problem can be solved without pictures, but a picture on paper or in your mind may help. ∎

For a linear program with many variables we can imagine the feasible region and the set of optimal solutions as *convex* sets in a high-dimensional space. We will make this more precise later, after we define what a convex set is (see §3 of Chapter 4).

A linear constraint $c_1x_1 + \cdots + c_nx_n = a$ gives a *hyperplane* (when not all $c_i = 0$) or the whole space (when it is of the form $0 = 0$), or the empty set (when all $c_i = 0$, but $a \neq 0$). A linear constraint $c_1x_1 + \cdots + c_nx_n \leq a$ gives either a *half-space* (when not all $c_i = 0$), or the whole space (when all $c_i = 0$ and $a \geq 0$), or the empty set (when all $c_i = 0$ but $a < 0$).

With computers, we can use 3-D (three-dimensional) graphics more efficiently than with paper. Paper is not a good media for working with high-dimensional pictures, but our imagination is.

Exercises

1–3. Try your hand at these exercises involving the digits a_i of your Social Security Number $a_1\, a_2\, a_3 \quad a_4\, a_5 \quad a_6\, a_7\, a_8\, a_9$. Solve the following problems:

1.
$$\begin{cases} \text{Maximize} & (a_1 - a_2)x \\ \text{subject to} & (a_3 + a_4)x \leq a_5, \\ & (a_6 + a_7)x \geq -a_8, \\ & (a_9 + 2)x \leq 10. \end{cases}$$

2.
$$\begin{cases} \text{Minimize} & f = a_1\, x - (a_1 + a_2)\, y \\ \text{subject to} & |(9 - a_3)x + a_4 y| \leq 10 + a_4, \\ & |a_5 x + (1 + a_6)y| \leq 8, \\ & |x + y| \leq a_7 + a_9 + 1. \end{cases}$$

3.
$$\begin{cases} \text{Minimize} & f = a_1\, x + a_2\, y \\ \text{subject to} & |(9 + a_3)x + a_4 y| \leq 10, \\ & |a_5 x + (9 + a_6)y| \leq 10, \\ & |x + y| \leq a_7 + a_8 + a_9. \end{cases}$$

4.
$$\begin{cases} \text{Maximize (over } x) & bx \\ \text{subject to} & |2x + 4| \leq 10, \\ & |x + 3| \leq 5. \end{cases}$$

where:

(*i*) $b = 7$

(*ii*) $b = -9$

(*iii*) b is any given number.

5.
$$\begin{cases} \text{Minimize} & f = x + 8y \\ \text{subject to} & |x| \leq 9, \\ & |y| \leq 9, \\ & |x + y| \leq 9. \end{cases}$$

6. Minimize x/y subject to $x \geq y$.

7. $\begin{cases} \text{Minimize} & f = x \cdot y \\ \text{subject to} & |x| + |y| \le 1. \end{cases}$

8. $\begin{cases} \text{Maximize} & w = x + y + z \\ \text{subject to} & |x| \le 1, \\ & |y| \le 1, \\ & |z| \le 1, \\ & x + 2y + 3z = 6. \end{cases}$

9. Maximize $\dfrac{1}{1 + x^2 + y^4}$.

10. Solve for b the linear equation $bx = c$, where x and c are given numbers.

11. $\begin{cases} \text{Maximize} & f = x + 9y \\ \text{subject to} & x \ge 0,\ y \ge 0, \\ & x - y \le 3, \\ & x - 3y \ge -5, \\ & 5x + 7y \le 35, \\ & x \text{ and } y \text{ integers.} \end{cases}$

12. Solve the diet problem involving three ingredients, energy, vitamin B_1 (thiamin), and vitamin B_2 (riboflavin) and two foods, almonds (nuts; refuse: shells, 60%) and blueberries (raw; refuse: 2%). Here are contents per 100 g of edible portion (excluding refuse) and prices in dollars per pound (not per 100 g!).

Food	Energy in kcal	B_1 in mg	B_2 in mg	Price
Almonds	578	0.24	0.81	2
Blueberries	56	0.05	0.05	3
RDA	2000	1.1	1.1	

13. Maximize $x^3 + y^3$ subject to $x + y \le 5$.

14. Maximize $|x| + y^2$ subject to $|x + y + 2| + |x - y + 3| \le 5$.

15. Maximize $x + 2y + 3z$ subject to $x, y, z \ge 0, x + y + y = 1$.

Chapter 2

Background

§4. Logic

Logical reasoning is as important in linear programming as it is in many other areas of science. It is only in college that problems may be presented to you in the form that you would immediately recognize as a linear program. On the contrary, you will usually be given data, some of which may even be extraneous, and it will be up to you to synthesize and collate the data until you recognize the kind of problem you want to solve. Think back to the examples we presented to you in §2 of Chapter 1. In no case did we start a problem by request that you maximize or minimize a linear form subject to certain linear constraints. Instead, we asked you to analyze a real-life situation, albeit a simple one, in order to come up with a detailed plan for activity. Taking the words and extracting from them the particular linear programming problem demand logical reasoning.

In order to develop your logical reasoning, you must be able to understand the mathematical meaning of certain words that might be different from the common English usage. For example, if we ask you, "Do you want coffee or tea?" we are using the word *or* to indicate that we expect you to choose between two beverages. However, if we told you, "the number -2 or 3 is a solution to the polynomial equation $x^2 - x - 6 = 0$," you would understand that both numbers are candidates for the solution. Mathematically speaking, you are not being asked to choose between them. In other words, *or* in mathematics usually means the inclusive disjunction. Notice that it is also the case in certain common everyday situations. If you were asked whether you wanted cream or sugar in your coffee, you would not be expected to choose only one. Some mathematical symbols including *or* are $\geq$, $\leq$, $\pm$.

Another seemingly simple word that has a more precise mathematical definition than it does in English is the word *and*. Think of

34

the various ways in which you use this word. You use it when you are listing a series of objects: "I bought a skirt, a sweater, and a dress." You use this word as a conjunction: "I washed my hair and I brushed my teeth." However, we use the word *and* in the mathematical sense when we want to indicate that two (or more) statements must be satisfied simultaneously. For instance, the use of the word *and* in the statement,

> "b is a positive integer
> and b is a solution of the equation $x^2 - x - 6 = 0$"

indicates that b must equal 3. Do you see why?

Your understanding of mathematics will be enhanced if you gain facility in using the following words in their mathematical sense and become familiar with their logical symbols:

- and (logical symbol: $\wedge$)
- or (logical symbol: $\vee$)
- implies (logical symbol: $\Rightarrow$)
- follows from (logical symbol: $\Leftarrow$)

Sometimes, the symbols $\Rightarrow$ and $\Leftarrow$ are combined into one symbol $\Leftrightarrow$ which means the same as "if and only if" (*iff* for short), "that is," "i.e.," "is equivalent," "means," "$\Rightarrow \wedge \Leftarrow$."

There are other equivalence relations in mathematics and elsewhere besides the logical equivalence. In saying "these theorems are equivalent," we often try to say something different from "either these theorems are both true or both false."

Usually, the comma means "and." For example, $x > 0, y > 0$ means $x > 0$ and $y > 0$. This can be also written as $x, y > 0$. Constraints in linear programs are often connected by commas, which means we want to satisfy all of them. However, in some situations the comma means something else. For example,

$$x = 0, 1, \text{or } 2$$

means that x takes one of these three values. In the example

$$x = 2 + 3, \text{ or } x = 5$$

the comma changes the meaning of *or* to *i.e.*

We often have at our disposal many words that have the same meaning mathematically. For instance, in the following true statement

$$x \geq 0 \text{ if } x = 1 \tag{4.1}$$

we can replace *if* by any of the following words or expressions:
   if, when, since, provided that, whenever, is weaker than,
   follows from, is a consequence of, because, is implied by, $\Leftarrow$.

In particular, we can write (4.1) as follows: $x \geq 0 \Leftarrow x = 1$. We can also rewrite (4.1) as

$$\text{If } x = 1, \text{ then } x \geq 0$$

or, equivalently,

$$x = 1 \text{ implies } x \geq 0. \tag{4.2}$$

We can replace *implies* in (4.2) by any of the following expressions:
   implies, is stronger than, only if, so, hence, results in,
   forces, gives, whence, therefore, thus, consequently, $\Rightarrow$.

Example 4.3. The following seventeen sentences have the same mathematical meaning:

- $x \geq 0$, because $x \geq 2$.   • $x \geq 0$, if $x \geq 2$.
- $x \geq 2$ only if $x \geq 0$.   • If $x \geq 2$, then $x \geq 0$.
- The bound $x \geq 2$ is sharper $x \geq 0$.
- Given $x \geq 2$, we conclude that $x \geq 0$.
- The bound $x \geq 2$ is better than $x \geq 0$.
- The constraint $x \geq 0$ is less tight than $x \geq 2$.
- The bound $x \geq 2$ is more precise than $x \geq 0$.
- In the view of condition $x \geq 2$, we have $x \geq 0$.
- The constraint $x \geq 0$ is less severe than $x \geq 2$.
- The constraint $x \geq 0$ is less strict than $x \geq 2$.
- The condition $x \geq 2$ implies the constraint $x \geq 0$.
- The constraint $x \geq 0$ is less stringent than $x \geq 2$.
- The constraint $x \geq 0$ is less demanding than $x \geq 2$.
- The condition $x \geq 2$ is sufficient to conclude that $x \geq 0$.
- The linear constraint $x \geq 0$ follows from the condition $x \geq 2$.

Remark. In definitions, *if* is used sometimes for "if and only if."

Many statements or conditions in this book are constraints on variables. Every constraint or a system of constraints gives a set, the feasible region. The feasible region corresponding to a stronger system is a part of the feasible region corresponding to a weaker system. Adding new constrains reduces the feasible region. For example, the point $x = 1$ belongs to the ray $x \geq 0$, which means that the condition $x = 1$ implies that $x \geq 0$. Equivalent systems of constraints imply each other, and they have the same feasible region. In the terms of feasible regions, *and* means the intersection, while *or* means the union. For example the condition "$x = 0$ and $y = 0$" gives a point in plane, while the condition "$x = 0$ or $y = 0$" gives the union of two straight lines.

Here is a more sophisticated example. Consider the following statement: $0 = 1$ implies $0 = 0$. This statement is true because any false statement implies everything in general. Also, truth follows from everything in general. In particular, we do not need any conditions to conclude that $0 = 0$. Another way to see that the statement is true is the following argument: given $0 = 1$, we can conclude that $0 = 1 = 0$, hence $0 = 0$. Multiplication of the equation $0 = 1$ by 0 also gives the desired conclusion. Finally, in the terms of feasible sets our statement is as follows: Nothing is a part of everything.

Now we consider more complicated linear equations. Given two linear equations

$$\begin{cases} x + 3y = 3 \\ x + 2y = 5, \end{cases}$$

we can take their difference and conclude that $y = -2$. Thus, the equation $y = -2$ follows from the system. More generally, given m equations, $f_1 = b_1, \ldots, f_m = b_m$, and any numbers c_i, we can take *linear combination of the given equations with the coefficients* c_i :

$$c_1 f_1 + \cdots + c_m f_m = c_1 b_1 + \cdots + c_m b_m.$$

This equation follows logically from the system, which means that every solution of the system satisfies the equation. Once we know how to solve systems of linear equations, it is easy to show that the converse is true in the case when the system is consistent and all equations are linear equations.

Namely, given a system of linear equations $f_1 = b_1, \ldots, f_m = b_m$, and another linear equation $f_0 = b_0$ in standard form that follows from the system, then either the equation $f_0 = b_0$ or the equation $0 = 1$ is a linear combination of the equations in the system. See §6 of this chapter for the general case and Exercises 38–41 for particular cases. A similar statement is true for linear inequalities if we are careful with the signs of coefficients (see §15 of Chapter 5).

Let us return to our small talk about logic in general. There is no way to list all English equivalents of logical symbols and their uses and abuses. Here is a list of common fallacies:

- $x \geq 0 \Rightarrow x = 0$ or 1 (false dilemma).
- $0 = 0$ so $0 = 1$ (argument from ignorance).
- $x = 1/3$, or $x = 0.33333$ as decimal (slippery slope).
- It is not true that $1 = 1$ or $0 = 1$ (complex conclusion).
- $0 > 1$ because it is what my instructor teaches (appeal to force).
- $0 > 1$ because I spent all night to get this result (appeal to pity).
- The minus in your equation $x^2 = -1$ is a mistake, for otherwise solving it would be waste of time (appeal to consequences).
- Every reasonable person will agree that $1/3 = 0.3$ (prejudicial language).
- Everybody knows that $1/3 = 0.3$ so it is true (appeal to popularity).
- I cannot agree that $0 = 1$ is a linear constraint because you even cannot spell constraint (attacking the person).
- The linear equation $0 = 1$ has a real solution because the *New York Times* wrote so and experts agreed (appeal to authority and anonymous authorities).
- All examples in Chapter 1 have less than 10 constraints so every linear program has less that 10 constraints (hasty generalization).
- I had no difficulties in the first week of classes, so this course is a piece of cake, and I do not need to work hard to pass (unrepresentative sample).
- To solve systems of linear equations, I add and subtract equations, so I will use the same operations to simplify my system of linear constraints (false analogy; while the sum of equations follows from the equations, this is not the case with inequalities of different types).
- Every linear equation for one unknown has a solution (fallacy of exclusion).
- This section follows Chapter 1; therefore, we do not need any logic in Chapter 1 but we need to know what linear programming is to be logical (coincidental correlation).
- Since all constraints in my optimization problem are linear, the objective function should be affine (joint effect; the assumption and the conclusion are both true for linear programs).
- $x > 0$ causes $x > 1$ (wrong direction).
- Complex numbers are not really numbers (equivocation).
- There is no solution for the equation $0 = 1$ (amphiboly, i.e., two different meanings).

• In linear programming all numbers in data and solutions are real, so every linear program has a feasible solution (existential fallacy).

If your head does not spin yet, what do you think about the logic in this statement: "Since I really understood the diet problem with 10 cereals, I am going to eat only cereals from now on." For more examples, see http://www.intrepidsoftware.com/fallacy/toc.htm.

Learning logic is similar to learning to walk: It takes patience and practice. We conclude this section with quotations. It is up to the reader to judge whether there is any logic in them.

Jacques Hadamard (1865–1963):
Logic merely sanctions the conquests of the intuition.

Antoine Arnauld (1612–1694):
Common sense is not really so common.

Ludwig Wittgenstein (1889–1951)
There can never be surprises in logic.

Morris Kline (b. 1908):
Logic is the art of going wrong with confidence.

Lord Dunsany (1878–1957):
Logic, like whiskey, loses its beneficial effect when taken in too large quantities.

Oliver Heaviside (1850–1925):
Logic can be patient, for it is eternal.
Why should I refuse a good dinner simply because I don't understand the digestive processes involved?

G. K. Chesterton (1874–1936):
You can only find truth with logic if you have already found truth without it.

Hermann Weyl (1885–1955):
Logic is the hygiene the mathematician practices to keep his ideas healthy and strong.

Richard Feynman (1918–1988):
... mathematics is not just another language. ... it is a language plus logic. Mathematics is a tool for reasoning.

Jeremy Bentham (1748–1832):
O Logic: born gatekeeper to the Temple of Science, victim of capricious destiny: doomed hitherto to be the drudge of pedants: come to the aid of thy master, Legislation.

Jules Henri Poincaré (1854–1912):
It is by logic we prove, it is by intuition that we invent. Thus, be it understood, to demonstrate a theorem, it is neither necessary nor even advantageous to know what it means. The geometer might be replaced by the "logic piano" imagined by Stanley Jevons; or, if you choose, a machine might be imagined where the assumptions were put in at one end, while the theorems came out at the other, like the legendary Chicago machine where the pigs go in alive and come out transformed into hams and sausages. No more than these machines need the mathematician know what he does.

Bertrand Russell (1872–1970):
Ordinary language is totally unsuited for expressing what physics really asserts, since the words of everyday life are not sufficiently abstract. Only mathematics and mathematical logic can say as little as the physicist means to say.

David van Dantzig (1900–1959):
Neither in the subjective nor in the objective world can we find a criterion for the reality of the number concept, because the first contains no such concept, and the second contains nothing that is free from the concept. How then can we arrive at a criterion? Not by evidence, for the dice of evidence are loaded. Not by logic, for logic has no existence independent of mathematics: it is only one phase of this multiplied necessity that we call mathematics. How then shall mathematical concepts be judged? They shall not be judged. Mathematics is the supreme arbiter. From its decisions there is no appeal. We cannot change the rules of the game, we cannot ascertain whether the game is fair. We can only study the player at his game; not, however, with the detached attitude of a bystander, for we are watching our own minds at play.

Exercises

1–31. Now it is your turn! Determine the validity of the following 30 statements. Write down yes or true if you agree with the statement and write no or false otherwise. Explain your reasoning.

1. $|x| = 1$ only if $x \geq 0$.

2. $xy = 0$ only if $y = 0$.

3. $|x| \leq 1$ if $x \leq 1$.

4. If $|x| \leq 1$, then $x \geq -1$.

5. $x \neq 1$ unless $x \geq 0$.

6. $|x| > 1$ hence $x > 0$.

7. $x \geq 0$ provided that $x > 2$.

8. If $0 = 1$, then $2 = 5$.

9. $x \geq 0 \Leftarrow x > 2$.

10. $x^2 = 0 \iff x = 0$.

11. The condition $x = 1$ does not imply the condition $x \geq 1$.

12. The condition $|x| = 1$ is stronger than the condition $x \geq 0$.

13. The condition $x \geq 0$ follows from the condition $x = 5$.

14. $|x| \leq 1$ if and only if $x \leq 1$ *and* $x \geq -1$.

15. $|x| \leq 1$ if and only if $x \leq 1$ *or* $x \geq -1$.

16. The equation $x^2 = 0$ is equivalent to the constraint $x = 0$.

17. The equation $x = y$ is equivalent to the constraint $x - y = 0$.

18. The constraint $|x| \geq 1$ means that either $x \leq -1$ or $x \geq 1$.

19. $|x| \geq 1$ if and only if $x \leq -1$ and $x \geq 1$.

20. If $xy = 0$ then $x = 0$ or $y = 0$.

21. The second condition in the system $x \geq 1, x \geq 0$ is redundant.

22. Given $0 = 1$, we can conclude that $2 = 5$.

23. The condition $x > 10$ is sufficient for the conclusion $x \geq 0$.

24. The constraint $x = 3$ forces $x \geq 0$.

25. The condition $x^2 > 10$ makes x positive.

26. If $|x - 1| > 4$, then either $x > 5$ or $x < -3$.

27. If $x > y$ then $-x < -y$.

28. The condition $x > 0$ is necessary but not sufficient for $x > 2$.

29. The condition $x = 1$ is weaker than the condition $x \geq 1$.

30. The condition $x = 1$ is a consequence of the condition $x \geq 1$.

31. $|x| \leq 1$ if and only if $x \leq 1$ and $x \geq -1$.   ∎

32–39. Find all implications between the following four conditions:

32.
(i) $x = 2, y = 3$ (ii) $x \geq 0$
(iii) $y \geq 0$ (iv) $x + y = 5$

33.
(i) $x = 2, y = 3$ or $x = y = 0$ (ii) $x, y \geq 0$;
(iii) x, y are integers   (iv) $|x + y| \leq 5$

34.
(i) $0 = 1$ (ii) $0 = 0$
(iii) $1 = 2$ (iv) $x = y$

35.
(i) $x = 1, y = 3$ (ii) $x + y = 3, x - y = 1$
(iii) $x = 1$ or $y = 2$ (iv) $2x + 3y = 8$

36.
(i) $x + y \geq 1, x - y \geq 2$ (ii) $2x \geq 3$
(iii) $2y \geq -1$ (iv) $x, y \geq 0$

37.
(i) $x = 0$ (ii) $x^3 = 0$
(iii) $0 = 0$ (iv) $xy = 0$

38.
(i) $x^2 + y^2 = 1$ (ii) $x^2 + y^2 \leq 1$
(iii) $x \leq 1, y \leq 1$ (iv) $x + y \leq 2$

39.
(i) $|x| + |y| \leq 1$ (ii) $x^2 + y^2 \leq 1$
(iii) $|x| \leq 1, |y| \leq 1$ (iv) $x^4 + y^4 \leq 1$ ■

40. What are other possible replacements for *if* in (4.1)?

41. What are other possible replacements for *implies* in (4.2)?

42–48. Do you agree?

42. This optimization problem is linear, so the objective function must be linear.

43. If every constraint in a linear program is feasible, then the program is feasible.

44. If a linear problem is infeasible, then one of given constraints must be infeasible.

45. If $x = 3$ and $y = -1$, then x is closer to 0 than is y.

46. $x \geq 0$ because $x > y$ and $y > 3$.

47. The constraint $a + 2b + 3c \geq 2$ follows from the system
$$a + b + c \geq 1, b + 2c \geq 1.$$

48. The constraint $a + 2b + 3c \leq 3$ follows from the system
$$a + b + c \leq 1, b + 2c \leq 1.$$ ■

49–52. Check whether the third equation follows from the first two equations, (i.e., it is redundant in the system). If it is, write it as a linear combination of the first two equations.

49. $\begin{cases} x = 0, \\ 2y = 5, \\ x + y = 2. \end{cases}$ **50.** $\begin{cases} x + y + z = 1, \\ x + 2y + 3z = 3, \\ 2x + 3y + 4z = 4. \end{cases}$

51. $\begin{cases} a - b + c + d = 0, \\ 2a + b - 3c = 1, \\ 3b - 5c - 2d = 1. \end{cases}$ **52.** $\begin{cases} a - b + c + d = 0, \\ 2a + b - 3c = 1, \\ -a - 2b + 4c + d = -1. \end{cases}$

§5. Matrices

Matrices are used often in linear algebra and linear programming. They allow us to write down systems of linear equations and inequalities in an abbreviated form. We will start by defining matrices and then we will see how linear programming problems can be written in a kind of "shorthand" notation by using matrices.

Definition 5.1. A *matrix* is a rectangular array of entries, which can be numbers, variables, polynomials, functions, and so on.   ∎

In general, a matrix A can be written in the following form:

$$A = \begin{bmatrix} a_{11} & a_{12} & \cdots & a_{1n} \\ a_{21} & a_{22} & \cdots & a_{2n} \\ \vdots & \vdots & \ddots & \vdots \\ a_{m1} & a_{m2} & \cdots & a_{mn} \end{bmatrix},$$

where $[a_{11} \quad a_{12} \quad \cdots \quad a_{1n}]$ is referred to as the first row of the matrix, $[a_{21} \quad a_{22} \quad \cdots \quad a_{2n}]$ is the second row, and so forth. Similarly,

$$\begin{bmatrix} a_{11} \\ a_{21} \\ \vdots \\ a_{m1} \end{bmatrix}$$

is the first column,

$$\begin{bmatrix} a_{12} \\ a_{22} \\ \vdots \\ a_{m2} \end{bmatrix}$$

is the second column, and so forth. In this case we say that the matrix A has m rows and n columns. We refer to A as an $m \times n$ matrix. We denote by a_{ij} the entry in the ith row and the jth column.

When $m = 1$ (i.e., our matrix has only one row), we have a row matrix or just a row, also called a row vector or just vector. When $n = 1$, we have a column matrix or a column, also referred to as a column vector or just vector.

Remark. The subscript in a_{ij} means a pair of numbers rather than a product. Use $a_{12,3}$ or $a_{1,23}$ instead of a_{123} for large m, n. We often uses commas in row matrices to avoid mix-ups. Compare $[1\ 2\ 34]$ and $[1, 2, 34]$.   ∎

Example 5.2.

(i) $\quad A = \begin{bmatrix} 1 & -1 & 3 \\ \sqrt{2} & \pi & 6 \\ -4 & 56 & 20 \end{bmatrix}$

is a 3×3 matrix with real entries.

(ii) $\quad B = \begin{bmatrix} 1 & x & x^2 \\ 1 & y^2 & y^4 \end{bmatrix}$

is a 2×3 matrix with polynomial entries.

(iii) $\quad \begin{bmatrix} x \\ y^2 \end{bmatrix}$ is the second column of B.

(iv) $\quad C = \begin{bmatrix} \sin x & e^x & \ln x & x+1 \\ \sqrt{x^2 - \pi^2} & 3/4 & 1/x & 5 \\ e^{7\cos x} & \tan \pi x & 2 & 5 \end{bmatrix}$

is a 3×4 matrix with functional entries.

(v) $\quad \begin{bmatrix} \sqrt{x^2 - \pi^2} & \dfrac{3}{4} & \dfrac{1}{x} & 5 \end{bmatrix}$ is the second row of C.

(vi) If we denote the entries of the matrix C by c_{ij}, where $1 \leq i \leq 3$ and $1 \leq j \leq 4$, then, since $\tan \pi x$ is in the third row and the second column, it is the entry c_{32}.

Definition 5.3. Two matrices A and B are *equal* if they have the same size and each entry of A is equal to the corresponding entry of B. ∎

In linear programming we also use inequalities between rows (or columns) of the same size. Unless stated otherwise, $A \geq B$ for two vectors of the same size means that every entry of A is greater than or equal to the corresponding entry of B.

Recall your introduction to the set of rational or real numbers. After the set was defined, the operations of addition, subtraction, multiplication, and division and their properties were explained. For instance, both addition and multiplication of numbers are commutative and associative; we can link addition and multiplication via the property known as distributivity of multiplication over addition. Let us do something similar with our newly defined objects called matrices.

Definition 5.4. The operation of *addition* of two $m \times n$ matrices, A and B, is defined component-wise; that is, the entry in the ith row and the jth column of the $m \times n$ matrix $A + B$ is $a_{ij} + b_{ij}$. Subtraction is defined in a similar way. ∎

It is easy to verify that matrix addition is commutative and associative.

Definition 5.5. The *product* of matrices, A and B, is defined only when the number of columns of A equals the number of rows of B. If A is an $m \times n$ matrix and B is an $n \times p$ matrix, the product $A \cdot B$ (or just AB) is an $m \times p$ matrix, whose entry in the ith row and the jth column is given by

$$(A \cdot B)_{ij} = \sum_{k=1}^{n} a_{ik} b_{kj}$$ ∎

Because of the nature of the definition of the product of two matrices, this operation is not commutative. The product in the reverse order, $BA = B \cdot A$, may not be defined, and even if it is, it may differ from $A \cdot B$.

However, matrix multiplication is associative. What do we mean by that? Let A be an $m \times n$ matrix, B an $n \times p$ matrix, and C a $p \times q$ matrix. Then the products $A \cdot B$, $(A \cdot B) \cdot C$, $B \cdot C$, and $A \cdot (B \cdot C)$ are all defined and, moreover,

$$(A \cdot B) \cdot C = A \cdot (B \cdot C).$$

Recall from your knowledge of the properties of real numbers that distributivity of multiplication over addition gives us the equality

$$a \cdot (b + c) = a \cdot b + a \cdot c.$$

Distributivity of matrix multiplication over addition holds as well, provided that the appropriate sums and products of matrices are defined. For example, if A is an $m \times n$ matrix, and B and C are two $n \times p$ matrices, then, $A \cdot (B + C)$ yields the same $m \times p$ matrix as does $A \cdot B + A \cdot C$. Therefore,

$$A \cdot (B + C) = A \cdot B + A \cdot C.$$

Proofs of these claims can be found in any linear algebra textbook, so we will omit them. Here are some more examples.

Example 5.6.

Let A and B be the following matrices:

$$A = \begin{bmatrix} 1 & -1 & 3 \\ 2 & 1/2 & 6 \end{bmatrix}, \quad B = \begin{bmatrix} 1/2 & -3/4 & 7 \\ 2 & 3 & 8 \end{bmatrix}.$$

   (*i*)   Then their sum, $A + B$, and their difference, $A - B$, are given by

$$A + B = \begin{bmatrix} 3/2 & -7/4 & 10 \\ 4 & 7/2 & 14 \end{bmatrix} \text{ and } A - B = \begin{bmatrix} 1/2 & -1/4 & -4 \\ 0 & -5/2 & -2 \end{bmatrix}.$$

Example 5.7.

Let C and D be the following 2×3 and 3×3 matrices:

$$C = \begin{bmatrix} 1 & 2 & 3 \\ 4 & 5 & 6 \end{bmatrix} \text{ and } D = \begin{bmatrix} 9 & 8 & 7 \\ 6 & 5 & 4 \\ 3 & 2 & 1 \end{bmatrix}.$$

Then the product, $C \cdot D$, is the following 2×3 matrix:

$$\begin{bmatrix} 1\cdot 9 + 2\cdot 6 + 3\cdot 3 & 1\cdot 8 + 2\cdot 5 + 3\cdot 2 & 1\cdot 7 + 2\cdot 4 + 3\cdot 1 \\ 4\cdot 9 + 5\cdot 6 + 6\cdot 3 & 4\cdot 8 + 5\cdot 5 + 6\cdot 2 & 4\cdot 7 + 5\cdot 4 + 6\cdot 1 \end{bmatrix}$$

$$= \begin{bmatrix} 30 & 24 & 18 \\ 84 & 69 & 54 \end{bmatrix}$$

Notice that the product $D \cdot C$ is not defined. Do you see why?

Example 5.8..

Let A be the 2×3 matrix

$$A = \begin{bmatrix} 1 & -2 & 0 \\ -3 & 4 & 1 \end{bmatrix},$$

x the 3×1 column matrix

$$x = \begin{bmatrix} x_1 \\ x_2 \\ x_3 \end{bmatrix},$$

and b the 2×1 column matrix

$$b = \begin{bmatrix} -5 \\ 6 \end{bmatrix}.$$

   Then the product $A \cdot x$ is a well-defined 2×1 column matrix. Therefore, the matrix equality $A \cdot x = b$ makes sense, and, by performing these operations, we obtain the system of linear equations

$$\begin{cases} x_1 - 2x_2 = -5 \\ -3x_1 + 4x_2 + x_3 = 6. \end{cases}$$

Conversely, the prededing system of equations can be written in the matrix form $A \cdot x = b$. This observation leads the following elementary but important result:

Any system of linear equations can be written in the standard matrix form

$$Ax = b,$$

where x *is the column of distinct unknowns (variables),* A *is a given matrix (the coefficient matrix), and* b *is a column of given numbers (constant terms).*

Note that the number of equations need not be equal to the number of variables. In other words, the coefficient matrix need not be a square matrix.

Definition 5.9. The *transpose* matrix of an $m \times n$ matrix A is the $n \times m$ matrix whose entry in the ij-position is the entry in the ji-position of the original matrix A. We denote the transpose of A by A^T. Thus, $(A^T)_{ij} = a_{ji}$. ∎

Proposition. Let A be an $m \times n$ matrix and B an $n \times p$ matrix. Then $A \cdot B$ and $B^T \cdot A^T$ are defined and $(A \cdot B)^T = B^T \cdot A^T$.

Sketch of Proof

    (i) Compute the product $A \cdot B$.

    (ii) Find the transpose of the matrix $A \cdot B$.

    (iii) Calculate the product $B^T \cdot A^T$.

    (iv) Compare the entries of the matrices $(A \cdot B)^T$ and $B^T \cdot A^T$.

Example 5.10.

The transpose of the 3×4 matrix

$$C = \begin{bmatrix} \sin x & e^x & \ln x & x+1 \\ \sqrt{x^2 - \pi^2} & 3/4 & 1/x & 5 \\ e^{7\cos x} & \tan \pi x & 2 & 5 \end{bmatrix}$$

is the following 4×3 matrix:

$$C^T = \begin{bmatrix} \sin x & \sqrt{x^2 - \pi^2} & e^{7\cos x} \\ e^x & 3/4 & \tan \pi x \\ \ln x & 1/x & 2 \\ x+1 & 5 & 5 \end{bmatrix}.$$

∎

The real numbers 0 and 1 are distinguished elements for addition and multiplication, respectively, because adding 0 to a number does not affect the number and, similarly, multiplication of any number by 1 yields the same number with which we started. We define matrices with similar properties.

Definition 5.11. The zero matrix 0 is the $m \times n$ matrix with entries consisting of zeros only. ■

Remark. The size of a zero matrix is often clear from context and need not to be specified. For example, the first 0 in the inequality $[0, 1, 2, 3] \geq 0$ means a number, while the second zero can be understood as an 1×4 matrix. In the equality

$$[1, -1, 2] \begin{bmatrix} 3 \\ 5 \\ 1 \end{bmatrix} = 0$$

the 0 on the right-hand side means a 1×1 matrix or a number.   ■

The *additive inverse* of an $m \times n$ matrix, A, is the $m \times n$ matrix consisting of the additive inverses of the entries of A; we denote the additive inverse of A by $-A$. Thus, $A + (-A) = 0$ for a matrix A of any size. By contrast, not every matrix has a multiplicative inverse. Recall that, even for real numbers, the number 0 is not invertible.

Definition 5.12. For a positive integer n we define the $n \times n$ identity matrix, 1_n, as follows: Its diagonal entries are ones and the other entries are zeros.   ■

Recall that the diagonal entries of a matrix $[a_{ij}]$ are a_{ii}. The identity matrix 1_n is a square diagonal matrix. In general, a *diagonal matrix* is defined as a matrix with all nondiagonal entries $= 0$.

Here is an example of a diagonal 2×3 matrix:

$$\begin{bmatrix} 1 & 0 & 0 \\ 0 & 1 & 0 \end{bmatrix}.$$

Here are two matrices that are not diagonal (can you see why?):

$$\begin{bmatrix} 1 & 2 \\ 0 & 1 \\ 5 & 0 \end{bmatrix}, \quad \begin{bmatrix} 0 & 1 & -1 & \pi & 1/2 \\ 0 & 0 & 1 & 3 & 4 \\ 0 & 0 & 0 & 0 & 1 \\ 0 & 0 & 0 & 0 & 0 \end{bmatrix}.$$

The 2×2 identity matrix 1_2 is the following matrix:

$$I_2 = \begin{bmatrix} 1 & 0 \\ 0 & 1 \end{bmatrix}.$$

Here is the 5×5 identity matrix:

$$I_5 = \begin{bmatrix} 1 & 0 & 0 & 0 & 0 \\ 0 & 1 & 0 & 0 & 0 \\ 0 & 0 & 1 & 0 & 0 \\ 0 & 0 & 0 & 1 & 0 \\ 0 & 0 & 0 & 0 & 1 \end{bmatrix}.$$

Definition 5.13. An $n \times n$ matrix, A, is said to be *invertible* if there exists an $n \times n$ matrix B such that $A \cdot B = B \cdot A = I_n$.

Proposition 5.14. Let A be an invertible $n \times n$ matrix. Then the inverse of A is unique.

Proof. Suppose that we have two inverses, B and C, for the invertible matrix A. By the definition, we have $A \cdot B = I_n$ and $C \cdot A = I_n$. Therefore, since, $C = C \cdot I_n$ and $A \cdot B = I_n$, we obtain $C = C \cdot (A \cdot B)$. Since matrix multiplication is associative, the latter product equals $(C \cdot A) \cdot B$ and this equals $I_n \cdot B = B$, because $C \cdot A = I_n$. Combining these equalities, we obtain that $C = B$. ∎

From now on, we will use the notation A^{-1} to denote the unique inverse of the $n \times n$ invertible matrix A. Before we continue this presentation of matrices, here are some questions worth thinking about:

(i) How can we calculate the inverse of a matrix?

(ii) Does every nonzero $n \times n$ matrix, A, have an inverse?

(iii) If not, what conditions should the $n \times n$ matrix, A, satisfy in order to have an inverse?

In order to answer these questions, we need to define some operations on matrices that do not have an analog to operations in the set of real numbers.

Definition 5.15. Let A be an $m \times n$ matrix. An *elementary row (column) operation* on A is one of the following procedures, performed on the rows (columns) of A:

(i)   Interchange two rows (columns) of A.

(ii)   Multiply a row (column) of A by a nonzero number.

(iii)   Add a multiple of a row (column) of A by a number to another row (column). ∎

Example 5.16. The result of interchanging the first and the third rows of

$$B = \begin{bmatrix} 1 & 2 \\ 3 & 4 \\ 5 & 6 \end{bmatrix}$$

is the matrix

$$B = \begin{bmatrix} 5 & 6 \\ 3 & 4 \\ 1 & 2 \end{bmatrix}.$$

Note that $B = PA$ and $A = PB$, where

$$P = \begin{bmatrix} 0 & 0 & 1 \\ 0 & 1 & 0 \\ 0 & 0 & 1 \end{bmatrix}$$

is a permutation matrix. The same permutation operation takes us from B back to A.

Example 5.17. By multiplying the second row of

$$A = \begin{bmatrix} 1 & 2 & 3 \\ 4 & 5 & 6 \end{bmatrix}$$

by $\sqrt{2}$, we obtain the matrix

$$B = \begin{bmatrix} 1 & 2 & 3 \\ 4\sqrt{2} & 5\sqrt{2} & 6\sqrt{2} \end{bmatrix}.$$

Note that $B = DA$ and $A = D^{-1}B$, where

$$D = \begin{bmatrix} 1 & 0 \\ 0 & 4\sqrt{2} \end{bmatrix}, \quad D^{-1} = \begin{bmatrix} 1 & 0 \\ 0 & \sqrt{2}/8 \end{bmatrix}$$

are diagonal matrices that differ from the identity matrix 1_2 in one diagonal entry. Another row multiplication operation takes us from B back to A.

Example 5.18. The matrix

$$B = \begin{bmatrix} 1 & 2 \\ -12 & -14 \\ 5 & 6 \end{bmatrix}$$

is obtained by replacing the second row of

$$A = \begin{bmatrix} 1 & 2 \\ 3 & 4 \\ 5 & 6 \end{bmatrix}$$

by the sum of its second row and -3 times its third row. A similar row addition operation, with -3 replaced by 3, takes us from B back to A. This row addition operation corresponds to multiplication by an elementary matrix on the left: $B = EA$ and $A = E^{-1}B$, where

$$E = \begin{bmatrix} 1 & 0 & 0 \\ 0 & 1 & 0 \\ 0 & -3 & 1 \end{bmatrix}, \quad E^{-1} = \begin{bmatrix} 1 & 0 & 0 \\ 0 & 1 & 0 \\ 0 & 3 & 1 \end{bmatrix}.$$

This elementary matrix differs from the identity matrix in one off-diagonal entry.
∎

In general, row (column) elementary operations correspond to multiplication by invertible matrices from the left (right).

Remark 5.19. For an invertible matrix A, the linear system $Ax = b$ has exactly one solution $x = A^{-1}b$. In general, solving the system $Ax = b$ is related with inverting a submatrix A of maximal possible size.

Remark 5.20. Sometimes numbers are called *scalars* to distinguish them from vectors. However, numbers can be also considered as 1×1 matrices; hence they can be considered as both row and column vectors. Addition and multiplication of numbers agrees with matrix addition and multiplication. However, it is tricky to interpret multiplication of a matrix by a scalar in terms of matrix multiplication. An interpretation involves *scalar matrices,* which are square diagonal matrices with the same diagonal entries.

Vectors with n components, written in a row or column, can be thought of as points or arrows (directed line segments) sticking out of the origin in an n-dimensional space. This language came from the cases $n = 1, 2, 3$, where those objects appear in geometry and mechanics. Matrices appear in connection with some geometric transformations.
∎

In conclusion, here is a quotation about matrices by Irving Kaplansky (from *Paul Halmos: Celebrating 50 Years of Mathematics,* New York: Springer-Verlag, 1991):

> We [he and Halmos] share a philosophy about linear algebra: we think basis-free, we write basis-free, but when the chips are down we close the office door and compute with matrices like fury.

Exercises

1–8. Let A equal the 1×4 matrix $[1, 2, 0, -3]$ and B equal the 1×4 matrix $[0, -1, -2, 4]$. Compute

1. $2A + 3B$ **2.** AB^T
3. BA^T **4.** $A^T B$
5. $B^T A$ **6.** $(A^T B)^2$
7. $(A^T B)^3$ **8.** $(A^T B)^{1000}$

9. Could you compute the products AB or BA? Why or why not?

10. Find a square matrix A such that $A \neq 0$ but $A^2 = 0$.

11. Find two matrices, A and B, such that both matrix products AB and BA are defined, $AB = 0$ and $BA \neq 0$.

12–14. Write the system of equations in a matrix form. *Hint :* some of these equations are not linear equations in standard form in the sense of Definition 1.3.

12. $5a + 2b + 3c - d = 1, -b + a - 3c = 2, 6c + a - 3c + 1 = a$.
13. $x + 2b + 3c - y = 1, -b + a - 3c = 2y, x + a - 3c + 1 = a$.
14. $x + 2b + 3c - y = 1, -b + a - 3c = 2y + d, x + 6a - y + 1 = a$.

15–17. Solve the systems 12–14.

18–25. Do Exercises 1–8 for matrices

$$A = [0, 1, -1, 0, 0, 2] \text{ and } B = [-2, 1, 0, 0, 3, 2].$$

26–33. Do Exercises 1–8 for matrices

$$A = [1, 1, -1, 0, 0, 2, 0] \text{ and } B = [-1, 1, 0, 3, 3, 2, -1].$$

34. Do Exercises 2–7 for matrices A, B in Example 5.6.

35. For matrix C in Example 5.7, compute $E_1 C$ and $E_2 C$, where $E_1 = \begin{bmatrix} 3 & 0 \\ 0 & -2 \end{bmatrix}$ (a diagonal matrix), $E_2 = \begin{bmatrix} 1 & 5 \\ 0 & 1 \end{bmatrix}$ (an elementary matrix). Also compute $(E_1)^2, (E_1)^9, (E_2)^{40}$.

36. For matrices C, D in Example 5.7, compute $CE_1, CE_2, DE_1,$ $DE_2, E_1 E_2, E_2 E_1, (E_1)^2, (E_1)^9, (E_2)^{40}$, where $E_1 = \begin{bmatrix} 2 & 0 & 0 \\ 0 & 3 & 0 \\ 0 & 0 & 4 \end{bmatrix}$ (a diagonal matrix), $E_2 = \begin{bmatrix} 1 & 0 & 0 \\ 0 & 1 & 0 \\ -3 & 0 & 1 \end{bmatrix}$ (an elementary matrix).

The point of Exercises 35–36 was to realize how to multiply by elementary and diagonal matrices (definition will be reminded in the next section). In general, zeros in a matrix help when you multiply by this matrix.

37. Let α be an invertible $m \times m$ matrix, β a $m \times n$ matrix, γ a $n \times m$ matrix, and δ a $n \times n$ matrix. Compute

$$\begin{bmatrix} 1_m & 0 \\ -\gamma\alpha^{-1} & 1_n \end{bmatrix} \begin{bmatrix} \alpha & \beta \\ \gamma & \delta \end{bmatrix} \begin{bmatrix} 1_m & -\alpha^{-1}\beta \\ 0 & 1_n \end{bmatrix}.$$

38. A matrix $A = [a_{i,j}]$ is called *diagonal* if $a_{i,j} = 0$ whenever $i \neq j$. Show that the sum of two diagonal matrices of the same size is a diagonal matrix. Show that the product of two diagonal matrices, when defined, is a diagonal matrix. Show that $AB = BA$ for two $n \times n$ diagonal matrices A, B.

39. A matrix $A = [a_{i,j}]$ is called *upper triangular* if $a_{i,j} = 0$ whenever $i > j$. Show that the sum of two upper triangular matrices of the same size is an upper triangular matrix. Show that the product of two upper triangular matrices, when defined, is an upper triangular matrix. Show that $AB \neq BA$ for some $n \times n$ upper triangular matrices A, B.

40. A matrix $A = [a_{i,j}]$ is called *lower triangular* if $a_{i,j} = 0$ whenever $i < j$. Show that the sum of two lower triangular matrices of the same size is a lower triangular matrix. Show that the product of two lower triangular matrices, when defined, is a lower triangular matrix. Show that $AB \neq BA$ for some $n \times n$ lower triangular matrices A, B.
41–44. Obtain a diagonal matrix by row and column elementary operations with the matrix A. *Remark*: The diagonal matrix in the answer is not unique, but the number of zero columns in it is unique. In many textbooks on linear algebra it is proved that this number equals the dimension of the space of solutions of $Ax = 0$.

41. $A = \begin{bmatrix} 1 & -2 & -1 & 0 \\ 5 & 1 & 3 & 1 \end{bmatrix}$

42. $A = \begin{bmatrix} 1 & 0 & -1 \\ -5 & 1 & 3 \\ 2 & 4 & 5 \\ 8 & -2 & -1 \end{bmatrix}$

43. $A = \begin{bmatrix} 1 & -2 & -1 \\ 5 & 1 & 3 \\ 2 & 4 & 5 \\ -1 & -2 & 0 \\ 8 & -2 & -1 \end{bmatrix}$

44. $A = \begin{bmatrix} 1 & -2 & -1 & 0 \\ 5 & 1 & 3 & 1 \\ 5 & 6 & 0 & 7 \end{bmatrix}$!!

§6. Systems of Linear Equations

The main object of this section is to recall how to solve systems of linear equations. To save paper we use matrices.

One of the main applications of elementary operations is solving system of linear equations. Every system of linear equations can be written in a matrix form $Ax = b$, where A is a given matrix, x is a column of distinct variables (unknowns), and b is a column of given numbers. The coefficient matrix A and the column b can be put together as the augmented matrix $[A|b]$.

Example 6.1.

The system $\begin{cases} x - 2z = -3 \\ 2x + 5y - 4z = 5 \end{cases}$ can be written as

$$\begin{bmatrix} 1 & 0 & -2 \\ 2 & 5 & -4 \end{bmatrix} \begin{bmatrix} x \\ y \\ z \end{bmatrix} = \begin{bmatrix} -3 \\ 5 \end{bmatrix}.$$

The augmented matrix is $\begin{bmatrix} 1 & 0 & -2 & -3 \\ 2 & 5 & -4 & 5 \end{bmatrix}.$ ∎

A row multiplication operation with the augmented matrix corresponds to multiplication of an equation in the system by a nonzero number. A row addition operation corresponds to adding a multiple of an equation to another equation. Interchange of two rows corresponds to interchange of two equations. Interchange of two columns correspond to interchange of two variables.

These operations with equations take the system to an equivalent system (with the same set of variables and the same solution set). Therefore, elementary row operations, applied to the augmented matrix to bring it to a "simplified" form, can be used to solve an arbitrary system of linear equations. For example, if the coefficient matrix is diagonal (see Exercise 38 of the previous section), then the system splits into independent set of linear equations in one variable each and hence can be solved easily.

The other column elementary operations correspond to changing of variables. They are not needed to solve systems of linear equations where the coefficients are numbers (rather than functions, differential operators, etc.) but could be useful for this purpose and for other purposes. If they are used for solving a system of linear equation, additional computations should be done to keep track of changes in variables, so in the end we can write the final answer in terms of the original variables. To keep track of those changes, we usually augment the augmented matrix $[A|b]$ by the identity matrix 1_n and start with the matrix

$$\begin{bmatrix} A & b \\ 1_n & \end{bmatrix},$$

where n is the number of variables. By row and column addition operations the coefficient matrix can be taken to diagonal form. Then, using multiplication operations, it is easy to finish solving our system in new variables. Then the matrix in the place of the additional 1_n is used to write a final answer in terms of original variables.

Since changes of variables may result in additional computations, many textbooks avoid them and work only with row operations. By row operations, every coefficient matrix can be brought to the so-called echelon form and then to a reduced echelon form. Then it is not so difficult to finish solving the system. An echelon form is an upper triangular matrix (see Exercise 39 of the previous section) with special properties.

In this section we explain how to use row elementary operations and column interchange operations to make any matrix diagonal. In this way, we keep the same set of variables, and all linear systems on the way from the original one to the final answer are equivalent in the sense that they have the same solution set.

A well-known Gauss-Jordan elimination first uses forward substitutions ("going down" row addition operations) to get an upper diagonal matrix U (sometimes column permutations are needed, and all zeros on the main diagonal of U are in the end), and then it uses backward substitutions ("going up" row addition operations) to get a diagonal matrix (zero rows in the coefficient matrix may give some complications preventing us from obtaining a diagonal matrix). So this method uses column permutations but no column addition or multiplication operations. An interpretation of this method is that we write our coefficient matrix A of size $m \times n$ as $A = LUP$, where L is a lower triangular $m \times m$ matrix with ones along the main diagonal, P an $n \times n$ permutation matrix, and U an upper triangular $m \times n$ matrix.

Actually, use of column operations and the corresponding permutations of variables can be minimized as follows. By row addition operations we can always bring any coefficient matrix of any size $m \times n$ to an upper triangular form (the entries below the main diagonal are zeros). A way to do so is as follows. First we work on the first column and make, if possible, its first entry nonzero. Then we use this entry as a pivot entry to kill (eliminate, i.e., make zero) all other entries. Then we do the same with the submatrix obtained by ignoring the first row, and so on.

Now we consider separately the following four cases:

Case 1: The diagonal entries of this upper triangular matrix U are all nonzero and the matrix is square (that is, the number n of unknowns equals the number of equations m).

Case 2: The diagonal entries of U are all nonzero and $n > m$.

Case 3: The diagonal entries of U are all nonzero and $n < m$.

Case 4: A diagonal entry of U is zero.

In Case 1, by n row multiplication operations we can make all n diagonal entries equal to one. Then by a few [at most $n(n-1)/2$] "going up" row addition operations, we make the coefficient matrix to be 1_n. Thus, the augmented matrix becomes $[1_n|b']$. The system is now solved, and the only solution is $x = b'$. In fact, we just dealt with the case when the coefficient matrix A is invertible, and we described a way to find the answer $x = A^{-1}b = b'$.

In Case 2, we transform as before the submatrix of U consisting of the first m rows and columns to 1_m. The augmented matrix becomes $[1_m, c|b']$ and our final answer is $y = cz + b'$ with arbitrary z, where y is the first m variables in x and z is the rest of variables.

In Case 3, the last $m-n$ rows of the coefficient matrix are zeros. If the same is true for the last $m-n$ rows of the augmented matrix, these rows say $0 = 0$, and they are redundant. Dropping them, we are reduced to Case 1, so we can find the unique solution by n row multiplication operations and a few row addition operations. On the other hand, if not all last $m-n$ entries in the last column of the augmented matrix are zeros, then we have a linear equation of the type $0 = $ nonzero number; hence our system has no solutions.

Finally, in Case 4, we start to use column permutation operations. To keep track of them, we write variables on top of the coefficient matrix, and when we permute columns we permute the corresponding labels on the top. By row addition and the column permutation operations, we bring the augmented matrix to the form

$$
\begin{array}{cc}
y^T & z^T \\
\end{array}
$$
$$
\left[\begin{array}{cc|c}
U & c & b' \\
0 & 0 & b''
\end{array}\right],
$$

where U is a upper triangular square matrix with nonzero diagonal entries, y is some variables in x labeling U, z is the rest of variables, and $[0, 0]$ stands for a zero row or several zero rows in the new coefficient matrix. If $b'' \neq 0$, (i.e., an entry of b'' is nonzero), then we have an equation of the type $0 = $ nonzero number; hence our system has no solutions. Otherwise, we drop the zero rows of the augmented matrix and we are reduced to Case 2. ∎

Thus, a complete answer to solving a system $Ax = b$ of linear equations has one of the following standard forms:
- $0 = 1$ (i.e., the system has no solutions)
- $x = d$ (the system has exactly one solution)
- $z = Cy + d$ (the system has infinitely many solutions), where the column z consists of some variables in x, the column y consists of the rest of variables.

In the last case C is a constant $k \times l$ matrix, where k is the number of variables in z and l is the number of variables in y, and d is a constant column with k entries.

The number k is known as the rank of A. The number l is called the *dimension of the set of solutions*. The variables in y take arbitrary values, and those values determine the values of variables in z. In the second case, when there is only one solution, we say that the dimension of the solution set is 0.

It is important to understand that in all three cases the equation or system of equations in the answer is equivalent to the original system $Ax = b$ in the sense that these two systems have exactly the same set (or space) of solutions. It may happen that the system $Ax = b$ can be solved for different subsets of variables y, but the numbers k, l are the same for all correct answers.

Remark 6.2. Solving a system of linear equations usually means a complete description of all solutions (or showing that they do not exist). In contrast, solving an optimization problem usually means finding an optimal solution and the optimal value (or showing that they do not exist).

Remark 6.3. In some textbooks, the final answer for solving a system $Ax = b$ is given in the form $x = Ct + d$, where t is a column of l variables (parameters) distinct from all variables in x. In this case, instead of discussing what is the equivalence of two systems with different sets of variables, it is better to talk about a 1-1 correspondence between two solution spaces given in a certain explicit way. Transformations of the form $t \mapsto Ct + d$ are known as *affine transformations*. Thus, the solution space (when nonempty) can be identified with all l-tuples of numbers by an affine transformation. ∎

We described the process of solving system of linear equations without appealing to the concepts of vector space and dimension.

Humans were solving those systems for at least 2000 years before these concepts and the word *algebra* appeared. By the way, the original meaning of this word was "reduction," and its main subject was solving system of equations by elimination, reducing the number of variables.

However these concepts do give an additional insight. A linear combination of vectors (say, rows) $v_1, \ldots, v_l$ with coefficients $c_1, \ldots, c_l$ is $c_1 v_1 + \ldots + c_l v_l$. Vectors are called linearly dependent if one of them is a linear combination of others. The rank of a matrix is the maximal number of its linearly independent rows (or columns). A system of $Ax = b$ of linear equations has a solution if and only if the rank of the coefficient matrix A equals the rank of the augmented matrix $[A|b]$. (This is a very easy exercise.) If the ranks are the same, then the dimension of the solution set equals the number of columns in A minus the rank.   ■

Now we give a few examples with solutions.

Problem 6.4. Solve the system of linear equations for x, y, z in Example 6.1.

Solution. We start with the augmented matrix

$$\begin{bmatrix} 1 & 0 & -2 & -3 \\ 2 & 5 & -4 & 5 \end{bmatrix}.$$

Adding the first row to the second row with coefficient -2 (i.e., replacing the second row by the second row minus 2 time the first row), we obtain a upper triangular matrix

$$\begin{bmatrix} 1 & 0 & -2 & -3 \\ 0 & 5 & 0 & 11 \end{bmatrix}$$

with nonzero diagonal entries 1, 5. So we are in Case 2. Multiplying the second row by $1/5$, we obtain the final matrix

$$\begin{bmatrix} 1 & 0 & -2 & -3 \\ 0 & 1 & 0 & 11/5 \end{bmatrix}.$$

Now we can write our answer with names of variables : $x = -3 + 2z, y = 11/5 = 2.2, z$ arbitrary. We did not write the names of variables on the top of the coefficient matrix because we did not change them. However, in case we need to be reminded how the augmented matrix is related to the system of equations, we can decorate the matrix with additional information:

$$\begin{array}{ccc} x & y & z \end{array} \;|\; =$$
$$\begin{bmatrix} 1 & 0 & -2 & | & -3 \\ 0 & 1 & 0 & | & 11/5 \end{bmatrix}.$$

Problem 6.5. Solve the system $x + 2y = 1, 3x + 6y = 2$.

Solution. Here is the augmented matrix:

$$\begin{bmatrix} 1 & 2 & 1 \\ 3 & 6 & 2 \end{bmatrix}.$$

Adding the first row to the second row with coefficient -3, we obtain a upper triangular matrix

$$\begin{bmatrix} 1 & 2 & 1 \\ 0 & 0 & -1 \end{bmatrix}.$$

Looking at the second row, we conclude the system has no solutions. A shorter way to write down the answer is $0 = 1$.

Problem 6.6. Solve

$$\begin{cases} x + 2y = 1, \\ 3x + 6y = 3. \end{cases}$$

Solution. Doing the same row addition as in Problem 6.5, we obtain a zero row in the augmented matrix. Dropping this row, we obtain the final augmented matrix

$$[1 \; 2 \; |1].$$

Answer: $x = -2y + 1, y$ arbitrary.

Problem 6.7. Solve $ax = b$ for x, where a, b are given numbers.

Answer. If $a \neq 0$, then $x = b/a$. If $a = b = 0$, then x is arbitrary. If $a = 0 \neq b$, there are no solutions.

Problem 6.8. Solve the system

$$\begin{cases} x + 2y + 3z = 4, \\ z = 2. \end{cases}$$

Solution. If we write the augmented matrix for x, y, z, namely,

$$\begin{bmatrix} 1 & 2 & 3 & 4 \\ 0 & 0 & 1 & 2 \end{bmatrix},$$

we find that we are in Case 4. Permuting y and z, we obtain the matrix

$$\begin{array}{ccc} x & z & y \end{array}$$
$$\left[\begin{array}{ccc|c} 1 & 3 & 2 & 4 \\ 0 & 1 & 0 & 2 \end{array}\right].$$

Adding the second row to the first one with coefficient -3, we obtain our final matrix:

$$\begin{array}{ccc} x & z & y \end{array}$$
$$\left[\begin{array}{ccc|c} 1 & 0 & 2 & -2 \\ 0 & 1 & 0 & 2 \end{array}\right].$$

Now we can write our final answer in the standard form

$$\begin{cases} x = -2y - 2, \\ z = 2 \end{cases}$$

where y is arbitrary. ■

Note that in this case passing to matrices did not result in saving time and room. The addition operation we used is the same as substitution of the second equation into the first one. However, for large systems with many nonzero coefficients, we save time and room when we use matrices to display information, by avoiding writing names of variables unnecessarily.

Problem 6.9. Solve for x_1, x_3, x_5:

$$\begin{cases} 2x_1 + 3x_2 + x_5 + 5 - 2x_2 = x_1 + x_4 + 1, \\ 2x_3 + x_5 + 2 = -3x_1 - 5x_2 + 3, \\ 3x_3 + 3x_2 - 1 = -x_5 - 6x_4 - 1. \end{cases}$$

Solution. The linear equations are not in standard form. We write them in standard form, and here is the augmented matrix:

$$\left[\begin{array}{ccccc|c} x_1 & x_3 & x_5 & x_2 & x_4 & = \\ 1 & 0 & 1 & 1 & -1 & -4 \\ 3 & 2 & 1 & 5 & 0 & 1 \\ 0 & 3 & 1 & 3 & 6 & 0 \end{array}\right].$$

By two downward addition operations, we make the matrix upper triangular:

$$-3 \downarrow \begin{array}{c} \leftarrow \\ \rightarrow \end{array} \left[\begin{array}{ccccc|c} 1 & 0 & 1 & 1 & -1 & -4 \\ 3 & 2 & 1 & 5 & 0 & 1 \\ 0 & 3 & 1 & 3 & 6 & 0 \end{array}\right]$$

$$\Downarrow$$

$$-3/2 \downarrow \begin{array}{c} \leftarrow \\ \rightarrow \end{array} \begin{bmatrix} 1 & 0 & 1 & 1 & -1 & | & -4 \\ 0 & 2 & -2 & 2 & 3 & | & 13 \\ 0 & 3 & 1 & 3 & 6 & | & 0 \end{bmatrix}$$

$$\Downarrow$$

$$\begin{bmatrix} 1 & 0 & 1 & 1 & -1 & | & -4 \\ 0 & 2 & -2 & 2 & 3 & | & 13 \\ 0 & 0 & 4 & 0 & 1.5 & | & -39/2 \end{bmatrix}.$$

Now we do two row multiplication operations to make the diagonal entries = 1:

$$\begin{array}{c} \\ -3/2 \cdot \\ 1/4 \cdot \end{array} \begin{bmatrix} 1 & 0 & 1 & 1 & -1 & | & -4 \\ 0 & 2 & -2 & 2 & 3 & | & 13 \\ 0 & 0 & 4 & 0 & 1.5 & | & -39/2 \end{bmatrix}$$

$$\Downarrow$$

$$\begin{bmatrix} 1 & 0 & 1 & 1 & -1 & | & -4 \\ 0 & 1 & -1 & 1 & 3/2 & | & 13/2 \\ 0 & 0 & 1 & 0 & 3/8 & | & -39/8 \end{bmatrix}.$$

Now we do two upward row addition operations (back substitution):

$$-1\uparrow \begin{array}{c} \longrightarrow \quad \longrightarrow \\ 1\uparrow \begin{array}{c} \rightarrow \\ \leftarrow \end{array} \begin{array}{c} \rightarrow \\ \leftarrow \end{array} \end{array} \begin{bmatrix} 1 & 0 & 1 & 1 & -1 & | & -4 \\ 0 & 1 & -1 & 1 & 3/2 & | & 13/2 \\ 0 & 0 & 1 & 0 & 3/8 & | & -39/8 \end{bmatrix}$$

$$\Downarrow$$

$$\begin{bmatrix} 1 & 0 & 0 & 1 & -11/8 & | & -71/8 \\ 0 & 1 & 0 & 1 & 15/8 & | & 13/8 \\ 0 & 0 & 1 & 0 & 3/8 & | & -39/8 \end{bmatrix}.$$

Now we can write the answer in standard form:

$$\begin{cases} x_1 = -x_2 + 11x_4/8 - 71/8 \\ x_3 = -x_2 - 15x_4/8 + 13/8 \\ x_5 = -3x_4/8 - 39/8. \end{cases}$$

Problem 6.10. Solve for x_1, x_2, x_3 the same system of linear equations (Problem 6.9).

Solution. We consider the last augmented matrix with columns permuted:

$$\begin{array}{cccccc} x_1 & x_2 & x_3 & x_5 & x_4 & = \\ \left[\begin{array}{ccccc|c} 1 & 1 & 0 & 0 & -11/8 & -71/8 \\ 0 & 1 & 1 & 0 & 15/8 & 13/8 \\ 0 & 0 & 0 & 1 & 3/8 & -39/8 \end{array}\right]. \end{array}$$

It is clear that we cannot to write the answer in the form

$$\begin{cases} x_1 = & \text{an affine function of } x_4, x_5 \\ x_2 = & \text{an affine function of } x_4, x_5 \\ x_3 = & \text{an affine function of } x_4, x_5. \end{cases}$$

So it is not clear what the problem means exactly. However, if it means that we have to solve the system for unknown x_1, x_2, x_3 with given numbers x_4, x_5, then here is a way to solve the problem.

If $x_5 + 3x_4/8 \neq -39/8$, then there are no solutions. Otherwise, we drop the last row in the matrix, switch the x_2-column with the x_3-column, and get the final matrix

$$\begin{array}{cccccc} x_1 & x_3 & x_2 & x_5 & x_4 & = \\ \left[\begin{array}{ccccc|c} 1 & 0 & 1 & 0 & -11/8 & -71/8 \\ 0 & 1 & 1 & 0 & 15/8 & 13/8 \end{array}\right]. \end{array}$$

Therefore, in the case $x_5 + 3x_4/8 = -39/8$, our answer in standard form is

$$\begin{cases} x_1 = -x_2 + 11x_4/8 - 7/8 \\ x_3 = -x_2 - 15x_4/8 + 13/8 \end{cases}$$

with an arbitrary x_2. ∎

We call two systems of equations equivalent if they have the same solutions. When we perform row elementary operations with an augmented matrix, we change the system of linear equations to an equivalent system. Is the converse true? See Exercises 40 and 41 at the end of this section. How about equivalence of systems involving different sets of variables? We leave this tricky question to the reader as a brain teaser. We do not need to address this issue if we do not change variables and do not consider elimination of variables as passing to an independent system with a smaller number of variables.

In §4 we stated the following fact:

Theorem 6.11. Given a system of linear equations $Ax = b$ and another linear equation $f_0 = b_0$ in standard form that follows from the system, then either the equation $f_0 = b_0$ or the equation $0 = 1$ is a linear combination of the equations in the system.

Proof. Note that if we do an elementary row operation, then all equations in the new system are linear combinations of old ones. In fact, only one of them is new, and it is a linear combination of an old equation in the case of multiplicative operation, and a linear combination of two old equations in the case of addition operation. Since linear combinations of linear combinations are linear combinations, after any number of elementary row operations, every new equation is a linear combination of the original equations.

If the system has no solutions, then we saw that we obtain the equation of the form $0 = c$ with $c \neq 0$, and a multiplication operation makes it $0 = 1$.

In the case when the system has exactly one solution $x = d$, the Gauss-Jordan method gives every equation $x_i = d_i$ as a linear combination of the original equations. Since $x = d$ satisfies the linear equation $f_0 = cx = b_0$, we have $cd = b_0$. Now we combine the equations $x = d$ with the coefficients c_i and obtain $cx = cd = b_0$.

Finally, assume that the system has infinitely many solutions. Then the Gauss-Jordan method gives the system of the form $z - Cy = d$ (see the standard answer on page 57). Substitution of this into $f_0 = cx = b_0$ gives a relation among C, d, c, b. Namely, writing $f_0 = cx = c'y + c''z$, the relation is $c'(Cz + d) + c''z = 0$ for all z; hence $c'C + c'' = 0$ and $c'd = 0$. Now taking the linear combination $c''(z - Cy) = cd$ of the equations $z - Cy = d$, we obtain the equation $f_0 = b_0$.

Remark 6.12. Can we solve systems of linear inequalities eliminating one variable after another like we did for systems of linear equations? Fourier asked this question and answered positively about 200 years ago. See Section A10 in the Appendix for the Fourier-Motzkin method. However, no one was able to make the method practical enough to compete with other methods. The problem with the method is that the number of constraints may grow very fast. Here is an example of when the elimination method works well.

Problem 6.13. Find a feasible solution for the system

$$\begin{cases} x_1 + x_2 + x_3 + x_4 \leq 3, \\ x_1 + 2x_2 + 3x_3 \geq 1, \\ \text{all } x_i \geq 0. \end{cases} \qquad (6.14)$$

Solution. We write down all two constraints involving x_4 in the form

$$0 \leq x_4 \leq 3 - x_1 - x_2 - x_3 \qquad (i)$$

and eliminate x_4 from the system. So we obtain the following system

$$\begin{cases} 0 \leq 3 - x_1 - x_2 - x_3 \\ x_1 + 2x_2 + 3x_3 \geq 1, \\ \text{all } x_i \geq 0 \end{cases}$$

for three variables. Now we write down all three constraints involving x_3 in the form

$$0, (1 - x_1 - 2x_2)/3 \leq x_3 \leq 3 - x_1 - x_2 \qquad (ii)$$

and eliminate x_3 from the system. So we obtain the system

$$\begin{cases} 0, (1 - x_1 - 2x_2)/3 \leq 3 - x_1 - x_2 \\ x_1, x_2 \geq 0 \end{cases}$$

for two variables. Finally, to finish our forward substitution, we write down all three constraints involving x_2 in the form

$$0 \leq x_2 \leq 3 - x_1, 8 - 2x_1 \qquad (iii)$$

and eliminate x_2 from the system. So we obtain the system

$$0 \leq 3 - x_1, 8 - 2x_1; \ x_1 \geq 0$$

for one variable. The last system is equivalent to $0 \leq x_1 \leq 3$. Taking any feasible value for x_1, we find one after another feasible values for x_2, x_3, x_4. By this back substitution we can find all feasible solutions to the original system (6.14).

For example, we start with $x_1 = 2$. The constraints (iii) for x_3 become $0 \leq x_2 \leq 1$. Let us pick $x_2 = 0$. The constraints (ii) are now $0 \leq x_3 \leq 1$. We pick $x_3 = 0$. The constraints (i) are now $0 \leq x_4 \leq 1$.

Let us pick $x_4 = 1$, and we are done. A feasible solution for (6.14) is found.

Remark 6.15. Most linear systems in real life have rational data. Therefore the answer involves only rational numbers. However, the rational numbers in the answer could have very large numerators and denominators, so their computation may take too much time. Usually systems are solved on a computer with limited precision. Approximate solutions are usually sufficient for practical applications. Uncertainty in data gives a good excuse for not finding exact solutions.

Exercises

1–8. Find whether the matrix A is invertible, reducing it to a diagonal matrix by row and column addition operations. *Remark*: The diagonal matrix in the answer is not unique, but the product of the diagonal entries in it is unique. In many textbooks on linear algebra it is proved that this product equals the *determinant* of A.

1. $A = \begin{bmatrix} 1 & 2 \\ 6 & 8 \end{bmatrix}$

2. $A = \begin{bmatrix} a & b \\ c & d \end{bmatrix}$

3. $A = \begin{bmatrix} a & 0 & 0 \\ 0 & b & 1 \\ 0 & 0 & c \end{bmatrix}$

4. $A = \begin{bmatrix} 0 & 0 & 1 \\ 0 & 1 & 0 \\ 1 & 0 & 0 \end{bmatrix}$

5. $A = \begin{bmatrix} 0 & 0 & -1 & 0 \\ 0 & 1 & 3 & 1 \\ 0 & -1 & 0 & 1 \\ -1 & 1 & 2 & 1 \end{bmatrix}$

6. $A = \begin{bmatrix} 1 & 0 & -1 \\ 5 & 1 & 3 \\ 2 & 4 & 5 \end{bmatrix}$

7. $A = \begin{bmatrix} 1 & -2 & -1 \\ 5 & 1 & 3 \\ 2 & 4 & 5 \end{bmatrix}$

8. $A = \begin{bmatrix} 1 & -2 & -1 & 0 \\ 5 & 1 & 3 & 1 \\ 0 & 1 & 0 & 1 \\ 5 & 6 & 0 & 7 \end{bmatrix}$

9–16. Solve for x, y. *Hints*: Equations could be given not in the standard form for linear equations. Treat b, t, u, z as given numbers.

9. $\begin{cases} x + 2y = 1 \\ 2x + 4y = 3 \end{cases}$

10. $\begin{cases} x + 2y = 3 \\ 2x + 4y = 6 \end{cases}$

11. $\begin{cases} 2x + 3y + 5z = 2, \\ 3x + 5y + 8z = b \end{cases}$

12. $\begin{cases} 2x + 3y + 1 = 2 + x - y, \\ 3x + 5y + 8 = 2x \end{cases}$

13. $\begin{cases} x + 2y = 3 + u \\ 2x + 4y = t \end{cases}$

14. $\begin{cases} x + ty = 3 \\ 2x + 4y = 1 \end{cases}$

15. $\begin{cases} x + t^2 y = 1 \\ x + y = t \end{cases}$ **16.** $\begin{cases} x + ty = t^2 \\ 2x + uy = 1 \end{cases}$

17. Solve the system in Exercise 11 for y and z.

18. Solve the system in Exercise 11 for x and z.

19. Is there a system of linear equations with exactly three solutions?

20–23. Find A^{-1} (if it exists) for the matrix A in Exercises 5–8.

24–27. Write the matrix A in Exercises 1–4 as $A = LU$ with an upper triangular matrix U and a lower triangular matrix L, if possible.

28–31. Write the matrix A in Exercises 5–8 as $A = LU$ with an upper triangular matrix U and a lower triangular matrix L, if possible. Then compute the matrix UL.

32–35. Solve each system for x, y, z:

32. $\begin{cases} x + y + z = a + b^2, \\ x + 2y + 3z = c^3, \\ x + 3y + 4z = d. \end{cases}$ **33.** $\begin{cases} x + 5y + z = y, \\ x + 2y + 3 = z, \\ x + 3y - 4z = d. \end{cases}$

34. $\begin{cases} x + y + z = u_1, \\ x + 2y + 3z = u_2, \\ x + 3y + 4z = u_3. \end{cases}$ **35.** $\begin{cases} u + y + z = x, \\ x + 2u + 3z = y, \\ x + 3y + 4z = v. \end{cases}$

36–38. Find a feasible solution for the system (or prove that the system is infeasible).

36. $x,\ y \geq 0,\ x + y \leq 2,\ z = 3x - 4y + 5$.

37. $x,\ y,\ z \geq 0,\ x + y + z \leq 2,\ x + 2y + 3z \geq 3$.

38. $x,\ y,\ z \geq 0,\ x + y - z \leq 2,\ x + 2y + 3z \geq 3$.

39. $x - 2y + 3z \geq 5,\ 3x + 4y - z \leq 8,\ x - 2y \geq -3$.

40. Given two systems of linear equations $Ax = b$ and $A'x = b'$ with the same column of n distinct variables x, the same nonempty solution set, and the same number m of equations, show that we can obtain the augmented matrix $[A'|b']$ from $[A|b]$ by row addition operations and a row multiplication operation. In particular, there is an invertible $m \times m$ matrix C such that $CA = A', Cb = b'$.

41. Given two systems of linear equations $Ax = b$ and $A'x = b'$ with the same column of n distinct variables x, the same nonempty solution set, and different numbers $m > m'$ of equations, show that we can obtain the matrix $\begin{bmatrix} A' & b' \\ 0 & 0 \end{bmatrix}$ of size $m \times n$, with last $m - m'$ rows being zero rows, from the matrix $[A|b]$ by row addition operations. In particular, there is an $m \times m'$ matrix C such that $CA = A', Cb = b'$. Show also that there is an $m' \times m$ matrix C' such that $C'A' = A, C'b' = b$.

Chapter 3

Tableaux and Pivoting

§7. Standard and Canonical Forms for Linear Programs

To save paper, we will write linear programs using matrix notation. Recall that a linear form $c_1x_1 + c_2x_2 + \cdots + c_nx_n$ in n variables $x_1, \ldots, x_n$ can be written as the matrix product cx, where $c = [c_1, \ldots, c_n]$ is a row of coefficients and $x = [x_1, \ldots, x_n]^T$ is the column of variables. If it is understood from the problem that the set of variables is given by the list $x_1, x_2, \ldots, x_n$, then we see that a linear form is uniquely determined by the row c; that is, if $z = cx$ and $\hat{z} = \hat{c}x$, then $z = \hat{z}$ for all x if and only if $c = \hat{c}$.

Recall that linear constraints were defined in §1 of Chapter 1 to be relations that can be written in the standard form

$$ax = b, \text{ or } ax \leq b, \text{ or } ax \geq b,$$

where ax is a linear form in the entries of x.

Given a system of inequalities of the same type, we can write all of them in matrix form $Ax \leq b$ or $Ax \geq b$. It is a worthwhile exercise to verify that the foregoing is true. Note that $b \leq b'$ for two columns b, b' of the same size means that every entry of b is less than or equal to the corresponding entry of b'. We use a similar convention for rows. When b is a column or a row, $b \geq 0$ means that all entries of b are nonnegative.

Example. The set of linear constraints

$$\begin{cases} .3x_a & + & .35x_b & + & .5x_c & + & .4x_d & \geq & .4 \\ .6x_a & + & .35x_b & + & .5x_c & + & .45x_d & \geq & .5 \\ .1x_a & + & .3x_b & & & + & .15x_d & \geq & .1 \end{cases}$$

can be written in matrix form as follows:

$$\begin{bmatrix} .3 & .35 & .5 & .4 \\ .6 & .35 & .5 & .45 \\ .1 & .3 & 0 & .15 \end{bmatrix} \begin{bmatrix} x_a \\ x_b \\ x_c \\ x_d \end{bmatrix} \geq \begin{bmatrix} .4 \\ .5 \\ .1 \end{bmatrix}. \qquad \blacksquare$$

The setting of a linear program has been given as follows:

maximize (or minimize) an affine function (a linear form plus a constant) subject to a finite number of linear constraints.

The main goal of this section is to develop a uniform way of writing linear programs, so that any linear program can be rewritten in this uniform way. The importance of this uniformity is easily seen when we want to give a method of solving of linear programs or state a theorem in linear programming. In its absence, we would have to consider all possible ways to write a linear program, making the writing of methods or proofs very cumbersome. Solving linear programs often involves using computers, so it is very important to be able to transform a linear program to a form acceptable for a given software.

In the literature, different models of uniformity are often called standard, canonical, and normal forms. They do not necessary mean the same thing in different publications. In this section, we first define standard and canonical forms for linear programs, and then we will see how *any* linear program can be written in canonical form and in standard form.

Definition 7.1. A linear program is said to be in *standard form* if it is a minimization problem, all variables are required to be non-negative, and all other constraints are linear equations.   ∎

In other words, using matrix notations, here is the standard form:

$$\text{minimize } cx + d, \text{ subject to } Ax = b, \ x \geq 0,$$

where c is a given row, d a given number, A a given matrix, x a column of distinct variables, b a given column.

We can write this even shorter as follows:

$$cx + d \to \min, \ Ax = b, \ x \geq 0.$$

Note that $Ax = b$ here is the standard form for a system of linear equations.

Example. $2x + 3y - z \to \min$, $x, y, z \geq 0, x + y + z = 1$ is a linear program in standard form.

Definition 7.2. The *canonical form* of a linear program is the following: Minimize $cx + d$, subject to $Ax \leq b$, $x \geq 0$,

where A, b, c, d, x are as before.   ∎

In other words, it is a minimization problem, all variables are required to be nonnegative, and all other constraints are of type $\leq$.

Example. $2x + 3y - z \to \min$, $x, y, z \geq 0, z \leq 2$ is a linear program in canonical form.   ∎

In different books on linear programming, standard and canonical (as well as normal) forms for linear programs are different. The *sign restrictions* $x \geq 0$ are common for all of them, but some of them have max instead of min or (and) $Ax \geq b$ instead of $Ax \leq b$ or $Ax = b$. Also different software packages use different input forms for linear programs. Some books and some software do not allow the constant d in the objective function.

So it is important to know how to go from one form to another. We will give a few little tricks that allow us to convert any form to any other form.

Using some elementary algebra, we can write any linear programming problem in a standard form as well as in a canonical form. We will use some elementary facts, summarized in the next lemma.

Lemma 7.3   **1.** Let a and b be real numbers. Then

(a) $a \leq b$ if and only if $-a \geq -b$.

(b) $a = b$ if and only if $a \leq b$ and $-a \leq -b$.

2. Minimization of a function f is equivalent to the maximization of the function $-f$, under the same constraints. That is, both optimization problems have the same feasible solutions and the same optimal solutions. The optimal values differ by a sign; that is, the minimum value of f equals the negative of the maximum value of $-f$.

3. Minimization of a function f is equivalent to the minimization of the function $f + d$, under the same constraints. That is, both optimization problems have the same feasible solutions and the same optimal solutions. The optimal values differ by d; that is, the minimum of $f + d$ equals the minimum of f plus d.

Proof

(1.a) $a \leq b$ if and only if $a - b \leq 0$ if and only if $-(a - b) \geq 0$ if and only if $b - a \geq 0$ if and only if $-a \geq -b$.

(1.b) If $a = b$, then certainly, $a \leq b$. Also, $-a = -b$ and from here it follows that $-a \leq -b$. For the converse, $a \leq b$ and $-a \leq -b$ are equivalent, by (1.a), to $a \leq b$ and $a \geq b$; it now follows that $a = b$.

2. Let c be a point where f attains its minimum; then $f(c) \leq f(x)$ for any x. Using (1.a), $-f(c) \geq -f(x)$ for any x, which means that $-f$ attains its maximum value at c.

3. An exercise for the reader.   ∎

Using this lemma we can rewrite any linear programming problem in its canonical or the standard form using the following "tricks" (here f represents a linear form and c is a constant).

Trick 7.4. By (1.a) from Lemma 7.3, the inequalities $f \leq c$ and $-f \geq -c$ are equivalent.   ∎

Thus, by rewriting the inequalities of the form $f \geq c$, we can make some or all the linear inequalities in our linear program to be of the form $f \leq c$. Also, if it is desirable, we can arrange them to be of the form $f \geq c$.

Trick 7.5. By (1.b) of Lemma 7.3, the equality $f = c$ is equivalent to the system of two inequalities $f \leq c$, $-f \leq -c$.   ∎

So we can replace any linear equation by two inequalities. We see now that, without loss of generality, we can assume that all the linear constraints in a linear programming problem are of the form $f \leq c$ or of the form $f \geq c$.

Trick 7.6. By (2) of Lemma 7.3, we can convert a minimization problem to a maximization problem and vice versa by multiplying the objective function by -1.   ∎

Do not forget that the optimal value changes its sign as well!

Using Tricks 7.4–7.6, we can write any linear program where all variables are required to be ≥ 0 in a canonical form without changing the set of variables, the feasible region, and the optimality region. Namely, if our problem is a maximization problem, we convert it to a minimization problem by Trick 7.6. By Trick 7.5 we can convert the equality constraints (if any) to inequalities. Finally, by Trick 7,4 we can convert the $\geq$ constraints besides the sign restrictions (if any) to $\leq$ constraints. In particular, the standard form $cx + d \rightarrow$ min, $Ax = b$, $x \geq 0$, can be converted to an equivalent the canonical form: $cx + d \rightarrow$ min, $Ax \leq b, -Ax \leq -b, x \geq 0$,

Trick 7.7. A constraint $f \leq c$ can be written as two constraints $f + s = c$ and $s \geq 0$ with an additional (*slack*) variable s.   ∎

Thus, we can replace any constraint given by an inequality by a pair of constraints, one given by an equality and the second being a sign restriction. The slack variable s should be different from other variables in the problem. This trick can be inverted, and the converse trick saves (eliminate) a variable s. This s should be eliminated from all constraints and from the objective function to obtain a problem with one variable less.

Combined with Trick 7.4, Trick 7.7 can be used for $\geq$ constraints: $f \geq c$ can be written as $f - s = c$, $s \geq 0$. In this case the new variable s is often called a *surplus* or *excess* variable, but we will also call it a slack variable. Spell ?

Using Trick 7.7 (and, if necessary, Thick 7.6), we can write any LP where all variables are required to be ≥ 0 in a standard form. For example, the canonical form $cx + d \to \min, Ax \leq b, \ x \geq 0$, can be written as the standard form $cx + d \to \min, Ax + u = b, \ x \geq 0, xu \geq 0$, where u is a column of slack variables (one slack variable for each inequality in the system $Ax \leq b$). This standard form is special because its system of linear equations is solved: $u = -Ax + b$ is a standard form for the answer.

Conversely, given any LP in standard form $cx + d \to \min, Ax = b, \ x \geq 0$, we can convert it to a canonical form by solving the system $Ax = b$ of linear equations. (If the system is inconsistent then the program is infeasible and there is no need to write it in any form.) In this way, we reduce the number of constraints and variables rather than doubling the number of constraint by Trick 7.5.

Allowing Trick 7.7 we may change the set of variables in our program adding slack variables or eliminating some variables, so we cannot say that we do not change the feasible region. However, the change is simple and explicit even after repeated use of the trick: The feasible region S of the original program P and the feasible region S' of the new program P$'$ are in 1-1 correspondence under *affine transformations*

$$x = Cx' + B, x' = C'x + B',$$

where x is the column of n variables of P, x' the column of n' variables in P$'$, and C, B, C', B' are constant matrices of sizes $n \times n'$, $n \times 1$, $n' \times n$, and $n' \times 1$, respectively. Moreover, the optimality regions are transformed to each other. So we could call these two programs *affinely equivalent*.

For example, the canonical form $cx + d \to \min, Ax \leq b, \ x \geq 0$, and the standard form $cx + d \to \min, Ax + u = b, \ x \geq 0, xu \geq 0$ above are connected by the affine transformations

$$x = [1_n \ 0] \begin{bmatrix} x \\ u \end{bmatrix}, \qquad \begin{bmatrix} x \\ u \end{bmatrix} = \begin{bmatrix} 1_n \\ -A \end{bmatrix} + \begin{bmatrix} 0 \\ b \end{bmatrix},$$

where n in the number of entries in the column x and 0 stands for zero matrices of appropriate sizes.

Trick 7.8. We can write any variable x in our linear program as the difference $x' - x''$ of two additional (artificial) variables, x' and x'', subject to $x' \geq 0, \ x'' \geq 0$. ∎

The purpose of this trick is to rid off a variable that could be negative. If we know that $x \leq a$, then a simpler substitution $x = a - x''$ with $x'' \geq 0$ works. If we know that $x \geq a$, the substitution $x = x' + a$ with $x' \geq 0$ works. If we know any upper or lower bonds on x but x occurs (with nonzero coefficient) in a linear equation, we can solve this equation for x and eliminate x completely from our linear program together with this equation. Otherwise, we still can exclude x from our constraints by Fourier-Motzkin elimination (see the end of §6), but if x occurs in many constraints, this method could be less practical than the standard Trick 7.8. If x does not occur in our constraints but occurs in the objective function, our linear problem is infeasible or unbounded.

Trick 7.8 is even dirtier (worse) than Trick 7.7 because not only do we change the set of variables, but we lose the affine equivalence. Although the transformation $x = x' - x''$ is affine, there is no affine transformation going the other way unless x is bounded (in which case we can fix x' or x''). There is a transformation given by

$$[x', x''] = \begin{cases} [x, \ 0] & \text{when } x \geq 0, \\ [0, -x] & \text{when } x \leq 0, \end{cases}$$

but this is not an affine transformation.

Using the preeding five tricks any number of times results in reduction of our linear program P to another linear program P′ such that

- every feasible solution to P gives easily a feasible solution to P′, and vice versa,

and

- every optimal solution to P gives easily an optimal solution to P′, and vice versa.

Combining Tricks 7.6, 7.7, and 7.8, we see that

any linear program can be written in standard form.

Combining Tricks 7.4, 7.5, 7.6, and 7.8, we see that

any linear program can be written in canonical form.

In some books and some software, the objective function of a linear program must be linear. By (3) of Lemma 7.3, this can be arranged. Do not forget that this changes the optimal value! Sometimes computer software is doing some tricks for us. Even tricks increasing the number of variables and constraints are widely used in linear programming software when the size of the problem is of no concern.

We allow the constant term in the objective function, because even if this term is not present in the beginning, it may appear in the process of solving the problem by simplex or by any other method. Here are some specific examples.

Problem 7.9. Convert $2x + 3y - z \to \min$, $x, y, z \geq 0$, $x + y + z = 1$ to a canonical form.

Solution. By the standard tricks, we obtain the canonical form $2x + 3y - z \to \min$, $x, y, z \geq 0$, $x + y + z \leq 1$, $-x - y - z \leq -1$. ∎

Another way to do Problem 7.9 is to solve the equation for, say, z. We obtain $z = 1 - x - y$. Substituting this in the linear program we obtain the canonical form $x + 2y - 1 \to \min$, $x, y \geq 0$, $x + y \leq 1$. Thus, we get a smaller canonical form, which can be easily solved: $\min = -1$ at $x = 0, y = 0, z = 1$.

Problem 7.10. Convert $f = 2x + 3y - z \to \max$, $x, y \geq 0$, $x + y + z \geq 1$ to a canonical form.

Solution. We multiply the objective function f and the constraint $x + y + z \geq 1$ by -1 and obtain $-f = -2x - 3y + z \to \min$, $x, y \geq 0$, $-x - y - z \geq 1$. Then we set $z = z' - z''$ with $z', z'' \geq 0$ and obtain a canonical form $-f = -2x - 3y + z' - z'' \to \min$, $x, y \geq 0$, $-x - y - z' + z'' \geq 1$. ∎

Note that $z = z' - z''$ is not a part of the original problem or the canonical form but a way to connect them. In general, before changing the objective function, you should name it. In the case when you are required to solve a linear program, the final answer should be given in terms of the original problem.

Rewriting a mathematical problem in one form or another may help to solve it or prepare the problem to be dealt with using sophisticated mathematical tools or computers. For example, to solve an equation $4 + x - 5x = 3 - x$ for x, we transform it first to a standard form $-3x = -1$ using addition and subtraction and then solve it using division.

Problem 7.11. Write the diet problem (Example 2.1) in canonical and standard forms.

Solution. To obtain a canonical form, all we need to do is to multiply three constraints of type $\geq$ by -1:

$$
\begin{cases}
10a & + & 15b & + & 5c & + & 60d & + & 8e & \to & \min \\
-0.3a & - & 1.2b & - & 0.7c & - & 3.5d & - & 5.5e & \leq & -50 \\
-73a & - & 96b & - & 20253c & - & 890d & - & 279e & \leq & -4000 \\
-9.6a & - & 7b & - & 19c & - & 57d & - & 22e & \leq & -1000 \\
a, & & b, & & c, & & d, & & e & \geq & 0.
\end{cases}
$$

To obtain a standard form, we add three slack variables corresponding to protein, vitamin A, and calcium:

$$\begin{cases} 10a + 15b + 5c + 60d + 8e \to \min \\ -0.3a - 1.2b - 0.7c - 3.5d - 5.5e + u_1 = -50 \\ -73a - 96b - 20253c - 890d - 279e + u_2 = -4000 \\ -9.6a - 7b - 19c - 57d - 22e + u_3 = -1000 \\ a, b, c, d, e, u_1, u_2, u_3 \geq 0. \end{cases}$$

Problem 7.12. Write the blending problem (Example 2.2) in canonical and standard forms.

Solution. The problem already has a standard form:

$$\begin{cases} 1.2a &+& 1.4b &+& 1.7c &+& 1.9d &\to& \min \\ .9a &+& .8b &+& .7c &+& .6d &=& .75 \\ .1a &+& .2b &+& .3c &+& .4d &=& .25 \\ a, && b, && c, && d &\geq& 0. \end{cases}$$

We do not need to include explicitly the redundant constraint $a + b + c + d = 1$. Using matrices, the problem can be written as

$$[1.2, 1.4, 1, 7, 1.9] \begin{bmatrix} a \\ b \\ c \\ d \end{bmatrix} \to \min,$$

$$\begin{bmatrix} 0.9 & 0.8 & 0.7 & 0.6 \\ 0.1 & 0.2 & 0.3 & 0.4 \end{bmatrix} \begin{bmatrix} a \\ b \\ c \\ d \end{bmatrix} = \begin{bmatrix} 0.75 \\ 0.25 \end{bmatrix}$$

$$[a, b, c, d] \geq 0.$$

To obtain a canonical form, we can replace every equation by two inequalities:

$$\begin{cases} 1.2a + 1.4b + 1.7c + 1.9d \to \min \\ .9a + .8b + .7c + .6d \leq .75 \\ -.9a - .8b - .7c - .6d \leq -.75 \\ .1a + .2b + .3c + .4d \leq .25 \\ -.1a - .2b - .3c - .4d \leq -.25 \\ a, b, c, d \geq 0. \end{cases}$$

Another way to obtain a canonical form is to solve the system of two linear equations for a, b and exclude a, b from the linear program.

Then we obtain a canonical form with only two variables, c, d, and this linear program can be easily solved graphically.

Remark. Gauss-Jordan elimination for systems of linear equations (see §6 above) also can be considered as reduction of a system to systems with smaller sets of variables with affine transformations between the solution sets giving a 1-1 correspondence.

However, in the Fourier-Motzkin elimination method for systems of inequalities (see §6), we have a somewhat more complicated relationship between the systems. In Chapter 7 we will use a substitution of variables that is not affine. Some recent method, like Karmirkar's method, use transformations that are not affine.

Exercises

1–11. Rewrite the following optimization problems as a linear program in a standard form and as a linear program in a canonical form. The answer may be given using matrices. You are not required to solve the linear programs. *Hint*: To get smaller forms, solve linear equations and eliminate variables rather than adding new variables.

1. $\begin{cases} \text{Maximize } 2x + 3y \\ \text{subject to } x \geq 1, \ y \geq -1, x + y \leq 5. \end{cases}$

2. $x \to \max, y = x + 1, x + y \leq 9, y \geq 1$

3. $x_1 - 5x_2 + x_3 + 3x_4 \to \min,$
$2x_2 + x_3 + 3x_4 = 3, \ x_1 - x_2 + 3x_4 \geq 3, \ \text{all } x_i \geq 0$

4. $x_1 - 5x_2 + x_3 + 3x_4 \to \max,$
$2x_2 + x_3 + 3x_4 = 3, \ x_1 - x_2 + 3x_4 \geq 3, \ x_1, x_2, x_3 \geq 0$

5–8. The linear programs in Examples 1.9–1.12.

9. $\begin{bmatrix} 1 & x & 2 \\ 0 & 5 & -1 \end{bmatrix} \begin{bmatrix} y \\ -1 \\ x+z \end{bmatrix} = \begin{bmatrix} x \\ 0 \end{bmatrix}, \begin{bmatrix} x \\ y \\ z \end{bmatrix} \geq 0, x + y + z \to \min$

10. $\begin{bmatrix} 1 & -2 & 2 \\ 0 & 5 & -1 \end{bmatrix} \begin{bmatrix} y \\ -1 \end{bmatrix} \geq \begin{bmatrix} x \\ 0 \end{bmatrix}, \begin{bmatrix} x \\ y \\ z \end{bmatrix} \geq 0, x + y + z \to \min$

11.
$3x_1 - x_2 + x_3 + 3x_4 + x_5 - 5x_6 + x_7 + 3x_8 + x_9 \to \min,$
$x_1 - 5x_2 + 2x_3 + 2x_4 - x_5 - x_6 - 2x_7 + 3x_8 + x_9 \geq 3,$
$-x_1 + x_2 + x_3 + x_4 + x_5 - x_6 - 2x_7 + 3x_8 + x_9 \leq -1,$
$2x_1 - 2x_2 - 2x_3 + 2x_4 + 3x_5 - x_6 - 2x_7 + x_8 + x_9 = 2,$
$x_1 + 3x_5 - x_6 - 2x_7 - x_9 = 0, \text{ all } x_i \geq 0$

§8. Pivoting Tableaux

Consider the following system of linear equations:

$$\begin{cases} 2x_1 + 3x_2 = 4 \\ 5x_1 + 6x_2 = 7 \end{cases} \tag{8.1}$$

In matrix notation, the system can be written in standard form $Ax = b$:

$$\begin{bmatrix} 2 & 3 \\ 5 & 6 \end{bmatrix} \begin{bmatrix} x_1 \\ x_2 \end{bmatrix} = \begin{bmatrix} 4 \\ 7 \end{bmatrix}.$$

Using *row tableaux*, this system can be rewritten as

$$\begin{matrix} x_1 & x_2 \\ \begin{bmatrix} 2 & 3 \\ 5 & 6 \end{bmatrix} & \begin{matrix} = 4 \\ = 7 \end{matrix} \end{matrix} \tag{8.2}$$

or as

$$\begin{matrix} x_1 & x_2 & -1 \\ \begin{bmatrix} 2 & 3 & 4 \\ 5 & 6 & 7 \end{bmatrix} & \begin{matrix} = 0 \\ = 0 \end{matrix} \end{matrix}$$

The rows in both tableaux correspond to the linear equations. The matrix in the first tableau is the coefficient matrix A, and the matrix in the second tableau is the augmented matrix $[A|b]$. We wrote the names of variables at the top margin for both tableaux.

Thus, a *tableau* is a matrix decorated (or marked) with additional information at the margins. Examples of the information we put at the margins are names of variables and constants.

Now we write the same system in *column* tableaux representing linear equations by columns and putting the names of variables at the left margin:

$$\begin{matrix} x_1 \\ x_2 \end{matrix} \begin{bmatrix} 2 & 5 \\ 3 & 6 \end{bmatrix} \\ \begin{matrix} = 4 & = 7 \end{matrix} \qquad \text{or} \qquad \begin{matrix} x_1 \\ x_2 \\ 1 \end{matrix} \begin{bmatrix} 2 & 5 \\ 3 & 6 \\ -4 & -7 \end{bmatrix} \\ \begin{matrix} = 0 & = 0. \end{matrix}$$

As you can see, tableaux provide us with another way of representing systems of linear equations. They can also be used to write out and handle the data of linear programs. The main advantage of using a tableau is that the variables need only be written out once, at the margin of the tableau. Tableaux are a "short-hand" way to handle linear programs, saving writing, paper, and time. With practice, they are easier to read.

In this respect tableaux are very similar to matrices used to solve system of linear equations. The columns of the coefficient matrix correspond to variables. Unless we switch variables or do other column operations corresponding to changes of variables, we do not need to write the names of variables at the top margin. In linear programming we put labels at margins of matrices to avoid confusion and mix-ups.

We will use tableaux to explain the way the simplex method works. Tableaux can be used to solve, by hand, linear programs with a small number of variables or constraints. When the matrix of data is sparse (i.e., it has many zero entries), other methods of handling data may be better than tableaux.

Example 8.3.
The standard form of a linear program (see Definition 7.1) can be written as a row tableau in the following way:

$$x^T$$
$$\begin{bmatrix} A \\ c \end{bmatrix} \begin{matrix} = \\ \rightarrow \end{matrix} \begin{matrix} b \\ \min, \end{matrix} \quad x \geq 0.$$

Here A is the matrix of coefficients, b is a column matrix, and c is a row matrix. Note that the variables on the top are written as a row matrix (x^T denotes the transpose of the column matrix of variables). The coefficients of the objective function, c, are written in the last row of the tableau. The nonnegativity constraint, $x \geq 0$, is written outside the tableau.

Example 8.4.
The canonical form of a linear program (see Definition 7.2) can be written as a tableau as follows:

$$\begin{matrix} x \\ 1 \end{matrix} \begin{bmatrix} -A^T & c^T \\ b^T & d \end{bmatrix}$$
$$\downarrow$$
$$\geq 0 \quad \min, \quad x \geq 0, \ y \geq 0$$

or

$$\begin{matrix} x \\ 1 \end{matrix} \begin{bmatrix} -A^T & c^T \\ b^T & d \end{bmatrix}$$
$$\downarrow$$
$$= y \quad \min, \quad x \geq 0, \ y \geq 0.$$

Here we used column tableaux. We have introduced new (slack) variables, y. ∎

One step of Gauss-Jordan elimination (see §6) can be explained as follows. To solve a system of linear equations, we first solve one of equations (the pivot equation) with respect to one of variables (the pivot variable) and then we eliminate this variable from the other equations by substitution.

Thus, we obtain a smaller system, one variable and one equation less. These steps are repeated (forward substitution), until no equations or no variables are left; see §6 for more details.

Now we explain how this step, the *pivot step* works in terms of tableaux. We start with a small example, the system of linear equations (8.1) and the same system written in a row tableau (8.2).

A step of the Gauss-Jordan elimination goes as follows:

1. Solve one of the equations, say, $2x_1 + 3x_2 = 4$ for one of the unknowns, say x_1:

$$x_1 = 4/2 - (3/2)x_2.$$

2. Substitute this expression for x_1 into the second equation: $5x_1 + 6x_2 = 7$ that becomes

$5(4/2 - (3/2)x_2) + 6x_2 = 7$, or $5 \cdot 4/2 + (6 - 5 \cdot 3/2)x_2 = 7$.

The new tableau is

$$\begin{matrix} & 4 & x_2 & \\ \begin{bmatrix} 1/2 & -3/2 \\ 5/2 & 7 - 2 \cdot 5/2 \end{bmatrix} & & = x_1 \\ & & = 7 \end{matrix} \qquad (8.5)$$

Now we replace the numbers 2, 3, 4, 5, 6, 7 in (8.1) and (8.2) by arbitrary numbers $\alpha, \beta, u, \gamma, \delta, v$ and rename the variables x_1, x_2 as x, y. That is, we consider an arbitrary system of two linear equations for x, y:

$$\begin{cases} \alpha x + \beta y = u \\ \gamma x + \delta y = v \end{cases} \qquad (8.6)$$

and the corresponding tableau

$$\begin{matrix} x & y & \\ \begin{bmatrix} \alpha^* & \beta \\ \gamma & \delta \end{bmatrix} & = u \\ = v \end{matrix} \qquad (8.7)$$

We assume now that the entry α marked by a star and called the *pivot entry* is not zero. Solving as before the first equation for x and substituting into the second equation, we obtain the tableau

$$\begin{matrix} & u & y \\ \begin{bmatrix} 1/\alpha & -\beta/\alpha \\ \gamma/\alpha & \delta - \beta\gamma/\alpha \end{bmatrix} & \begin{matrix} = x \\ = v \end{matrix} \end{matrix}. \tag{8.8}$$

Going from tableau (8.7) to tableau (8.8) is called the *pivot step*. So here is the pivot step for 2-by-2 tableaux:

$$\begin{matrix} x & y \\ \begin{bmatrix} \alpha^* & \beta \\ \gamma & \delta \end{bmatrix} & \begin{matrix} = u \\ = v \end{matrix} \end{matrix} \quad \mapsto \quad \begin{matrix} u & y \\ \begin{bmatrix} 1/\alpha & -\beta/\alpha \\ \gamma/\alpha & \delta - \beta\gamma/\alpha \end{bmatrix} & \begin{matrix} = x \\ = v \end{matrix} \end{matrix}. \tag{8.9}$$

Pivoting tableaux of an arbitrary size works similarly. Here are examples of pivot steps with tableaux of smaller size:

$$\begin{matrix} x \\ \begin{bmatrix} \alpha^* \end{bmatrix} & = u \end{matrix} \quad \rightarrow \quad \begin{matrix} u \\ \begin{bmatrix} 1/\alpha \end{bmatrix} & = x \end{matrix},$$

$$\begin{matrix} x & y \\ \begin{bmatrix} \alpha^* & \beta \end{bmatrix} & = u \end{matrix} \quad \mapsto \quad \begin{matrix} u & y \\ \begin{bmatrix} 1/\alpha & -\beta/\alpha \end{bmatrix} & = x \end{matrix}.$$

Here is an example of a bigger size:

$$\begin{matrix} x & y \\ \begin{bmatrix} \gamma_1 & \delta_1 \\ \alpha^* & \beta \\ \gamma_2 & \delta_2 \end{bmatrix} & \begin{matrix} = v_1 \\ = u \\ = v_2 \end{matrix} \end{matrix} \quad \mapsto \quad \begin{matrix} u & y \\ \begin{bmatrix} \gamma_1/\alpha & \delta_1 - \beta\gamma_1/\alpha \\ 1/\alpha & -\beta/\alpha \\ \gamma_2/\alpha & \delta_2 - \beta\gamma_2/\alpha \end{bmatrix} & \begin{matrix} = v_1 \\ = x \\ = v_2 \end{matrix} \end{matrix}.$$

These examples indicate that for a tableau of any size the pivot step looks as follows:

Pivot Rules

- Switch the labels of the pivot row and column ($x \leftrightarrow u$; this is the only change on the top and left margins).
- Replace the pivot entry $\alpha \neq 0$ by its multiplicative inverse $\alpha' = 1/\alpha$ ($\alpha \mapsto 1/\alpha$).
- Divide every entry β in the pivot row which is not in the pivot column by $-\alpha$ (i.e., replace every such β by $\beta' = -\alpha'\beta$).
- Replace every entry δ outside the pivot column and pivot row by $\delta - \beta\gamma/\alpha = \delta + \beta'\gamma$, where β is in the pivot row and in the same column as δ (above or below δ) and γ is in the pivot column and in the same row as δ (on left or right of δ).

● Divide every entry β in the pivot column that is not in the pivot row by the pivot entry α (i.e., replace every such β by $\beta' = \alpha'\beta$). ∎

So a pivot step on a tableau of size $m \times n$ can be done in one switch of labels, one division, $mn - 1$ multiplications, $mn - m - n + 1$ additions, and $n - 1$ sign changes.

As an exercise, let us do a pivot step on the tableau (8.5) indicating the pivot entry $-3/2$ in the second row and second column by *:

$$7 \leftrightarrow x_2$$

$$\alpha = -3/2 \mapsto \alpha' = 1/\alpha = -2/3$$

$$\beta = 5/2 \mapsto \beta' = -\beta/\alpha = -\alpha'\beta = 5/3$$

$$\delta = 1/2 \mapsto = \delta - \beta\gamma/\alpha$$

$$= \delta + \beta'\gamma = 1/2 + (5/3)(-3/2) = 1/2 - 5/2 = -2$$

$$\gamma \mapsto \gamma/\alpha = \gamma\alpha' = -1/17$$

$$
\begin{array}{cc}
4 & x_2 \\
\end{array}
\begin{bmatrix} 1/2 & -3/2 \\ 5/2 & -3/2^* \end{bmatrix}
\begin{array}{l} = x_1 \\ = 7 \end{array}
\quad\mapsto\quad
\begin{array}{cc}
4 & 7 \\
\end{array}
\begin{bmatrix} -2 & 1 \\ 5/3 & -2/3 \end{bmatrix}
\begin{array}{l} = x_1 \\ = x_2 \end{array}.
$$

Thus, we solved the system (8.1):

$$x_1 = 4 \cdot (-2) + 7 \cdot 1 = -1,$$

$$x_2 = 4 \cdot (5/3) + 7 \cdot (-2/3) = 2.$$

In fact, we solved a more general system with constant terms 3, 5 replaced by arbitrary numbers u, v. The solution is

$$x_1 = -2u + v,$$

$$x_2 = 5u/3 - 2v/3.$$

In other words, we computed

$$\begin{bmatrix} 2 & 3 \\ 5 & 6 \end{bmatrix}^{-1} = \begin{bmatrix} -2 & 1 \\ 5/3 & -2/3 \end{bmatrix}.$$

Now we will verify the pivot rules for a tableau of arbitrary size. We consider a system of linear equations $AX = b$ given by a row tableau:

$$X^T$$
$$[\;A\;]\quad = b$$

where $X^T = [x_1 \quad \cdots \quad x_n]$ is a row of variables, b is a column constants (they also can be considered as variables), $b^T = [b_1, \ldots, b_m]$, and A is the $m \times n$ coefficient matrix

$$A = \begin{bmatrix} a_{11} & a_{12} & \cdots & a_{1n} \\ a_{21} & a_{22} & \cdots & a_{2n} \\ \vdots & \vdots & \ddots & \vdots \\ a_{m1} & a_{m2} & \cdots & a_{mn} \end{bmatrix}.$$

Recall that a long way of writing the system $AX = b$ is

$$a_{11}x_1 + a_{12}x_2 + \cdots + a_{1n}x_n = b_1$$
$$a_{21}x_1 + a_{22}x_2 + \cdots + a_{2n}x_n = b_2$$
$$\ldots$$
$$\ldots$$
$$\ldots$$
$$a_{m1}x_1 + a_{m2}x_2 + \cdots + a_{mn}x_n = b_m$$

Any nonzero entry in the matrix of coefficients, A, can be chosen as a pivot entry (the reason for choosing a nonzero entry is because we need to divide by this number in some of the steps of the pivoting that we are about to describe). This element is used to obtain a new tableau by pivoting as described next. We remark that the new tableau obtained by pivoting represents a system of equations equivalent to the one we started out with; that is, any solution of the initial system is a solution of the new system and vice versa.

Let $\alpha = a_{i,j}$ be the pivot entry (so $\alpha \neq 0$). Let $x = x_j$ be the entry in the top margin of the same column (pivot column) as α and $u = b_i$ the entry at the right margin in the same row (pivot row) as α. The following tableau describes this situation:

$$\begin{array}{ccc} * & x & * \\ \begin{bmatrix} * & * & * \\ * & \alpha & * \\ * & * & * \\ \vdots & \vdots & \vdots \end{bmatrix} & \begin{array}{l} = * \\ = u \\ = * \\ \vdots \end{array} \end{array}$$

In particular, we have an equation of the form $\cdots + \alpha x + \cdots = u$ in our system of linear equations. We use this equation to switch x and u at the margin by solving this equation for x. We observe that the number of x_is on the top is reduced by 1 and we get closer to solving system of equations for X. All entries of the coefficient matrix will undergo some changes, so we will obtain a new matrix of coefficients.

To fix ideas, we assume that α is in the first row and the first column (so $x = x_1, u = u_1$), and we concentrate our attention on the second row and the second column, setting $y = x_2, v = b_2$ (the case when we have only one column or only one row is simpler, and we leave it as an exercise for the reader):

$$
\begin{array}{cc}
x & y & \cdots \\
\begin{bmatrix} \alpha & \beta & \cdots \\ \gamma & \delta & \cdots \\ \vdots & \vdots & \vdots \end{bmatrix} & \begin{array}{c} = u \\ = v \\ \vdots \end{array}
\end{array}
$$

First we solve the equation $\alpha x + \beta y + \cdots = u$ for x. So $x = \dfrac{u}{\alpha} - \dfrac{\beta}{\alpha} y - \cdots$. Then we substitute this expression for x into all the other equations of the system. In particular, the second equation, $\gamma x + \delta y + \cdots = v$ takes the form $\gamma(\dfrac{u}{\alpha} - \dfrac{\beta}{\alpha} y - \cdots) + \delta y + \cdots = v.$ Thus, we obtain the following tableau:

$$
\begin{array}{cc}
u & y & \cdots \\
\begin{bmatrix} 1/\alpha & -\beta/\alpha & \cdots \\ \gamma/\alpha & \delta - \beta\gamma/\alpha & \cdots \\ \vdots & \vdots & \vdots \end{bmatrix} & \begin{array}{c} = x \\ = v \\ \vdots \end{array}
\end{array}
$$

This confirms the pivot rules given previously.

Computationally, pivoting tableaux, as just done, to solve systems of linear equations is not more efficient than using augmented matrices, as in §6. We will pivot tableaux to solve linear programs.

In this section, to practice pivoting, we will solve some systems of linear equations.

Example 8.10. Consider the system of linear equations:

$$
\begin{cases} x + 2y = 3 \\ 4x + 7y = 5. \end{cases}
$$

We can rewrite the system as the following tableau:

$$\begin{array}{cc} x & y \end{array}$$
$$\begin{bmatrix} 1 & 2 \\ 4 & 7 \end{bmatrix} \begin{array}{l} = 3 \\ = 5. \end{array}$$

Let us pick up the entry 1, located in the first row and the first column, as the pivot entry. It is an appropriate choice because $1 \neq 0$. It is also a good choice because we avoid creating fractions as we execute the steps listed. We mark the pivot entry by a superscript $*$:

$$\begin{array}{cc} x & y \end{array}$$
$$\begin{bmatrix} 1^* & 2 \\ 4 & 7 \end{bmatrix} \begin{array}{l} = 3 \\ = 5. \end{array}$$

Step 1: The entries at the margins are switched:

$$\begin{array}{cc} 3 & y \end{array}$$
$$\begin{bmatrix} 1^* & 2 \\ 4 & 7 \end{bmatrix} \begin{array}{l} = x \\ = 5 \end{array}$$

Step 2: The pivot entry 1 is replaced by $1/1=1$:

$$\begin{array}{cc} 3 & y \end{array}$$
$$\begin{bmatrix} 1^* & 2 \\ 4 & 7 \end{bmatrix} \begin{array}{l} = x \\ = 5 \end{array}$$

Step 3: The entry 2 appearing in the pivot row is replaced by $-2/1 = -2$:

$$\begin{array}{cc} 3 & y \end{array}$$
$$\begin{bmatrix} 1^* & -2 \\ 4 & 7 \end{bmatrix} \begin{array}{l} = x \\ = 5. \end{array}$$

Step 4: The entry 4 that appears in the pivot column is replaced by $4/1 = 4$:

$$\begin{array}{cc} 3 & y \end{array}$$
$$\begin{bmatrix} 1^* & -2 \\ 4 & 7 \end{bmatrix} \begin{array}{l} = x \\ = 5. \end{array}$$

Step 5: The entry 7, which is the only entry outside the pivot row or the pivot column, is replaced by $7 - 2 \cdot 4 = -1$:

$$\begin{array}{cc} 3 & y \end{array}$$
$$\begin{bmatrix} 1^* & -2 \\ 4 & -1 \end{bmatrix} \begin{array}{l} = x \\ = 5 \end{array}$$

Notice that the new system of equations is

$$\begin{cases} 3 - 2y = x \\ 4 \cdot 3 - y = 5 \end{cases}$$

(compare with the steps of the substitution method).

Now we pick -1 as the pivot entry. Again, we mark the pivot entry by a superscript $*$:

$$\begin{array}{cc} 3 & y \end{array}$$
$$\begin{bmatrix} 1 & -2 \\ 4 & -1^* \end{bmatrix} \begin{array}{l} = x \\ = 5. \end{array}$$

By pivoting, step by step, as we did before, we obtain the following tableau:

$$\begin{array}{cc} 3 & 5 \end{array}$$
$$\begin{bmatrix} -7 & 2 \\ 4 & -1 \end{bmatrix} \begin{array}{l} = x \\ = y. \end{array}$$

Here we make the following two remarks:

(a) The matrix $\begin{bmatrix} -7 & 2 \\ 4 & -1 \end{bmatrix}$ is the inverse matrix of the matrix of coefficients $\begin{bmatrix} 1 & 2 \\ 4 & 7 \end{bmatrix}$; that is,

$$\begin{bmatrix} 1 & 2 \\ 4 & 7 \end{bmatrix}^{-1} = \begin{bmatrix} -7 & 2 \\ 4 & -1 \end{bmatrix}.$$

(b) We obtain a tableau that has no variables on the top margin. Therefore, we can combine the two columns into a single column, having the constant 1 in the top margin, as follows: $3 \cdot -7 + 5 \cdot 2 = -11$; $3 \cdot 4 + 5 \cdot -1 = 7$. The final tableau gives the solution of the linear system of equations:

$$\begin{array}{c} 1 \end{array}$$
$$\begin{bmatrix} -11 \\ 7 \end{bmatrix} \begin{array}{l} = x \\ = y. \end{array}$$

It follows that the solution is $x = -11 \cdot 1 = -11$; $\quad y = 7 \cdot 1 = 7$.

Thus, we can use pivoting to invert matrices (if the matrix is invertible) and to solve linear systems of equations. This may not be the best or most efficient method to find the inverse of a matrix or to solve a linear system of equations, but it works. Here are more details about solving an arbitrary system of linear equations, $AX = b$, by pivoting. We write the system as a row tableau

$$\begin{array}{c} X^T \\ \begin{bmatrix} A \end{bmatrix} = b. \end{array}$$

Then by repeated pivoting, we switch constants at the right margin with variables on the top until we cannot do so anymore.

After this, we combine all constant columns into the last column with a coefficient of 1 in the top margin; if there are no constant columns, we add one, consisting of a coefficient of 1 in the top margin and zeros in the remaining entries. Before looking at the final tableau, let us see how this constant column is formed.

Suppose that, after switching constants in the right margin with variables on the top, we have the following tableau:

$$
\begin{array}{cccccc}
y_1 & c_1 & y_2 & c_2 & \cdots & \\
\left[\begin{array}{ccccc}
a_{11} & b_{11} & a_{12} & b_{12} & \cdots \\
a_{21} & b_{21} & a_{22} & b_{22} & \cdots \\
\vdots & \vdots & \vdots & \vdots & \vdots
\end{array}\right] & & & & & \begin{array}{l} = z_1 \\ = z_2 \\ \vdots, \end{array}
\end{array}
$$

where y_i represents a row matrix of variables, c_i represents a constant, a_{ij} represents a matrix and b_{ij} represents a column matrix. The columns having the constants c_is in the top margin represent the constant columns. Thus, in the first equation of the new linear system of equations, the constant term is given by

$$c_1 \cdot b_{11} + c_2 \cdot b_{12} + \cdots = b_1'.$$

Similarly, the constant term in the second equation is

$$c_1 \cdot b_{21} + c_2 \cdot b_{22} + \cdots = b_2',$$

and so on. Writing this column as the last, with coefficient 1 in the top margin, we obtain our final tableau:

$$
\begin{array}{cc}
y & 1 \\
\left[\begin{array}{cc}
a & b' \\
c & d
\end{array}\right] & \begin{array}{l} = z \\ = e. \end{array}
\end{array}
$$

In this final tableau, the letters y and z represent a row matrix of variables and a column matrix of variables, respectively; a and c represent matrices; and $d, e,$ and b' represent column matrices with constant entries.

Note that all entries of the matrix c are zeros. The reason is the following: If c has a nonzero entry, we pivot on that entry to switch the corresponding variable from y and the corresponding entry in e. But since we are in the final tableau, no more of these switches is possible. Therefore, c has no nonzero entries. Notice also that c may not be present in the final tableau (see Example 8.10).

If $d \neq e$, then our system has no solutions (it is an *inconsistent* system). Otherwise, we can disregard the bottom portion (which yields the equality $d = e$,) and now the system takes the form

$$
\begin{array}{cc} y & 1 \end{array}
$$
$$
\begin{bmatrix} a & b' \end{bmatrix} \quad = z,
$$

that is, $z = b' + ay$. So the variables in y (if any) can take arbitrary values and then the variables in z (if any) take the corresponding values $z = b' + ay$.

Definition 8.11.   The number of variables in y is called the *dimension* of the set of solutions; that is, the dimension of the set of solutions is the number of columns in the matrix a. ∎

Thus, any linear system can be solved by pivot steps. Can you see that the total number of pivot steps needed is at most the number of variables in the system?

Remark 8.12. It is possible to generalize our pivot step by replacing a nonzero pivot entry α by an invertible submatrix α. Taking care to write matrices in matrix products in correct order, we obtain the following "generalized pivot step":

$$
\begin{array}{cc} x & y \end{array} \qquad\qquad\qquad \begin{array}{cc} u^T & y \end{array}
$$
$$
\begin{bmatrix} \alpha^* & \beta \\ \gamma & \delta \end{bmatrix} \begin{array}{c} = u \\ = v \end{array} \quad \mapsto \quad \begin{bmatrix} \alpha^{-1} & -\alpha^{-1}\beta \\ \gamma\alpha^{-1} & \delta - \gamma\alpha^{-1}\beta \end{bmatrix} \begin{array}{c} = x^T \\ = v \end{array}.
$$

This step can be replaced by k ordinary pivot steps and a permutation of rows or columns of the tableau, where $k \times k$ is the size of matrix α.

Exercises

1–2. Using a row tableau, write

1. The diet problem (Example 2.1).

2. The blending problem (Example 2.2). ∎

3. Read the following tableau and rewrite it as a system of linear equations $Ax = b$, for x', u, y, z with $x = [x', u, y, z]^T$:

$$
\begin{array}{cccc} x' & z & y & 1 \end{array}
$$
$$
\begin{bmatrix} 3 & 2 & 1 & y \\ -1 & 3 & 0 & 1 \\ 0 & -2 & 2 & 0 \\ 0 & 0 & 1 & 2 \end{bmatrix} \begin{array}{c} = u \\ = z \\ = x' \\ = z. \end{array}
$$

4–7. In the following optimization problems it is given that all variables $x, y, z \geq 0$, and the rest of data are given by tableaux. Rewrite the problems as linear programs in standard and canonical forms. You are not required to solve the optimization problems.

4.
$$
\begin{array}{cccc}
x & y & z & 1 \\
\end{array}
$$
$$
\left[
\begin{array}{cccc}
3 & 2 & 1 & 3 \\
-1 & 3 & 0 & 1 \\
0 & -2 & 2 & 0 \\
0 & 0 & 1 & 2 \\
\end{array}
\right]
\begin{array}{l}
\geq 0 \\
= 3 \\
\leq x \\
\to \min
\end{array}
$$

5.
$$
\begin{array}{cccc}
x & 3 & y - 5z & 1 \\
\end{array}
$$
$$
\left[
\begin{array}{cccc}
3 & 2 & 1 & 0 \\
-1 & 3 & 0 & 1 \\
0 & -2 & 2 & 0 \\
0 & 0 & 1 & 2 \\
\end{array}
\right]
\begin{array}{l}
\geq 3 \\
= y \\
\geq 1 \\
\to \min
\end{array}
$$

6.
$$
\begin{array}{cccc}
x & -y & z & -3 \\
\end{array}
$$
$$
\left[
\begin{array}{cccc}
3 & 2 & 1 & x \\
-1 & 3 & 0 & 1 \\
0 & -2 & 2 & 0 \\
0 & 0 & 1 & 2 \\
\end{array}
\right]
\begin{array}{l}
= 0 \\
\geq 0 \\
\leq 0 \\
\to \max
\end{array}
$$

7.
$$
\begin{array}{cccc}
x & y & z & 1 \\
\end{array}
$$
$$
\left[
\begin{array}{cccc}
3 & 2 & 1 & 2 \\
-1 & 3 & 0 & 1 \\
0 & -2 & 2 & 0 \\
-1 & 0 & 1 & 2 \\
\end{array}
\right]
\begin{array}{l}
\geq 0 \\
\geq 0 \\
\geq 0 \\
= \to \min
\end{array}
\qquad \blacksquare
$$

8. In the following tableau perform the steps of pivoting at the entry whose value is 1.

$$
\begin{array}{cccc}
x' & x'' & y & 1 \\
\end{array}
$$
$$
\left[
\begin{array}{cccc}
3 & 2 & 1 & b \\
-1 & 3 & 0 & 1 \\
0 & -2 & 2 & 0 \\
0 & 0 & 1 & 2 \\
\end{array}
\right]
\begin{array}{l}
= u \\
= a \\
= x' \\
= z
\end{array}
$$

(a) In the first row (b) In the last row

9. Invert the matrix of Exercise 1 in four pivoting steps.

10–17. In each of the following tableaux, perform the pivoting procedure in the entry marked with a superscript *.

10.
$$\begin{matrix} & 1 & a & 3 & x \\ & \begin{bmatrix} 1^* & 0 & b & a \\ -1 & 2 & 3 & 1 \end{bmatrix} & & & & \begin{matrix} = y \\ = z \end{matrix} \end{matrix}$$

11.
$$\begin{matrix} & 1 & a & 3 & x \\ & \begin{bmatrix} 1 & 0 & b & a \\ -1^* & 2 & 3 & 1 \end{bmatrix} & & & & \begin{matrix} = y \\ = z \end{matrix} \end{matrix}$$

12.
$$\begin{matrix} & 1 & a & 3 & x \\ & \begin{bmatrix} 1 & 0 & b & a \\ -1 & 2 & 3 & 1^* \end{bmatrix} & & & & \begin{matrix} = y \\ = z \end{matrix} \end{matrix}$$

13.
$$\begin{matrix} x \\ \begin{bmatrix} 5^* \end{bmatrix} & = 2 \end{matrix}$$

14.
$$\begin{matrix} & 1 & a & 0 & x & x \\ & \begin{bmatrix} 1 & 0 & b & a & -3 \\ -1 & 2^* & 3 & 1 & 0 \end{bmatrix} & & & & & \begin{matrix} = y \\ = z \end{matrix} \end{matrix}$$

15.
$$\begin{matrix} & 1 & 0 & 0 & x & 1 \\ & \begin{bmatrix} 1 & 0 & b & a & -3 \\ -1 & 2 & 3 & 1 & 0 \\ -1 & 2^* & 3 & 1 & 1 \\ -1 & 2 & 3 & 1 & 0 \end{bmatrix} & & & & & \begin{matrix} = y \\ = z \\ = u \\ = v \end{matrix} \end{matrix}$$

16.
$$\begin{matrix} & x_1 & x_2 & x_3 & x_4 & 1 \\ & \begin{bmatrix} 1 & 0 & 1 & 1 & -3 \\ -1 & 0 & 2 & 1 & 0 \\ -1 & 2 & 3 & 1 & 1 \\ -1 & 2 & 3^* & 1 & 0 \\ -1 & 1 & 0 & 1 & 1 \end{bmatrix} & & & & & \begin{matrix} = x_5 \\ = x_6 \\ = x_7 \\ = x_8 \\ = v \end{matrix} \end{matrix}$$

17.
$$\begin{matrix} & x_1 & x_2 & x_3 & x_4 & x_5 & x_6 & 1 \\ & \begin{bmatrix} 1 & 0 & 1 & 1 & -3^* & 1 & 0 \\ -1 & 0 & 2 & 1 & 0 & 1 & -2 \\ -1 & 2 & 3 & 1 & 1 & 0 & 0 \\ -1 & 2 & 3 & 1 & 0 & 1 & 1 \\ -1 & 1 & 0 & 1 & 1 & 2 & 3 \end{bmatrix} & & & & & & \begin{matrix} = x_7 \\ = x_8 \\ = x_9 \\ = x_{10} \\ = v \end{matrix} \end{matrix}$$

∎

18. Suppose two systems of equations are given by two tableaux

$$\begin{matrix} y \\ \begin{bmatrix} A \end{bmatrix} & = z \end{matrix} \quad \text{and} \quad \begin{matrix} y \\ \begin{bmatrix} A' \end{bmatrix} & = z \end{matrix}$$

with the same variables y on the top and the same variables z at the right margin. Assume also that the systems are equivalent (i.e., have the same sets of solutions). Show that the matrices A and A' are equal.

§9. Standard Row Tableaux

In this section we write linear programs in tableau form and then use pivoting to solve them. We would like to find a method of choosing pivot entries that works for any linear program; that is, this method, when applied to a linear programming problem, would give the optimal value and an optimal solution (if they exist). This method of choosing pivot entries is, roughly speaking, what we mean when we use the phrase *simplex method*.

Using the algebraic tricks of §7, we can arrange that all the variables in our linear program are nonnegative, with the possible exception of the objective variable. We will not write these constraints in our tableaux, but they must not be forgotten (write them near your tableau).

Definition 9.1   A standard row tableau has the form

$$
\begin{matrix} x & 1 \end{matrix}
$$
$$
\begin{bmatrix} A & b \\ c & d \end{bmatrix} \begin{matrix} = u \\ = z \end{matrix} \mapsto \min, \quad x \geq 0, \quad u \geq 0,
$$

where A is the matrix of coefficients, b is a column matrix representing the right-hand side of the linear constraints, c is the row matrix representing the coefficients of the objective function, d is a given number, x is a row matrix of variables, u is a column of variables and z, which may or may not be written, is the objective variable; all variables in x, u, z are distinct.

More precisely, a tableau is *standard* if

• The coefficient matrix does not contain any unknowns (only given numbers or, sometimes, parameters).

• Every row represents a linear equation (rather than an inequality) with the possible exception of the last row.

• The last row represents the objective function to be minimized.

• All variables at the margins, with the possible exception of the objective variable, are nonnegative. (Recall that we will write these nonnegativity constraints near the tableau, but not as part of it.)

• All variables at the margins are distinct independent variables.

• There is exactly one constant at the top margin; it is 1 and it appears over the last column.   ∎

As we mentioned in Definition 9.1, the objective variable need not be written out in a standard tableau.

Thus, solving a linear program by a simplex method, after stating the problem precisely and collecting data, involves the following steps:

1. Write the problem as a standard tableau (here, if necessary, we change variables and include in the tableau only nonnegative variables, with the possible exception of the objective variable).

2. Apply pivoting until we either obtain an optimal solution or find out that there are no solutions (how we choose pivot elements depends on the method; later we will discuss how to make these choices).

3. Write the answer in terms of the original variables.

4. Give an interpretation of the answer in terms of the original word problem. Sometimes, if we find that the answer is not satisfactory, we adjust the model, collect more data, solve the corresponding linear programming problem, and so on.

Example 9.2. Suppose that our linear program in variables x, y, z, w is given by the following tableau (without any other constraints):

$$
\begin{array}{c}
\quad 2 \quad -x \quad y \quad -1 \\
\begin{bmatrix}
z & 0 & 1 & 2 \\
0 & 0 & 1 & 0 \\
z & 0 & 0 & 0 \\
1 & 2 & 3 & 4
\end{bmatrix}
\begin{array}{l}
= w \to \max \\
\geq 0 \\
\geq 1 \\
= y
\end{array}
\end{array}
$$

This tableau is not standard. In fact, every rule we have just discussed is broken. Namely,

• We do not have the constraint $x \geq 0$

• The variable z is in the matrix (in standard tableaux, it belongs in the margins)

• The second row and the third row of the tableau have the $\geq$ sign;

• The objective function is represented by the first row of the matrix rather than the last one

• We have a maximization problem instead of minimization

• The variable y shows up twice; the variable x occurs with coefficient -1 rather than 1

- There are too many constants at the margin
- The constant over the last column of the matrix is -1 rather than 1.

Can you see how to put this tableau in standard form?

Actually, there are many ways to rewrite this linear programming problem using a standard row tableau. For example, we read the tableau and obtain

$$\begin{cases} 2z + y - 2 = w \to \max \\ y \geq 0 \\ 2z \geq 1 \\ 2 - 2x + 3y - 4 = y, \end{cases}$$

or

$$\begin{cases} -2z - y + 2 = -w \to \min \\ y \geq 0 \\ z \geq 1/2 \\ x = y - 1. \end{cases}$$

We introduce a *slack* variable u to write $z - 1/2 = u \geq 0$ instead of $z \geq 1/2$. Now we write our problem in standard tableau form:

$$\begin{array}{ccc} y & z & 1 \end{array}$$
$$\begin{bmatrix} 0 & 1 & -1/2 \\ -1 & -2 & 2 \end{bmatrix} \begin{array}{l} = u \\ = w \to \min \end{array}$$

with $y \geq 0, z \geq 0, u \geq 0, x = y - 1$.

Note that the tableau does not include the variable x and that the additional information ($\to \min$, $y \geq 0$, $z \geq 0$, $u \geq 0$, $x = y - 1$) is written outside the tableau.

Example 9.3.

Suppose our linear problem is given as follows:

$$\begin{array}{cccccc} v & 1 & x & y & x & -2 \end{array}$$
$$\begin{bmatrix} 0 & 1 & 2 & 3 & 4 & 5 \\ 0 & 0 & -7 & 8 & -9 & 1 \\ 0 & 0 & 0 & 1 & 0 & 0 \\ 0 & 3 & 1 & 0 & 2 & 0 \\ 0 & 3 & 1 & 0 & 2 & 0 \\ 1 & 0 & 0 & 2 & 1 & 0 \\ 0 & 0 & 1 & 0 & 2 & 0 \\ 1 & 0 & 0 & 0 & -1 & 0 \end{bmatrix} \begin{array}{l} = 0 \\ \geq 0 \\ \geq 0 \\ \geq 0 \\ \leq 0 \\ = z \to \max \\ = u \\ = u. \end{array}$$

Let us bring this tableau to the standard form. We start by combining the constant columns (that is, the columns marked with 1 and -2 at the top margin) to get the column

$$1 \cdot 1 + (-2 \cdot 5) = -9$$
$$1 \cdot 0 + (-2 \cdot 1) = -2$$
$$1 \cdot 0 + (-2 \cdot 0) = 0$$
$$1 \cdot 3 + (-2 \cdot 0) = 3$$
$$1 \cdot 3 + (-2 \cdot 0) = 3$$
$$1 \cdot 0 + (-2 \cdot 0) = 0$$
$$1 \cdot 0 + (-2 \cdot 0) = 0$$
$$1 \cdot 0 + (-2 \cdot 0) = 0.$$

We will write this constant column as the last column in the tableau, marked with a coefficient equal to 1 on the top.

$$
\begin{array}{ccccc}
v & x & y & x & 1 \\
\end{array}
$$
$$
\begin{bmatrix}
0 & 2 & 3 & 4 & -9 \\
0 & -7 & 8 & -9 & -2 \\
0 & 0 & 1 & 0 & 0 \\
0 & 1 & 0 & 2 & 3 \\
0 & 1 & 0 & 2 & 3 \\
1 & 0 & 2 & 1 & 0 \\
0 & 1 & 0 & 2 & 0 \\
1 & 0 & 0 & -1 & 0
\end{bmatrix}
\begin{array}{l}
= 0 \\
\geq 0 \\
\geq 0 \\
\geq 0 \\
\leq 0 \\
= z \rightarrow \max \\
= u \\
= u.
\end{array}
$$

Then we can combine the two columns marked with x to obtain a single column marked with x; the column in the matrix of coefficients is obtained by adding the corresponding entries of the two columns

$$2 + 4 = 6$$
$$-7 - 9 = -16$$
$$0 + 0 = 0$$
$$1 + 2 = 3$$
$$1 + 2 = 3$$
$$0 + 1 = 1$$
$$1 + 2 = 3$$
$$0 - 1 = -1.$$

We obtain the following tableau:

$$
\begin{array}{cccc}
v & y & x & 1 \\
\end{array}
$$

$$
\left[\begin{array}{cccc}
0 & 3 & 6 & -9 \\
0 & 8 & -16 & -2 \\
0 & 1 & 0 & 0 \\
0 & 0 & 3 & 3 \\
0 & 0 & 3 & 3 \\
1 & 2 & 1 & 0 \\
0 & 0 & 3 & 0 \\
1 & 0 & -1 & 0
\end{array}\right]
\begin{array}{l}
= 0 \\
\geq 0 \\
\geq 0 \\
\geq 0 \\
\leq 0 \\
= z \to \max \\
= u \\
= u.
\end{array}
$$

Next, we write the coefficients of the objective function in the last row of the matrix:

$$
\begin{array}{cccc}
v & y & x & 1 \\
\end{array}
$$

$$
\left[\begin{array}{cccc}
0 & 3 & 6 & -9 \\
0 & 8 & -16 & -2 \\
0 & 1 & 0 & 0 \\
0 & 0 & 3 & 3 \\
0 & 0 & 3 & 3 \\
0 & 0 & 3 & 0 \\
1 & 0 & -1 & 0 \\
1 & 2 & 1 & 0
\end{array}\right]
\begin{array}{l}
= 0 \\
\geq 0 \\
\geq 0 \\
\geq 0 \\
\leq 0 \\
= u \\
= u \\
= z \to \max.
\end{array}
$$

We can also subtract one u-row from the other. This gives us two rows − one will be one of the original rows decorated with u at the right margin and the other row will be marked with 0 at the right margin. We do this is because we want distinct variables at the margins:

$$
\begin{array}{cccc}
v & y & x & 1 \\
\end{array}
$$

$$
\left[\begin{array}{cccc}
0 & 3 & 6 & -9 \\
0 & 8 & -16 & -2 \\
0 & 1 & 0 & 0 \\
0 & 0 & 3 & 3 \\
0 & 0 & 3 & 3 \\
0 & 0 & 3 & 0 \\
1 & 0 & -4 & 0 \\
1 & 2 & 1 & 0
\end{array}\right]
\begin{array}{l}
= 0 \\
\geq 0 \\
\geq 0 \\
\geq 0 \\
\leq 0 \\
= u \\
= 0 \\
= z \to \max.
\end{array}
$$

Now, since we do not want constants at the right margin, we introduce new non-negative (slack or surplus) variables w_1 and w_2 for the inequalities in the second and fourth rows of the preceding matrix. Note that since the inequality in the fifth row reads $3x + 3 \leq 0$, we replace it by the equivalent inequality $-3x - 3 \geq 0$ and then introduce the nonnegative slack variable w_3, so that we have the equality $-3x - 3 = w_3$. Since the third row, $y \geq 0$, is just the nonnegativity constraint for the variable y, we write it separately near the tableau. We also swap the columns labeled with x and y. The new tableau is

$$
\begin{array}{cccc}
v & x & y & 1 \\
\end{array}
$$
$$
\left[
\begin{array}{cccc}
0 & 6 & 3 & -9 \\
0 & -16 & 8 & -2 \\
0 & 3 & 0 & 3 \\
0 & -3 & 0 & -3 \\
0 & 3 & 0 & 0 \\
1 & -4 & 0 & 0 \\
1 & 1 & 2 & 0 \\
\end{array}
\right]
\begin{array}{l}
= 0 \\
= w_1 \\
= w_2 \\
= w_3 \\
= u \\
= 0 \\
= z \to \max
\end{array}
$$

with $y \geq 0, w_1 \geq 0, w_2 \geq 0, w_3 \geq 0$.

Now we take the constraint $u = 3x$ (fifth row of the matrix) out of the tableau and write it separately, near the tableau. We also permute the rows so that the rows marked with a constant at the right margin appear first:

$$
\begin{array}{cccc}
v & x & y & 1 \\
\end{array}
$$
$$
\left[
\begin{array}{cccc}
0 & 6 & 3 & -9 \\
1 & -4 & 0 & 0 \\
0 & -16 & 8 & -2 \\
0 & 3 & 0 & 3 \\
0 & -3 & 0 & -3 \\
1 & 1 & 2 & 0 \\
\end{array}
\right]
\begin{array}{l}
= 0 \\
= 0 \\
= w_1 \\
= w_2 \\
= w_3 \\
= z \to \max
\end{array}
$$

with $y \geq 0, w_1 \geq 0, w_2 \geq 0, w_3 \geq 0, u = 3x$.

We still have two constants at the right margin. We can fix this by pivoting at 1 in the second row and 6 in the first row, so that the constants will appear at the top margin:

$$
\begin{array}{cccc}
0 & 0 & y & 1 \\
\end{array}
$$
$$
\left[
\begin{array}{cccc}
0 & 1/6 & -1/2 & 3/2 \\
1 & 2/3 & -2 & 6 \\
0 & -8/3 & 16 & -26 \\
0 & 1/2 & -3/2 & 15/2 \\
0 & -1/2 & 3/2 & -15/2 \\
1 & 5/6 & -1/2 & 15/2 \\
\end{array}
\right]
\begin{array}{l}
= x \\
= v \\
= w_1 \\
= w_2 \\
= w_3 \\
= z \to \max
\end{array}
$$

with $y \geq 0, w_1 \geq 0, w_2 \geq 0, w_3 \geq 0, u = 3x.$

In order to have a minimization problem instead of a maximization problem, we multiply the last row by -1 :

$$
\begin{array}{cccc}
0 & 0 & y & 1
\end{array}
$$
$$
\begin{bmatrix}
0 & 1/6 & -1/2 & 3/2 \\
1 & 2/3 & -2 & 6 \\
0 & -8/3 & 16 & -26 \\
0 & 1/2 & -3/2 & 15/2 \\
0 & -1/2 & 3/2 & -15/2 \\
-1 & -5/6 & 1/2 & -15/2
\end{bmatrix}
\begin{array}{l}
= x \\
= v \\
= w_1 \\
= w_2 \\
= w_3 \\
= -z \to \min
\end{array}
$$

with $y \geq 0, w_1 \geq 0, w_2 \geq 0, w_3 \geq 0, u = 3x.$

The variables x and v are not required to be nonnegative, so they cannot be present in a standard tableau. On the other hand, we do not need to decide anything about these variables, so we put the first two constraints outside the tableau. The columns with 0 on the top margin can be dropped. The following is a standard tableau representing the linear program:

$$
\begin{array}{cc}
y & 1
\end{array}
$$
$$
\begin{bmatrix}
16 & -26 \\
-3/2 & 15/2 \\
3/2 & -15/2 \\
1/2 & -15/2
\end{bmatrix}
\begin{array}{l}
= w_1 \\
= w_2 \\
= w_3 \\
= -z \to \min,
\end{array}
$$

$y \geq 0, w_1 \geq 0, w_2 \geq 0, w_3 \geq 0, u = 3x, x = 3/2 - y/2, v = 6 - 2y.$ ∎

Going from General LP in Canonical Form to Standard Tableau and Back

Usually, it takes less effort to write a linear program in a standard tableau. For example, when our linear program is written in the canonical form

$$\text{minimize } cx + d, \text{ subject to } Ax \leq b, \ x \geq 0,$$

we proceed as follows:

1. Replace the constraints $Ax \leq b$ by the equivalent constraints $-Ax \geq -b$.

2. The constraints $-Ax \geq -b$ can be read as $-Ax + b \geq 0$. In order to have equations rather than inequalities, we introduce the

slack variables u, defined as $u = Ax + b$. Note that we now require $x \geq 0$ and $y \geq 0$. The standard row tableau can be written right away:

$$
\begin{array}{cc}
x^T & 1
\end{array}
$$
$$
\begin{bmatrix} -A & b \\ c & d \end{bmatrix}
\begin{array}{l} = \quad u \\ \rightarrow \quad \min, \quad x \geq 0, \quad y \geq 0 \end{array}
$$

where y is the column matrix of slack variables (all variables in y are distinct and different from the variables in the column matrix x).

It is also very easy to go from a standard tableau to a linear program in canonical form (we drop the variables at the right margin). So there is almost no difference between standard tableaux and canonical forms.

Going from General LP in Standard Form to Standard Tableau and Back

If our linear program is given in the standard form

$$\text{minimize } cx + d, \text{ subject to } Ax = b, x \geq 0:$$

then there are at least two ways to put it in a standard tableau.

(i) (artificial variables) Reduce the problem to a canonical form

$$\text{minimize } cx + d \text{ subject to } Ax \leq b, -Ax \leq -b, x \geq 0$$

and then proceed as before.

Thus, we obtain the standard tableau

$$
\begin{array}{cc}
x^T & 1
\end{array}
$$
$$
\begin{bmatrix} -A & b \\ A & -b \\ c & d \end{bmatrix}
\begin{array}{l} = \quad u \\ = \quad v \\ \rightarrow \quad \min, \quad x \geq 0, \quad y \geq 0 \end{array} .
$$

The variables in the columns u, v are called *artificial variables*. They are connected with artificial variables in Trick 7.7 by duality; see Chapter 4.

(ii) Solve the system $Ax = b$ (this can be done, for example, by pivoting, as we explained before). Then we either find that the system has no solutions, hence the linear program is infeasible, or we get an answer of the form $y = Dz + e$, where y are some variables in x and z is the rest of variables in x. Then we express our objective function $cx + d = c'z + d'$ in terms of z. Thus, we obtain the standard row tableau

$$\begin{array}{cc} z^T & 1 \end{array}$$
$$\begin{bmatrix} D & e \\ c' & d' \end{bmatrix} \quad \begin{array}{c} = \\ \to \end{array} \quad \begin{array}{c} y \\ \min, \end{array} \quad z \ge 0, \quad y \ge 0$$

It is even easier to go back from any standard tableau to the corresponding linear program in standard form: Just take the variables at the right margin to the left-hand sides of the corresponding equations and take the constant terms to the right-hand sides of those equations. That is, rewrite $Ax + b = u$ in Definition 9.1 as $Ax - u = b$. ∎

If we have a linear program with some variables that are not required to be nonnegative, sometimes we can easily eliminate them from the program to get a standard form.

Example 9.4. Write in a standard tableau:
$$x + y + z \to \min, \; 2x + 3y + 4z \le 5, \; x - y + 4z = 3, \; x \ge 0, y \ge 0.$$

Solution. We solve the equation for z and eliminate it from the problem: $z = (3 - x + y)/4$,
$$x + y + z = x + y + (3 - x + y)/4 = 3x/4 + 5y/4 + 3/4 \to \min,$$
$$2x + 3y + 4z = 2x + 3y + 4(3 - x + y)/4 = x + 4y + 3 \le 5, \; x \ge 0, y \ge 0,$$
or
$$3x/4 + 5y/4 + 3/4 \to \min, \; x + 4y \le 2, x \ge 0, y \ge 0.$$

The problem with two variables x, y is in normal form, so can be written easily in a standard tableau:

$$\begin{array}{ccc} x & y & 1 \end{array}$$
$$\begin{bmatrix} -1 & -4 & 2 \\ 3/4 & 5/4 & 3/4 \end{bmatrix} \quad \begin{array}{c} = u \\ \to \min \end{array}, \; x \ge 0, y \ge 0, z = (3 - x + y)/4.$$

We put $z = (3 - x + y)/4$ on side, so the value of z can be included in the final answer. Although it was not required, it is now clear that the optimization problem has exactly one optimal solution: $x = 0, y = 0$, min$= 3/4, z = 3/4$. ∎

Example 9.5. Write in a standard tableau:

x_1	x_2	x_3	x_4	x_5	x_6	3	
1	0	1	1	−3	1	0	$= x_7$
−1	0	2	1	0	1	−2	$= x_8$
−1	2	3	1	1	0	0	$= x_9$
−1	2	3	1	0	1	1	$= x_{10}$
−1	1	0	1	1	2	3	$= v \to \max$

$$x_1, x_2, x_3, x_4, x_5, x_6, x_7, x_{10} \ge 0.$$

Solution. We eliminate x_8, x_9 from the tableau, multiply the objective variable v by -1, and multiply the last column by 3:

$$
\begin{array}{ccccccc}
x_1 & x_2 & x_3 & x_4 & x_5 & x_6 & 1 \\
\left[\begin{array}{cccccc}
1 & 0 & 1 & 1 & -3 & 1 \\
-1 & 2 & 3 & 1 & 0 & 1 \\
1 & -1 & 0 & -1 & -1 & -2
\end{array}\right. & & & & & & \left.\begin{array}{c}
0 \\
3 \\
-9
\end{array}\right]
\end{array}
\begin{array}{l}
= x_7 \\
= x_{10} \\
= -v \to \min
\end{array},
$$

$$x_1, x_2, x_3, x_4, x_5, x_6 \geq 0,$$

$$x_8 = -x_1 + 2x_3 + x_4 + x_6 - 6, \quad x_9 = -x_1 + 2x_2 + 3x_3 + x_4 + x_5. \ \blacksquare$$

Exercises

1. Solve the linear program of Example 9.3 and give your answer in terms of the original variables x, u and v.

2–10. Rewrite the following optimization problems as linear programs using standard row tableaux. Optimization is not required. *Hint*: first write your problem as a linear program in canonical form.

2. $\begin{cases} \text{Maximize} & P = 2x + 3y \\ \text{subject to} & x \geq 0, \ y \geq 0 \\ & 4x + 5y \leq 7. \end{cases}$

3. $\begin{cases} \text{Maximize} & x \\ \text{subject to} & |x + y| \leq 1 \\ & |x - y| \leq 1. \end{cases}$

4.

$$
\begin{array}{ccccccc}
x_1 & x_2 & x_3 & x_4 & x_5 & x_6 & -1 \\
\left[\begin{array}{ccccccc}
1 & 0 & 1 & 1 & -3 & 1 & 0 \\
-1 & 0 & 2 & 1 & 0 & 1 & -2 \\
-1 & 2 & 3 & 1 & 1 & 0 & 0 \\
-1 & 1 & 0 & 1 & 1 & 2 & 3
\end{array}\right]
\end{array}
\begin{array}{l}
\geq 0 \\
= x_8 \\
= 0 \\
= v \to \min
\end{array}
$$

$$x_1, x_2, x_3, x_4, x_5, x_6, x_7, x_8, x_9, x_{10} \geq 0.$$

5.

$$
\begin{array}{ccccccc}
x_1 & x_2 & x_3 & x_4 & x_5 & x_6 & 1 \\
\left[\begin{array}{ccccccc}
1 & 0 & 1 & 1 & -3 & 1 & 0 \\
-1 & 0 & 2 & 1 & 0 & 1 & -2 \\
-1 & 2 & 3 & 1 & 1 & 0 & 0 \\
0 & 2 & 3 & 1 & 0 & 1 & 1 \\
-1 & 1 & 0 & 1 & 1 & 2 & 3
\end{array}\right]
\end{array}
\begin{array}{l}
= x_7 \\
= x_8 \\
= x_9 \\
= x_2 \\
= v \to \max
\end{array}
$$

$$x_1, x_2, x_3, x_4, x_5, x_6, x_7 \geq 0.$$

6.

$2x_1$	x_2	$-3x_1$	x_4	x_5	x_6	1	
1	0	1	1	-3	1	0	$= x_7$
-1	0	2	1	0	1	-2	$= x_7$
0	2	3	1	1	0	0	≤ 0
-1	2	3	1	0	1	1	$= x_3$
-1	1	0	1	1	2	3	$= v \to \min$

$x_1, x_2, x_3, x_4, x_5, x_6, x_7 \geq 0.$

7.

x_1	x_2	x_3	x_4	x_5	x_6	1	
1	0	1	1	-3	1	0	$= x_2$
-1	0	2	1	0	1	-2	$= x_4$
-1	2	3	1	1	0	0	$= v \to \min$

$x_1, x_2, x_3, x_4, x_5, x_6, x_7, x_8, x_9, x_{10} \geq 0.$

8.

x_1	x_2	x_3	x_4	x_5	x_6	-4	
1	0	1	1	-3	1	0	≥ 0
-1	0	2	1	0	1	-2	$= 1$
-1	2	3	1	1	0	0	$= 0$
-1	2	3	1	0	1	1	$= x_1$
-1	1	0	1	1	2	3	$= v \to \min$

$x_1, x_2, x_3, x_4, x_5, x_6 \geq 0.$

9.

x_7	x_2	x_3	x_4	x_5	x_6	1	
1	0	1	1	-3	1	0	$= 0$
-1	0	2	1	0	1	-2	$= 0$
-1	2	3	1	1	0	0	$= 0$
-1	2	3	1	0	1	1	$= x_1$
-1	1	0	1	1	2	3	$= v \to \min$

$x_1, x_2, x_3, x_4, x_5, x_6 \geq 0.$

10.

x_1	x_2	x_3	x_4	x_5	x_6	-1	
1	0	1	1	-3	1	0	$= x_1$
-1	0	2	1	0	1	-2	$= 2x_1$
-1	2	3	1	1	0	0	x_7
-1	2	3	1	0	1	1	$= x_2$
-1	1	0	1	1	2	3	$= v \to \max$

$x_1, x_2, x_3, x_4, x_5, x_6 \geq 0.$

Chapter 4

Simplex Method

§10. Simplex Method, Phase 2

The simplex method, described in terms of standard row tableaux, is a way of choosing pivot entries in order to reach a terminal tableau in finitely many pivot steps. Usually, a simplex method works in two stages, or phases. In Phase 1 our objective is to find a feasible solution, and in Phase 2 we try to improve our objective function, staying feasible until we reach an optimal solution. In this section we provide a precise description of Phase 2. Phase 1 is similar but somewhat more complicated and will be described in detail in the next section. There are tricks to reduce Phase 1 to Phase 2 as well as Phase 2 to Phase 1. The dual point of view switches the two phases (Chapter 5).

We start with an arbitrary linear program written in standard row tableau:

$$
\begin{array}{cc}
x & 1
\end{array}
$$
$$
\begin{bmatrix} A & b \\ c & d \end{bmatrix} \begin{array}{l} = u \\ = z \to \min, \quad x \geq 0, \quad u \geq 0 \end{array} \tag{10.1}
$$

Here A is a matrix of coefficients, b is a column representing the right-hand sides of linear constraints, c is a row representing the coefficients of the objective function, d is a given number, x is a row of variables (called *nonbasic variables*), u is a column of variables (called *basic variables*), and z is the objective variable; all variables in x, u, z are distinct.

In matrix notation, our linear program is

minimize $z = cx^T + d$, subject to $Ax^T + b = u \geq 0$, $x \geq 0$.

If we obtained this tableau from the linear program in canonical form

minimize $z = cx^T + d$, subject to $-Ax^T \geq b$, $x \geq 0$,

then the variables in u are slack variables. However, afte
steps, some slack variables can become nonbasic (i.e.,
to the top margin). In this case we say that these variables a....
corresponding inequalities in the normal form become *active*.

Remark. In general, an inequality constraint in an optimization
problem is called *active*, or *tight*, for a feasible solution if the inequality holds as equality. For comparison, a *binding* constraint
is a constraint whose removal changes the optimality region (in a
stronger version, the optimal value). For a linear program, any binding constraint is tight for an optimal solution. ∎

Note that the system of linear equations $A^T \cdot x + b = u$ has a
solution given by $x = 0$, $u = b$. Such a solution is called the *basic
solution* for the given standard row tableau. The corresponding value
for the objective function is d. This justifies the following definition:

Definition 10.2. A standard tableau [see (10.1)] is called *row feasible* if $b \geq 0$ (i.e., the basic solution is feasible). ∎

For example, the following two standard tableaux are feasible:

$$
\begin{array}{cc}
x & 1 \\
\begin{bmatrix} -1 & 1 \\ 0 & 0 \\ -2 & -3 \end{bmatrix} & \begin{array}{l} = u \\ = v \\ \rightarrow \min \end{array}
\end{array}
\quad \text{and} \quad
\begin{array}{c}
1 \\
\begin{bmatrix} 2 \\ 3 \\ -2 \end{bmatrix} \begin{array}{l} = x \\ = y \\ \rightarrow \min. \end{array}
\end{array}
\tag{10.3}
$$

The corresponding basic solutions are

$$x = 0, u = 1, v = 0 \text{ and } x = 2, y = 3.$$

They are both feasible. Here are two standard tableaux that are not
feasible:

$$
\begin{array}{cc}
x & 1 \\
\begin{bmatrix} -1 & 1 \\ 1 & -1 \\ -2 & -3 \end{bmatrix} & \begin{array}{l} = u \\ = v \\ \rightarrow \min \end{array}
\end{array}
\quad \text{and} \quad
\begin{array}{c}
1 \\
\begin{bmatrix} -2 \\ 3 \\ -2 \end{bmatrix} \begin{array}{l} = x \\ = y \\ \rightarrow \min. \end{array}
\end{array}
\tag{10.4}
$$

The corresponding basic solutions are

$$x = 0, u = 1, v = -1 \text{ and } x = -2, y = 3.$$

These solutions are not feasible.

Definition 10.5. The preceding tableau (10.1) is called *optimal* if
$b \geq 0$ and $c \geq 0$. ∎

In this case, the basic solution is an optimal solution and d is the optimal value. Indeed, on one hand the basic solution $x = 0$, $u = b$ is feasible (because $b \geq 0$) and the objective function z reaches the value d at this basic solution. On the other hand, since $c \geq 0$, we conclude that $z = c^T x + d \geq d$ for any row matrix $x \geq 0$, so we cannot get a smaller value for z (i.e., d is the minimum value that z can attain). For example, the first tableau in (10.3) is not optimal while the second one is, with min $= -2$.

In fact, any optimal tableau describes all the optimal solutions for our linear program by a system of linear constraints. All nonbasic variables corresponding to nonzero entries of c must be zero in any optimal solution. For other variables, the tableau gives a system of linear equations, which, together with the conditions of nonnegativity for the variables, are necessary and sufficient for optimality. Also an optimal solution gives other useful information about the basic optimal solution (cf. Chapter 5).

Thus, our linear program is solved if it can be described by an optimal tableau. A simplex method attempts to obtain an optimal tableau by a few pivoting steps, starting from an arbitrary (standard) tableau. We will never choose pivot entries in the last column (the column of constants with 1 in the top margin) or the last row (the row that represents the objective function), so our pivoting steps take standard tableaux into standard tableaux.

Phase 2 of our simplex method starts with a feasible tableau, and it will produce either an optimal tableau or a tableau with a *bad column*, which is defined as follows.

Definition 10.6. A *bad column* in a standard tableau is a column that is not the last column (labeled by 1 on top) with the last entry negative and all other entries nonnegative. ∎

For example, all tableaux in (10.3) and (10.4) have no bad columns, but the following two tableaux have one bad column each:

$$
\begin{array}{cc}
y & 1 \\
\left[\begin{array}{cc}
1 & 1 \\
1 & 0 \\
-3 & -1
\end{array}\right]
\begin{array}{l}
= u \\
= v \\
=\to \min,
\end{array}
\end{array}
\qquad
\begin{array}{ccc}
x & y & 1 \\
\left[\begin{array}{ccc}
0 & -1 & 1 \\
0 & 1 & -3 \\
-1 & -3 & -1
\end{array}\right]
\begin{array}{l}
= u \\
= v \\
=\to \min.
\end{array}
\end{array}
$$

Suppose we have a bad column. Then we can give the variable on the top larger and larger values and set all other variables on the top (if any exist) to be 0. In this way we obtain feasible solutions with smaller and smaller values for the objective function. Thus, we encounter an unbounded problem and there are no optimal solutions. In particular, we have no chance of obtaining an optimal tableau. A short-hand for "the problem is unbounded" is min $= -\infty$.

In this section we describe Phase 2 of the simplex method in detail and show that it always terminates in finitely many steps.

We stated that the simplex method is a way of choosing pivot entries to reach a terminal tableau in finitely many pivot steps. More specifically, in Phase 2 all tableaux are standard and feasible, the values of the objective function (on the basic solutions) improve or (in the degenerate case) they stay the same after each pivot step, and in finitely many pivot steps we obtain a terminal tableau, which is either optimal or has a bad column.

We start with a feasible standard tableau (10.1):

$$\begin{array}{cc} x & 1 \end{array}$$
$$\begin{bmatrix} A & b \\ c & d \end{bmatrix} \begin{array}{l} = u \\ = z \to \min, \quad x \geq 0, \quad u \geq 0. \end{array}$$

By Definition 10.2, the feasibility means that $b \geq 0$. We would like to obtain an optimal tableau by pivoting. The simplex method consists of checking for terminal tableaux (when we finish), a procedure to choose the pivot entries, and pivoting. We divide a loop containing one pivot step into four substeps.

Simplex Method, Phase 2

1. If $c \geq 0$, our feasible tableau is already *optimal*. So we write Answer: min $= d$ at $x = 0, u = b$.
We are done!

2. We check whether we have a bad column. That is, for each $c_j < 0$ we check whether all entries of the matrix A in this column are nonnegative. If we find a bad column, we write
Answer: min $= -\infty$, as $x_j \to \infty$, the other variables on the top margin, (nonbasic variables) x_i, $i \neq j$, are set equal to 0, and $u = A \cdot x + b$. We are done!

3. Pick some $c_j < 0$. Consider all $a_{ij} < 0$ in the jth column (which is going to be the pivot column). We compare b_i/a_{ij} for those a_{ij} and pick the one, $a_{i'j}$, with the maximal ratio (i.e., with the ratio closest to zero) as the pivot element.

4. Pivot and go to Substep **1.**   ■

By pivoting, we obtain a new standard tableau that is feasible. Indeed, $b_{i'} \to -b_{i'}/a_{i'j} \geq 0$. If $i \neq i'$, then $b_i \mapsto \bar{b}_i = b_i - a_{ij} \cdot b_{i'}/a_{i'j}$. If $a_{ij} \geq 0$, then $\bar{b}_i \geq b_i \geq 0$. If $a_{ij} < 0$, then $\bar{b}_i \geq 0$, because $b_i/a_{ij} \leq b_{i'}/a_{i'j}$. The value of the objective function, z, improved or stayed the same: It changed from d to $d - c_j b_i/a_{i'j} \leq d$, where $a_{i'j}$ is the pivot entry. It stays the same if and only if the last entry b_i in the pivot row is 0. In this case, our pivot step is *degenerate*. A degenerate pivot step does not change the basic feasible solution, but it does change the basis (the set of variables at the right margin).

Problems 10.7. Solve the following linear programs given by standard row tableaux:

(a)
$$\begin{bmatrix} x_1 & x_2 & x_3 & x_4 & 1 \\ 1 & -1 & 0 & 1 & 0 \\ -1 & 1 & -2 & -1 & -2 \\ 0 & 1 & 2 & -1 & -2 \end{bmatrix} \begin{matrix} = e \\ = f \\ \to \min \end{matrix}$$

(b)
$$\begin{bmatrix} x_1 & x_2 & x_3 & x_4 & 1 \\ 1 & -1 & 0 & 1 & 0 \\ -1 & 1 & -2 & -1 & 1 \\ 1 & 1 & 2 & 0 & -2 \end{bmatrix} \begin{matrix} = e \\ = f \\ \to \min \end{matrix}$$

(c)
$$\begin{bmatrix} x_1 & x_2 & x_3 & x_4 & 1 \\ 1 & 1 & 0 & 1 & 0 \\ -1 & 1 & -2 & -1 & 1 \\ 0 & -1 & 2 & 0 & 2 \end{bmatrix} \begin{matrix} = e \\ = f \\ \to \min \end{matrix}$$

(d)
$$\begin{bmatrix} x_1 & x_2 & x_3 & x_4 & 1 \\ 1 & 1 & 0 & 1 & 0 \\ -1 & -1 & -2 & -1 & 1 \\ 0 & -1 & 2 & 0 & 2 \end{bmatrix} \begin{matrix} = e \\ = f \\ \to \min \end{matrix}$$

Solutions

(a) The tableau is not feasible, so we cannot apply Phase 2 to it. We will solve this problem later.

(b) The tableau is optimal, so *min* $= -2$ at $x_1 = x_2 = x_3 = x_4 = 0, e = 0, f = 1$. If required, we can describe all optimal solutions: $x_1 = x_2 = x_3 = 0, 0 \leq x_4 \leq 1, e = x_4, f = 1 - x_4$.

(c) 1. The tableau is not optimal. 2. The x_2-column is bad, so the problem is unbounded (min $= -\infty$).

(d) 1. The tableau is not optimal. 2. There are no bad columns. 3. The pivot column isthe x_2-column. The pivot entry is -1 in the f-row.

$$\begin{array}{ccccc} x_1 & x_2 & x_3 & x_4 & 1 \\ \left[\begin{array}{ccccc} 1 & 1 & 0 & 1 & 0 \\ -1 & -1^* & -2 & -1 & 1 \\ 0 & -1 & 2 & 0 & 2 \end{array}\right] & & & & \begin{array}{l} = e \\ = f \\ \to \ min. \end{array} \end{array}$$

4. We compute first the new entries in the last row and column:

$$\begin{array}{ccccc} x_1 & f & x_3 & x_4 & 1 \\ \left[\begin{array}{ccccc} & & & & 1 \\ & & & & 1 \\ 1 & 1 & 4 & 1 & 1 \end{array}\right] & & & & \begin{array}{l} = e \\ = x_2 \\ \to \ min. \end{array} \end{array}$$

Now we see that the new tableau is optimal without computing the other entries. Also we can write the answer without other entries: min $= 1$ at $x_1 = f = x_3 = x_4 = 0, e = 1, x_2 = 1$. Notice that the pivot step kept the tableau standard and feasible, and it improved the objective function, as it should in Phase 2. ∎

Problem 10.8. Solve

$$\begin{array}{ccccc} x_1 & -x_2 & 2x_3 & x_4 & -2 \\ \left[\begin{array}{ccccc} 1 & 1 & 0 & 1 & 0 \\ -1 & -1 & -2 & -1 & 1 \\ 0 & -1 & 2 & 0 & 2 \end{array}\right] & & & & \begin{array}{l} = 3x_5 \\ = -x_6 \ , \\ \to \ min. \end{array} \end{array} \qquad \text{all } x_i \geq 0$$

Solution. The tableau is not standard. To make it standard, we multiply the second column by -1, the last column by $-2$and the second row by -1:

$$\begin{array}{ccccc} x_1 & x_2 & 2x_3 & x_4 & 1 \\ \left[\begin{array}{ccccc} 1 & -1 & 0 & 1 & 0 \\ 1 & -1 & 2 & 1 & 2 \\ 0 & 1 & 2 & 0 & -4 \end{array}\right] & & & & \begin{array}{l} = 3x_5 \\ = x_6 \ , \\ \to \ min. \end{array} \end{array} \qquad \text{all } x_i \geq 0$$

There is no need to scale the third column and the first row or replace $2x_2$ and $3x_5$ by other variables. The tableau is optimal, and we write the answer: $x_1 = x_2 = x_3 = x_4 = 0, x_5 = 0, x_6 = 2$, min$= -4$.

Problem 10.9. Solve

$$\begin{array}{ccccc} x_1 & x_2 & x_3 & x_4 & 1 \\ \left[\begin{array}{ccccc} 1 & 1 & 0 & 1 & 0 \\ -1 & -1 & -2 & -1 & 1 \\ 0 & -1 & 2 & 0 & 2 \end{array}\right] & & & & \begin{array}{l} = x_5 \\ = 0 \\ \to \ min. \end{array} \end{array} \qquad \text{all } x_i \geq 0$$

Solution. The tableau is not standard. To make it standard, we pivot on the first entry -1 in the second row. This pivot step switches x_1 and 0; only the first row in the matrix changes:

$$
\begin{array}{ccccc}
0 & x_2 & x_3 & x_4 & 1
\end{array}
$$
$$
\left[
\begin{array}{ccccc}
-1 & 0 & -2 & 0 & 1 \\
-1 & -1 & -2 & -1 & 1 \\
0 & -1 & 2 & 0 & 2
\end{array}
\right]
\begin{array}{l}
= x_5 \\
= x_1 \\
\to \min.
\end{array}
\qquad \text{all } x_i \geq 0
$$

Now we drop the first column to obtain a standard tableau:

$$
\begin{array}{cccc}
x_2 & x_3 & x_4 & 1
\end{array}
$$
$$
\left[
\begin{array}{cccc}
0 & -2 & 0 & 1 \\
-1 & -2 & -1 & 1 \\
-1 & 2 & 0 & 2
\end{array}
\right]
\begin{array}{l}
= x_5 \\
= x_1 \\
\to \min.
\end{array}
\qquad \text{all } x_i \geq 0
$$

Since the tableau is standard, we can use Phase 2 of the simplex method. In this case the simplex method gives us only one choice: Switch x_1 and x_2:

$$
\begin{array}{cccc}
x_2 & x_3 & x_4 & 1
\end{array}
\qquad\qquad
\begin{array}{cccc}
x_1 & x_3 & x_4 & 1
\end{array}
$$
$$
\left[
\begin{array}{cccc}
0 & -2 & 0 & 1 \\
-1^* & -2 & -1 & 1 \\
-1 & 2 & 0 & 2
\end{array}
\right]
\begin{array}{l}
= x_5 \\
= x_1 \\
\to \min
\end{array}
\;\mapsto\;
\left[
\begin{array}{cccc}
0 & -2 & 0 & 1 \\
-1 & -2 & -1 & 1 \\
1 & 0 & 1 & 1
\end{array}
\right]
\begin{array}{l}
= x_5 \\
= x_2 \\
\to \min.
\end{array}
$$

The tableau is optimal, so we write the answer: $x_1 = x_3 = x_4 = 0, x_2 = 1, x_5 = 1, \min = 1$. There are other optimal solutions, but we were not required to find them. ∎

Problem 10.10. Solve

$$
\begin{array}{ccccc}
x_1 & x_2 & x_3 & x_4 & 1
\end{array}
$$
$$
\left[
\begin{array}{ccccc}
1 & 1 & 0 & 1 & 0 \\
-1 & -1 & -2 & -1 & 1 \\
0 & -1 & 2 & 0 & 2
\end{array}
\right]
\begin{array}{l}
= -x_5 \\
= -x_6 \\
\to \max.
\end{array}
\qquad \text{all } x_i \geq 0
$$

Solution. The tableau is not standard. To make it standard, we multiply every row by -1. Since we change the objective function, let $f = -x_2 + 2x_3 + 2$. Then the new objective function is $-f$:

$$
\begin{array}{ccccc}
x_1 & x_2 & x_3 & x_4 & 1
\end{array}
$$
$$
\left[
\begin{array}{ccccc}
-1 & -1 & 0 & -1 & 0 \\
1 & 1 & 2 & 1 & -1 \\
0 & 1 & -2 & 0 & -2
\end{array}
\right]
\begin{array}{l}
= x_5 \\
= x_6 \\
= -f \to \min.
\end{array}
\qquad \text{all } x_i \geq 0
$$

Now the tableau is standard. Since it is not feasible, we cannot use Phase 2. Should we wait until we learn Phase 1? Not necessary. Let us look at the problem closely, trying to use common sense. The first row in the tableau is "almost bad." It says $-x_1 - x_2 - x_4 = x_5$. Since all $x_i \geq 0$, we conclude that $x_1 = x_2 = x_4 = 0 = x_5$. So it remains to determine x_3 and x_6 for which we have a smaller standard tableau:

$$\begin{array}{c} \quad x_3 \quad 1 \\ \begin{bmatrix} 2 & -1 \\ -2 & -2 \end{bmatrix} \begin{array}{l} = x_6 \\ = -f \to \min. \end{array} \end{array} \qquad x_1 = x_2 = x_4 = x_5 = 0; x_3, x_6 \geq 0$$

Still this tableau is not feasible. However, it has a bad column. Since we are not in Phase 2, we cannot conclude that the problem is unbounded. However, the same argument, which was used to prove the unboundness in the presence of a bad column in Phase 2, works in our case. Namely, we drive the objective function to $-\infty$ by sending x_3 to $-\infty$ and we are feasible for $x_2 \geq 1/2$. Thus, the problem is unbounded, $\min(-f) = -\infty$. In the terms of the original problem, it is also unbounded: $\max = \infty$. As an alternative, we could solve the optimization problem for x_3, x_6 by the graphical method.   ∎

The total number of possible tableaux we can obtain by the simplex method starting from an initial tableau is finite: There are only so many ways to put some variables on the top and the rest of variables at the right margin; after the positions of the variables are fixed, the tableau is determined. Remember that all our systems of linear equations written in those tableaux are equivalent (for the initial tableau as previosly, the system is $Ax^T = u$), so they have the same solutions.

In fact, if we work with m variables at the right margin (basic variables) and n variables on the top (nonbasic variables), then there are $(m+n)!$ ways to arrange $m+n$ variables at $m+n$ positions. After the positions for the variables are fixed, the whole $(m+1) \times (n+1)$ matrix is determined. Some of these tableaux are feasible. Some of these configurations could be impossible because our system of linear equations could be unsolvable for some sets of m variables. If we permute the basic variables or nonbasic variables, we do not change the basic solution. So the number of the basic solutions is at most $\frac{(m+n)!}{m!n!}$.

By the simplex method, Phase 2, we either obtain an optimal tableau in finitely many steps, or we obtain a tableau that occurred before. In the latter case we can get into a cycle passing through the same set of tableaux with the same entry d. Note that the last entry d in the last row—that is, the current value for the objective function—must be constant along the cycle because it cannot go up. Since d does not change, the last entry in the pivot row is zero along the cycle.

We call a pivot step *degenerate* if the last entry in the pivot row is 0. Equivalently, a degenerate pivot step does not change the basic solution (although it does change the set of basic variables).

A good way to avoid cycling is to choose i' in Phase 2 at random, whenever there is a choice. There are many ways to specify the choice of the pivot in Substep 3, so there are many simplex algorithms. None of them is shown to be the best for all data, but some of them allow us to avoid cycling.

Another good rule for choosing the pivot column is to pick the smallest c_j in Step 3 (this is known as the least-coefficient rule). However this rule needs to be refined for the case when there are several minimal entries to be an exact algorithm.

Or we can pick the pivot entry by computing the improvements in the objective function for all possible pivot entries that preserve the feasibility condition and choose one that gives the best improvement (the largest decrease rule). This rule, however, may contradict the previous rule.

In practical problems cycling happens so rarely that often nothing is done to prevent it. There are other, more serious problems with the simplex method, which we will discuss soon. However, for theoretical reasons it is important to show how to avoid cycling; that is, it is important to give a simplex method that works always (by "works" we mean that it terminates in finitely many steps). We will do so next.

The first way suggested to prevent cycling was *perturbation*, a small change of entries in the last column to avoid zeros. More precisely, we replace the column $b = [b_1, \ldots, b_m]^T$ by the perturbed column $b(\varepsilon) = [b_1 + \varepsilon, \ldots, b_n + \varepsilon^n]^T$. Formally, we work with *polynomials* in ε. Informally, $b(\varepsilon)$ is a small perturbation of b; ε is considered a small positive number. We can define how to compare these polynomials and then apply the simplex method to the perturbed matrix. All entries in the b-part stay nonzero, which makes cycling impossible. See Section A6 of the Appendix.

The perturbation requires additional computations and has very little practical value. But it showed that there was a way to avoid cycling.

In 1976, Robert G. Bland introduced cycle-proof rules that also make all choices in Step 3 unique. Here is a paraphrase of Bland's rule. First, before pivoting, list (sort) the names or labels of variables in some linear order (in a row or column). For example, if you were working with three variables, you could name them and order them as x, y, z or label them and order them as t_1, t_2, t_3. Then, whenever the simplex method gives you a choice, pick the variable that appears first in the list of variables, according to your ordering. When the variables are ordered by subscripts, this rule is known as the least subscript rule.

For example, if we have two negative entries in the last row, corresponding to x and y, then the pivot column is going to be the x-column, since in the ordering, x appears before y. Similarly, if there are two maximal ratios corresponding to x and y, the x-row is going to be the pivot row.

We now show that Bland's rule works.

Theorem 10.11. If we choose pivot entries in Phase 2 of the simplex method according to Bland's rule, cycling is impossible.

Proof. Let $x_1, \ldots, x_{n+m}$ be our variables; we use the least subscript rule. We work with $(m+1) \times (n+1)$ standard feasible row tableaux.

Suppose that we have a cycle. A *swing* variable is a variable that changes its position along the cycle. While we make the cycle, the swing variable labels (marks) the pivot column for a tableau and it labels the pivot row for another tableau.

Let x_t be the swing variable with the largest subscript. We consider a tableau

$$\begin{matrix} y & 1 \\ \begin{bmatrix} A & b \\ c & d \end{bmatrix} & \begin{matrix} = z \\ = f \end{matrix} \end{matrix} \quad \to \min, \qquad (10.12)$$

where x_t labels the pivot column (i.e., x_t is on the top, but it is at the side in the next tableau). Then the last entry c_t in the pivot column is negative, and $c_j \geq 0$ for the variables x_j on top with $j < t$, because otherwise we would choose another pivot column according to Bland's rule. Note that the last entries in the swing rows are all zeros.

Now we consider a tableau

$$\begin{matrix} y' & 1 \\ \begin{bmatrix} A' & b' \\ c' & d' \end{bmatrix} & \begin{matrix} = z' \\ = f \end{matrix} \end{matrix} \quad \to \min, \qquad (10.13)$$

where x_t marks the pivot row. That is, x_t is at the right margin (one of variables in z') but is about to move to the top. Note that $d = d'$ because the objective function cannot increase along the cycle.

Let

$$x_s$$

$$\begin{bmatrix} a \\ c_s \end{bmatrix}$$

be the pivot column in (10.13). So $c_s < 0$, $a_i \geq 0$ for the entry of a in the x_i-row when x_i is a swing variable at the right margin with $i \neq t$, and $a_t < 0$.

Now we define a solution h of the system $Ay^T = z$ or an equivalent system $A'y'^T = b'$, which is not feasible and need not be basic, as follows. We set $x_s = 1$ and set all other variables in y' to 0, hence $z' = a + b'$ and $f(h) = d + c_s < d$. In particular, for this solution, we have $x_j = h_j = 0$ for any variable x_j on top ($j \neq s$) and $x_i = h_i = a_i \geq 0$ for any swing x_i variable at the right in (10.13).

On the other hand, we can plug this solution into the tableau (10.12) and obtain that $f(h) = d + \sum c_j h_j$ where the sum is taken over all swing variables in y [recall that the values for nonswing variables on top in (10.12) and (10.14) are 0]. In this sum, $c_t h_t = c_t a_t > 0$ (since $c_t, a_t < 0$) and all other terms are ≥ 0 (because $c_j, h_j > 0$ for $j \neq t$). Thus, $f(h) > d$, which contradicts the preceding estimate. ∎

Exercises

1–3. Solve the linear programs given by standard row tableaux:

	a	b	c	d	1	
	.3	.35	.5	.4	.4	$= y_1$
	−.3	−.35	−.5	.4	.4	$= y_2$
	.6	.35	.5	.45	0	$= y_3$
1.	−.6	−.35	−.5	5	.5	$= y_4$
	.1	.3	0	.15	1	$= y_5$
	−.1	−.3	0	−.15	.1	$= y_6$
	11	12	16	14	0	→ min

	y_7	y_8	y_9	y_{10}	1	
	.3	.35	.5	.4	−.4	$= y_1$
	−.3	−.35	−.5	−.4	−.4	$= y_2$
	.6	.35	.5	.45	−.5	$= y_3$
2.	−.6	−.35	−.5	−.45	.5	$= y_4$
	.1	.3	0	.15	−.1	$= y_5$
	−.1	−.3	0	−.15	.1	$= y_6$
	11	12	16	−14	0	$= C$ → min

3.

$$\begin{bmatrix}
 & z_1 & z_2 & z_3 & z_4 & 1 \\
0 & .35 & .5 & .4 & .4 \\
3 & -.35 & -.5 & -.4 & .4 \\
.6 & .35 & .5 & .45 & 0 \\
.6 & -.35 & -.5 & -.45 & .5 \\
.1 & .3 & 0 & .15 & -.1 \\
0 & -.3 & 0 & -.15 & .1 \\
-11 & 12 & 16 & 14 & 0
\end{bmatrix}
\begin{matrix}
= y_1 \\
= y_2 \\
= y_3 \\
= y_4 \\
= y_5 \\
= y_6 \\
\rightarrow \min
\end{matrix}$$ ∎

4–6. State whether you agree or disagree.

4. A standard tableau is optimal if its basic solution is optimal.

5. Every standard tableau with one row is feasible.

6. If a standard tableau has only one row and no bad columns, then it is optimal.   ∎

7. Solve the linear program from Example 2.3.

8. Solve the linear program $x + z \rightarrow \min,\ y - u = 1, x \geq 0, y \geq 0$.

9. Solve $1.2a + 1.4b + 1.7c + 1.9d \rightarrow \min,\ a + b + c + d = 1,$ $0.1a + 0.2b + 0.3c + 0.4d = 0.25,\ [a, b, c, d] \geq 0.$

10. Solve

$$\begin{bmatrix}
x_1 & -x_2 & x_3 & x_4 & -1 \\
1 & 1 & 0 & 1 & -3 \\
-1 & -1 & -2 & -1 & 1 \\
0 & -1 & 2 & 0 & 2
\end{bmatrix}
\begin{matrix}
= x_5 \\
= -x_6 \\
\rightarrow \min.
\end{matrix}$$ all $x_i \geq 0$

11. Solve

$$\begin{bmatrix}
x_1 & -x_2 & x_3 & x_4 & -1 \\
1 & -1 & 0 & 1 & -3 \\
-1 & 1 & -2 & -1 & 1 \\
0 & -1 & 2 & 0 & 2
\end{bmatrix}
\begin{matrix}
= x_5 \\
= -x_6 \\
\rightarrow \max.
\end{matrix}$$ all $x_i \geq 0$

12. Solve

$$\begin{bmatrix}
x_1 & -x_2 & x_3 & x_4 & 1 \\
1 & -1 & 0 & 1 & -3 \\
-1 & 1 & -2 & -1 & 1 \\
0 & -1 & 2 & 0 & 2
\end{bmatrix}
\begin{matrix}
= x_5 \\
= -x_6 \\
\rightarrow \max.
\end{matrix}$$ all $x_i \geq 0$

13. Describe how Phase 2 works for an arbitrary standard tableau with only one row. *Hint*: Consider first a few examples.

14. Describe how Phase 2 works for an arbitrary feasible tableau with only one two columns (that is, only one variable on the top). *Hint*: Consider first a few examples and use the graphical method.

§11. Simplex Method, Phase 1

A good simplex method produces an optimal tableau for an arbitrary linear program that has an optimal solution. No method can produce an optimal tableau for a problem that has no optimal solutions. In Phase 1, our goal is to obtain a feasible tableau. Since there are infeasible linear programs, we have to allow other outcomes for simplex methods. Recall that the simplex method works with standard tableaux.

Definition 11.1. A *bad row* in a standard tableau is a row that is not the last row and whose last entry is strictly negative while the other entries are nonpositive.   ∎

In other words, it is a row $[A_i, b_i]$ in the matrix $[A, b]$ [see (10.1)] such that $A_i \leq 0$ and $b_i < 0$. The constraint corresponding to a bad row reads $A_i^T x + b_i = u_i$. Since the left-hand side here is strictly negative and the right-hand side is nonnegative, we obtain a contradiction. Thus, a linear program with a standard tableau having a bad row has no feasible solutions. This is because the bad row is inconsistent with the nonnegativity conditions.

For example, the x-row in the second tableau in (10.4) is bad, while the first tableau in (10.4) has no bad rows. A short-hand for "the problem is infeasible" is min $= \infty$.

The following are logical implications between properties of a linear program and properties of its standard tableaux that are obvious from our definitions:

$$\begin{array}{ccccc} \text{there is a} & & \text{the} & & \text{there is no} \\ \text{tableau with} & \Rightarrow & \text{program is} & \Rightarrow & \text{feasible} \\ \text{a bad row} & & \text{infeasible} & & \text{tableaux} \end{array} \qquad (11.2)$$

Note that the existence of an infeasible tableau does not imply that min $= \infty$ and that min $= \infty$ does not imply the existence of a bad column in every tableau.

Phase 1 of our simplex method will produce either a feasible tableau or a tableau with a bad row. So it shows that either our linear program is not feasible (the second case) or it yields a feasible solution (namely, the basic solution of the feasible tableau in the first case.) Given this, we have the converse implications of (11.2):

$$\begin{array}{ccccc} \text{there is a} & & \text{the} & & \text{there is no} \\ \text{tableau with} & \Leftrightarrow & \text{program is} & \Leftrightarrow & \text{feasible} \\ \text{a bad row} & & \text{infeasible} & & \text{tableaux} \end{array}$$

Given this, we can summarize this discussion in the following theorem:

Theorem 11.3. Our simplex method will have one and only one of the following three outcomes:

1. An optimal tableau; in this case we can read off the optimal solution(s) from the optimal tableau, as well as the optimal value.

2. A feasible tableau with a bad column; in this case the problem is unbounded, so there are no optimal solutions.

3. A tableau with a bad row; this gives us an infeasible problem (i.e., a problem without feasible solutions). ∎

So there are terminal tableaux of three different types for the simplex method. Including row feasible tableaux that are terminal for Phase 1, there are four different types. It is very important to recognize these terminal tableaux when it is time to stop and write the answer. To memorize terminal tableaux, the following sketch may help:

Row feasible tableau:
$$\begin{array}{cc} \oplus & 1 \\ \left[\begin{array}{cc} * & \oplus \\ * & * \end{array}\right] & \begin{array}{l} = \oplus \\ \to \min, \end{array} \end{array}$$
terminal for Phase 1, go to Phase 2

Bad row:
$$\begin{array}{cc} \oplus & 1 \\ \left[\begin{array}{cc} \ominus & - \end{array}\right] & = \oplus, \end{array}$$
a terminal tableau, $\min = \infty$

Optimal:
$$\begin{array}{cc} \oplus & 1 \\ \left[\begin{array}{cc} * & \oplus \\ \oplus & * \end{array}\right] & \begin{array}{l} = \oplus \\ \to \min, \end{array} \end{array}$$
a terminal tableau, write answer

Bad column:
$$\begin{array}{c} \oplus \\ \left[\begin{array}{c} \oplus \\ - \end{array}\right] & \begin{array}{l} = \oplus \\ \to \min. \end{array} \end{array}$$
terminal if in Phase 2

A feasible tableau with a bad column :
$$\begin{array}{c} \oplus \quad \oplus \quad \oplus \quad 1 \\ \left[\begin{array}{cccc} * & \oplus & * & \oplus \\ * & - & * & * \end{array}\right] & \begin{array}{l} = \oplus \\ \to \min. \end{array} \end{array}$$
terminal, $\min = -\infty$.

Here

$\oplus$ represents positive and zero entries, that is, nonnegative entries (≥ 0);

$\ominus$ represents negative and zero entries, that is, nonpositive entries, (≤ 0);

— represents a negative entry;

+ represents a positive entry;

* represents arbitrary numbers.

In §10 we presented a method to choose pivoting entries from a feasible standard tableau. However, what can we do when a given standard tableau is not feasible? We will present a method to make a standard tableau feasible or to show that this cannot be done. We start with a standard tableau:

$$
\begin{array}{c} x \quad\ 1 \\ \begin{bmatrix} A & b \\ c & d \end{bmatrix} \begin{array}{l} = u \\ = z \to \min, \quad x \geq 0, \quad u \geq 0 \end{array} \end{array}
$$

As in Phase 2, we split a loop involving a pivot step into four substeps.

Simplex Method, Phase 1

1. If $b \geq 0$, then our tableau is already feasible, so we are done with Phase 1 and we proceed to Phase 2.

2. Check whether we have a bad row in the tableau. That is, for each $b_i < 0$, check the sign of the entries a_{ij} in its row. If for one of these b_i, all the entries, a_{ij}, in the row are nonpositive, then we cannot satisfy this constraint. Therefore, there are no feasible solutions. We can write this as follows:

Answer: min $= \infty$.

3. Pick the row, i', closest to the top of the tableau, with $b_{i'} < 0$. This means that $b_i \geq 0$ for every index i for which b_i is above $b_{i'}$. Choose any positive entry, $a_{i'j'} > 0$, in this row. Select $i'' \leq i'$ such that $b_{i''}/a_{i''j'} = \max(b_{i'}/a_{i'j'}, b_i/a_{ij'})$, where the maximum is taken over all indexes $i \leq i'$ for which $a_{ij'} < 0$.

4. Pivot and go to Substep **1.**   ∎

Note that the b-entries above $b_{i'}$ remain nonnegative and $b_{i'}$ either maintains or increases its value. Indeed, b_i with $i \leq i'$ is replaced by $\bar{b}_i = b_i - a_{ij'} \cdot b_{i''}/a_{i''j'}$. If $a_{ij'} \geq 0$ (for example, when $i = i'$), then $\bar{b}_i \geq b_i$. If $a_{ij'} < 0$, then $\bar{b}_i \geq 0$.

The method just described produces either a tableau with a bad row, or a feasible tableau, or we get into a cycle. Cycling is possible only when there are zero entries in the b-column and those zeros are in the pivot rows (the degenerate case). To prevent cycling we can use Bland's rule (see §10) or the perturbation technique when choosing the pivot entries. Bland's rule works, because if we get into a cycle, then we get cycling for a linear program in Phase 2. This program is obtained by changing sign in the first row with the negative last entry in the standard tableau, treating it as the objective function, and ignoring all following rows.

Practical computations show that the number of pivoting steps in a good simplex method is proportional to $\min(m, n)$, where $m \times n$ is the size of the tableau. However, linear programs have been constructed that require a much larger number of pivoting steps for many common ways to specify the choice of pivot entries (including Bland's rule). Fortunately, such "pathological" problems do not occur in practice.

It is still an open problem whether there is a number t such that for any natural numbers m, n and any standard feasible tableau with $m + 1$ rows and $n + 1$ columns (so we have $m + n$ decision variables) there is a sequence with less than $(m + n)^t$ pivot steps that produces feasible tableaux (but not necessary follows the simplex method) and results in a terminal tableau. Without the restriction that the tableaux are feasible, take $t = 1$, because we can obtain a terminal tableau in $\leq \min(m, n)$ pivot steps.

If you use a computer, then besides storage room for data, you need about the same room for the current tableau. Since round-off errors accumulate, you need occasionally to recalculate the current tableau starting from the initial tableau (this is why you have to store an initial tableau). This requires $\leq \min(m, n)$ pivot steps. Otherwise the accumulation of round-off errors can take you far away from optimal and feasible solutions. In general, since time is money, it is a good idea to check occasionally whether the progress in computations (say, improvement in objective function) justifies the time and effort spent.

Large linear programs often have sparse tableaux; that is, most entries of the matrix of coefficients are equal to zero. We can save RAM (random access memory) by storing only nonzero entries and using multiplicative simplex methods that represent the current matrix as a product of a few elementary operations applied to the initial tableau.

Problem 11.4. Solve

$$
\begin{array}{cccc}
x & y & z & 1
\end{array}
$$
$$
\begin{bmatrix}
1 & -1 & 1 & 2 \\
0 & -1 & 0 & -1 \\
1 & 0 & 2 & 0
\end{bmatrix}
\begin{array}{l}
= u \\
= v \\
= w \to \min.
\end{array}
\qquad x, y, z, u, v \geq 0
$$

Solution. The tableau is standard but not feasible. The v-row is bad, so the problem is infeasible ($\min = \infty$ for short).

Problem 11.5. Solve

$$
\begin{array}{cccc}
x & y & z & 1
\end{array}
$$
$$
\begin{bmatrix}
1 & -1 & 1 & 2 \\
0 & 1 & 0 & -1 \\
1 & 0 & 2 & 0
\end{bmatrix}
\begin{array}{l}
= u \\
= v \\
= w \to \min.
\end{array}
\qquad x, y, z, u, v \geq 0
$$

Solution. The tableau is standard but not feasible. There are no bad rows. We look for a positive entry in the second row, and there is only one, in the y-column, which is going to be the pivot column. We compare $2/(-1) < (-1)/1 < 0$; hence the v-row is the pivot row. Now we pivot and obtain

$$
\begin{array}{cccc}
x & v & z & 1
\end{array}
$$
$$
\begin{bmatrix}
1 & -1 & 1 & 1 \\
0 & 1 & 0 & 1 \\
1 & 0 & 2 & 0
\end{bmatrix}
\begin{array}{l}
= u \\
= y \\
= w \to \min.
\end{array}
\qquad x, y, z, u, v \geq 0
$$

As it should be, the tableau stays standard, and the first entry in the last column stays ≥ 0. Moreover the tableau becomes optimal. Answer: $x = v = z = 0, y = u = 1$, min$= 0$. ∎

Problem 11.6. Solve the problem in 10.4 by the simplex method.

Solution. After we get the standard tableau

$$
\begin{array}{ccccc}
x_1 & x_2 & x_3 & x_4 & 1
\end{array}
$$
$$
\begin{bmatrix}
-1 & -1 & 0 & -1 & 0 \\
1 & 1 & 2 & 1 & -1 \\
0 & 1 & -2 & 0 & -2
\end{bmatrix}
\begin{array}{l}
= x_5 \\
= x_6 \\
= -f \to \min
\end{array}
\qquad \text{all } x_i \geq 0
$$

as before, we play stupid and follow Phase 2. Let us see whether we get the same answer.

There are four choices for the pivot column consistent with Phase 1. Only one of them, the x_3-column, gives a nondegenerate pivot step. So we pivot on 2 and obtain

$$
\begin{array}{ccccc}
x_1 & x_2 & x_6 & x_4 & 1
\end{array}
$$
$$
\begin{bmatrix}
-1 & -1 & 0 & -1 & 0 \\
-1/2 & -1/2 & 1/2 & -1/2 & 1/2 \\
1 & 2 & -1 & 1 & -3
\end{bmatrix}
\begin{array}{l}
= x_5 \\
= x_3 \\
= -f \to \min.
\end{array}
\qquad \text{all } x_i \geq 0
$$

This tableau is feasible, so we proceed with Phase 2. The tableau has a bad column, so the problem is unbounded. To get more practice, try other three choices for the pivot entry. Can you get into cycling? Fat chance!

Problem. Solve the linear program

$$
\begin{array}{ccc}
x & y & 1 \\
\end{array}
$$
$$
\begin{bmatrix}
-1 & 1 & 1 \\
2 & -1 & 0 \\
-1 & 0 & 1
\end{bmatrix}
\begin{array}{l}
\geq 0 \\
\leq 0 \\
= z \to \max \quad x, y \geq 0.
\end{array}
$$

Solution. The tableau is not standard. Multiplying the last two rows by -1 and introducing two slack variables u, v we make it standard:

$$
\begin{array}{ccc}
x & y & 1 \\
\end{array}
$$
$$
\begin{bmatrix}
-1 & 1 & 1 \\
-2 & 1 & 0 \\
1 & 0 & -1
\end{bmatrix}
\begin{array}{l}
= u \\
= v \\
= -z \to \min \quad x, y \geq 0, u, v \geq 0.
\end{array}
$$

The tableau is optimal. So the basic solution is optimal: $x = y = 0, u = 1, v = 0, \min = -1$. In the terms of the original problem, the answer is: $x = y = 0$, max= 1. All optimal solutions are given by $x = 0, y \leq 1$. ∎

Problem. Solve the linear program given by a standard row tableau

$$
\begin{array}{cccc}
x_1 & x_2 & x_3 & 1 \\
\end{array}
$$
$$
\begin{bmatrix}
1 & 0 & -1 & -1 \\
-1 & 1 & 1 & 0 \\
0 & -1 & 0 & 1 \\
1 & -1 & 1 & 0
\end{bmatrix}
\begin{array}{l}
= x_4 \\
= x_5 \\
= x_6 \\
\to \min .
\end{array}
$$

Solution. The tableau is not feasible. According to Phase 1, we have to switch x_1 and x_4. After the pivot step, we obtain

$$
\begin{array}{cccc}
x_4 & x_2 & x_3 & 1 \\
\end{array}
$$
$$
\begin{bmatrix}
1 & 0 & 1 & 1 \\
-1 & 1 & 0 & -1 \\
0 & -1 & 0 & 1 \\
1 & -1 & 2 & 1
\end{bmatrix}
\begin{array}{l}
= x_1 \\
= x_5 \\
= x_6 \\
\to \min .
\end{array}
$$

We switch x_2 and x_5 and obtain

$$
\begin{array}{cccc}
x_4 & x_5 & x_3 & 1 \\
\end{array}
$$
$$
\left[
\begin{array}{cccc}
1 & 0 & 1 & 1 \\
1 & 1 & 0 & 1 \\
-1 & -1 & 0 & 0 \\
0 & -1 & 2 & 0
\end{array}
\right]
\begin{array}{l}
= x_1 \\
= x_2 \\
= x_6 \\
\to \min
\end{array}
.
$$

Now the tableau is feasible, so we proceed with Phase 2. We switch x_5 and x_6 and after a degenerate pivot step obtain an optimal tableau:

$$
\begin{array}{cccc}
x_4 & x_6 & x_3 & 1 \\
\end{array}
$$
$$
\left[
\begin{array}{cccc}
* & * & * & 1 \\
* & * & * & 1 \\
* & * & * & 0 \\
1 & 1 & 2 & 0
\end{array}
\right]
\begin{array}{l}
= x_1 \\
= x_2 \\
= x_5 \\
\to \min .
\end{array}
$$

So $\min = 0$ at $x_4 = x_6 = x_3 = 0, x_4 = 1, x_2 = 1, x_5 = 0$.

Remark. Some textbooks and some computer implementations reduce Phase 1 to Phase 2 by formal tricks rather than by treating Phase 1 as an equal stage of the simplex method. Here we give such a trick. Consider a standard row tableau

$$
\begin{array}{cc}
x & 1 \\
\end{array}
$$
$$
\left[
\begin{array}{cc}
A & b \\
c & d
\end{array}
\right]
\begin{array}{l}
= u \\
= z \to \min, \quad x \geq 0, \quad u \geq 0
\end{array}
\qquad (11.7)
$$

that is not feasible (i.e., the minimal entry $-\mu$ of b is negative). We introduce a new variable $t \geq 0$ and consider the standard tableau

$$
\begin{array}{ccc}
x & t & 1 \\
\end{array}
$$
$$
\left[
\begin{array}{ccc}
A & I & b \\
0 & 1 & 0
\end{array}
\right]
\begin{array}{l}
= u \qquad x \geq 0, t \geq 0 \\
\to \min \qquad u \geq 0,
\end{array}
\qquad (11.8)
$$

where I is the column of ones. By one pivot step with the pivot entry in the t-column and the last entry in the pivot row being $-\mu$, we obtain a feasible tableau, so Phase 1 is done in one pivot step. If the optimal value for (11.8) is 0, then we can easily obtain a feasible tableau for (11.7) from an optimal tableau for (11.8).If the optimal value for (11.8) is not 0, then the problem (11.7) is infeasible.

A modification of this is the "big M method." We replace the last row $[0, 1, 0]$ in (11.8) by $[c, M, d]$ with large M. By the same pivot step, this M-tableau becomes feasible. If the M-program has an optimal solution with $t = 0$, then we obtain an optimal solution for (11.7) by dropping t.

There are other tricks to avoid or shorten Phase 1. Some of them involving introducing a new variable for each variable in u.

Remark. It is also possible to reduce Phase 2 to Phase 1. A simple-minded way to do this is to replace the objective function z be linear constraints $f \le M$ for various goals M. This approach could be computationally taxing. Another way, using duality, will be given in the next chapter.

Exercises

1. Solve
$$\begin{array}{cccc} x & y & z & 1 \end{array}$$
$$\begin{bmatrix} 1 & -1 & 1 & 2 \\ 0 & -1 & 0 & -1 \\ 1 & 0 & 2 & 0 \end{bmatrix} \begin{array}{l} = u \\ = v \\ = w \to \min. \end{array}$$
$$\begin{array}{l} x, y, z \ge 0 \\ u, v \ge 0 \end{array}$$

2. Solve
$$\begin{array}{cccc} x & y & z & 1 \end{array}$$
$$\begin{bmatrix} 1 & -1 & 1 & 2 \\ 0 & -1 & 0 & 0 \\ 1 & 0 & 2 & 0 \end{bmatrix} \begin{array}{l} = u \\ = v \\ = w \to \min \end{array}$$
with $x \ge 0, y \ge 0, z \ge 0, u \ge 0, v \ge 0$.

3. Solve
$$\begin{array}{cccc} x & y & z & 1 \end{array}$$
$$\begin{bmatrix} 1 & -1 & 1 & 2 \\ 0 & -1 & 0 & 0 \\ 1 & 0 & -2 & 0 \end{bmatrix} \begin{array}{l} = u \\ = v \\ = w \to \min \end{array}$$
with $x \ge 0, y \ge 0, z \ge 0, u \ge 0, v \ge 0$.

4–8. Solve the linear programs, where all the decision variables, a, b, c, d, e, are required to be nonnegative.

4.
$$\begin{array}{cccc} a & -b & c & -2 \end{array}$$
$$\begin{bmatrix} 1 & 0 & -1 & 2 \\ 2 & -1 & 0 & -3 \\ 0 & 2 & -1 & 0 \end{bmatrix} \begin{array}{l} = d \\ = e \\ \to \min \end{array}$$

5.
$$\begin{array}{cccc} a & b & c & 1 \end{array}$$
$$\begin{bmatrix} 1 & 2 & 3 & -1 \\ 2 & 0 & 1 & 3 \\ -1 & 1 & 0 & 0 \end{bmatrix} \begin{array}{l} = d \\ = e \\ \to \min \end{array}$$

6.
$$
\begin{array}{cccc}
a & b & c & -1 \\
\end{array}
$$
$$
\left[\begin{array}{cccc}
1 & 2 & 3 & -3 \\
2 & 1 & 0 & -1 \\
-3 & 0 & 1 & 0
\end{array}\right]
\begin{array}{l}
= d \\
= e \\
\rightarrow \min
\end{array}
$$

7.
$$
\begin{array}{cccc}
a & b & c & -1 \\
\end{array}
$$
$$
\left[\begin{array}{cccc}
1 & 0 & 1 & -1 \\
0 & 1 & 1 & 1 \\
1 & 2 & 3 & 0
\end{array}\right]
\begin{array}{l}
= c \\
= d \\
= f \rightarrow \min
\end{array}
$$

8.
$$
\begin{array}{cccc}
a & b & c & -1 \\
\end{array}
$$
$$
\left[\begin{array}{cccc}
1 & 0 & 1 & -1 \\
0 & 1 & 1 & 1 \\
1 & 2 & 3 & 0
\end{array}\right]
\begin{array}{l}
= e \\
= d \\
= f \rightarrow \min
\end{array}
$$

9–10. Do you agree or disagree with the following statements?

9. There is no standard feasible tableau with a bad row.

10. There is no standard tableau with has both a bad row and a bad column.

11–13. Solve the linear program given by a standard row tableau.

11.
$$
\begin{array}{ccccc}
x_1 & x_2 & x_3 & x_4 & 1 \\
\end{array}
$$
$$
\left[\begin{array}{ccccc}
1 & 0 & -2 & -3 & -1 \\
-1 & 1 & 1 & 1 & 1 \\
2 & -1 & 0 & 1 & 3 \\
1 & -1 & 1 & 0 & 2
\end{array}\right]
\begin{array}{l}
= x_5 \\
= x_6 \\
= x_7 \\
\rightarrow \min
\end{array}
$$

12.
$$
\begin{array}{cccc}
x_1 & x_2 & x_3 & 1 \\
\end{array}
$$
$$
\left[\begin{array}{cccc}
1 & 0 & -1 & -1 \\
-1 & 3 & 1 & 0 \\
3 & -1 & 2 & 1 \\
1 & -1 & 1 & 0 \\
1 & -1 & -1 & 0
\end{array}\right]
\begin{array}{l}
= x_4 \\
= x_5 \\
= x_6 \\
= x_7 \\
\rightarrow \min
\end{array}
$$

13.
$$
\begin{array}{cccccc}
x_1 & x_2 & x_3 & x_4 & x_5 & 1 \\
\end{array}
$$
$$
\left[\begin{array}{cccccc}
1 & 0 & -1 & -1 & 0 & 1 \\
-1 & 3 & 1 & 0 & -2 & -1 \\
3 & -1 & 2 & 1 & 2 & 1 \\
1 & -1 & 1 & 0 & -1 & 1 \\
0 & -1 & -1 & 0 & -1 & 2
\end{array}\right]
\begin{array}{l}
= x_6 \\
= x_7 \\
= x_8 \\
= x_9 \\
\rightarrow \min
\end{array}
$$

§12. Geometric Interpretation

In this section we will explain the geometric meaning of Phase 2 (the second stage) of the simplex method. The objective is to engage our geometric imagination. The price we have to pay is to learn some geometric language: n-tuples of numbers become points in R^n, certain linear combinations of two n-tuples become lines, feasible regions become subsets of R^n with interesting properties, Phase 2 of simplex method is a way to move in the feasible region, and so on. We will first look at the properties of the feasible region as a subset of the n-dimensional space R^n.

We saw in §3 (Chapter 1) that the feasible region of a linear programming problem in two or less variables is a set with very special properties. In fact, when we try to generalize rays, intervals, in one dimension to higher-dimensional spaces, we are led to the notion of convex sets, which we proceed to define.

Definition 12.1. A set S is called *convex* if $(1 - a)x + ay$ belongs to S whenever x and y belong to S and $0 \le a \le 1$.   ∎

For the definition of a convex set to make sense, $(1 - a)x + ay$ should be defined. This is the case when, for example, S is a set of rows $[a_1, a_2, \ldots, a_n]$ or a set of columns $[a_1, a_2, \ldots, a_n]^T$, where $n \ge 1$ is a fixed integer. Clearly, we can multiply these rows (columns) by numbers and add them to obtain a new row (column).

Definition 12.2. A point $(1 - a)x + ay$ with $0 \le a \le 1$ is called a *convex combination* or a *mixture* of x and y.   ∎

The set of all convex linear combinations of two distinct points x and y is called the *interval* or *line segment* connecting or joining x and y. It can be also called the convex hull of x and y. The points x and y are the *endpoints* of the segment. Sometimes these terms are used also in the case $x = y$ when the segment becomes a point.

Figure 12.3 shows three examples of convex sets.

Figure 12.3. Three convex sets in the plane:

They are

diamond $|x| + |y| \leq 1$,

disc $(x - 3)^2 + y^2 \leq 1$,

and

interval $(1 - \alpha)[5, -1] + \alpha[6, 1]$ $(0 \leq \alpha \leq 1)$ connecting the point $x = 5, y = -1$ with the point $x = 6, y = 1$.

Figure 12.4 shows three examples of nonconvex sets.

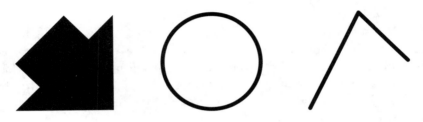

Figure 12.4. Three nonconvex sets in the plane:
the union of a square and a triangle;
circle $(x - 3)^2 + y^2 = 1$; the union of two intervals.

Problem. Show that the disc $x^2 + y^2 \leq 1$ in the (x, y)-plane is a convex set.

Solution. If (x_1, y_1) and (x_2, y_2) are two elements of the disc, then $x_1^2 + y_1^2 \leq 1$ and $x_2^2 + y_2^2 \leq 1$. Then for $0 \leq a \leq 1$, $(1 - a)(x_1, y_1) = ((1 - a)x_1, (1 - a)y_1)$ and $a(x_2, y_2) = (ax_2, ay_2)$. Therefore, the point $(1 - a)(x_1, y_1) + a(x_2, y_2)$ has coordinates $x = (1 - a)x_1 + ax_2$ and $y = (1 - a)y_1 + ay_2$ and, by substituting these expressions for x and y, we have to conclude that $x^2 + y^2 \leq 1$. To do this use the following hint: $(x_1 - x_2)^2 \geq 0$, hence $2x_1y_1 \leq x_1^2 + y_1^2$.

Proposition 12.5. The intersection of any collection of convex sets is also a convex set.

Proof. Let S_i, $i \in I$, be a collection of convex sets. We want to prove that the intersection $S = \bigcap_{i \in I} S_i$ is also convex.

If x and y are elements of S, then x and y belong to S_i for every $i \in I$. Therefore, by the definition of a convex set, $(1 - a)x + ay$ belongs to S_i for every $i \in I$. Thus, $(1 - a)x + ay \in S$, which proves the convexity of S. ∎

The importance of convex sets in linear programming is easily seen from the following two theorems.

Theorem 12.6. The feasible region S of any linear program is a convex set.

Proof. Since S is given by the *intersection* of a finite set of linear constraints, it suffices to show that the region given by one linear constraint is convex.

Let our constraint, for variables $z_1, \ldots, z_n$, be as follows:

$$a_1 z_1 + \cdots + a_n z_n \le c$$

(The cases when we have $=$ or $\ge$ instead of $\le$ can be dealt with similarly.) Take any a in the interval $0 \le a \le 1$. Let both $x = (x_1, \ldots, x_n)$ and $y = (y_1, \ldots, y_n)$ satisfy the linear constraint; that is,

$$a_1 x_1 + \ldots + a_n x_n \le c \text{ and } a_1 y_1 + \ldots + a_n y_n \le c.$$

Taking the linear combination of these inequalities with the coefficients a and $1 - a$, we obtain that

$$a_1 \left(a\, x_1 + (1 - a)\, y_1\right) + \ldots + a_n \left(a\, x_n + (1 - a)\, y_n\right) \le c;$$

that is, $a\, x + (1 - a)\, y$ satisfies the linear constraint. ∎

Since linear equations are linear constraints, Theorem 12.6 implies that the solution set for any system of linear equations is convex. In fact, more is true in this case: For any two distinct solutions x, y the whole straight line $\{(1 - a)x + ay\}$ passing through x, y consists of solutions, not only the line segment $\{(1 - a)x + ay : 0 \le a \le 1\}$. The point $(1 - a)x + ay$ is called an *affine combination* of x and y. ∎

As illustration of convexity, the n-dimensional unit simplex

$$x_1 + \cdots + x_{n+1} = 1, \text{ all } x_i \ge 0$$

is convex. When $n = 1$ this is just the unit interval $0 \le x_1 \le 1$, and the two-dimensional simplex is just a triangle.

Theorem 12.7. The optimality region for any linear program is a convex set.

Proof. This follows from Theorem 12.6 because the optimality region can be described by adding one additional linear constraint

$$\text{objective function} = \text{optimal value}$$

to the set of constraint for the feasible region. ∎

Besides the feasible set and the set of optimal solutions, there is a third convex set associated to any linear program, the set of feasible values. First we need a definition.

Definition. 12.8. Given an optimization problem, the values of the objective function on the feasible region are called *feasible values.* ∎

If the feasible region is empty (i.e., the system of constraints is inconsistent), then there are no feasible values.

Proposition 12.9.   The image of a convex set under an affine transformation is convex.

Proof.   Let S be a convex set in the space of all columns x with n entries and $x \mapsto Cx + d$ be an affine transformation, where C is a constant $m \times n$ matrix and d a constant column with m entries. Let $Cx + d$ and $Cy + d$ be two points in the image. Then $a(Cx + d) + (1 - a)(Cy + d) = C(ax + (1 - a)y) + d$ belongs to the image too, for any scalar a in the interval $0 \le a \le 1$. ∎

The point of the proof was the fact that under any affine transformation, straight lines go to straight lines and line segments go to line segments.

Corollary 12.10. The set of feasible values for any linear program is convex.

Proof. Apply Proposition 12.9 to the case when the convex set is the feasible set of a linear program and the affine transformation is the objective function of this program.

Definition 12.11. A *vertex* (or extreme point) of a convex set S is defined as a point of S that is *not* the half-sum $(y + z)/2$ of two distinct points y, z of S. ∎

In other words, a point is extreme if it is an endpoint of every line segment that lies wholly in S. Still another way to express this is that a point x is a vertex if and only if the set S stays convex after deleting x. See Exercise 14 for an alternative definition of a vertex.

Example 12.12. The interval $0 \le x \le 1$ has exactly two vertices, its ends $x = 0$ and $x = 1$. Indeed, if $0 = (y + z)/2$ with both y and z in the interval, then obviously $y = z = 0$; hence y and z are not distinct. Similarly, 1 is a vertex. On the other hand, if x is in the interval and $x \ne 0, 1$, then we can write $x = (y + z)/2$ with $y = x - \varepsilon$ and $z = x + \varepsilon$, where $\varepsilon = \min(x, 1 - x) \ne 0$, hence both y, z are in the interval, and they are distinct.

Example 12.13. The triangle $x \ge 0, y \ge 0, x + y \le 1$ has three vertices: $x = y = 0; x = 0, y = 1; x = 1, y = 0$. Prove this.

Example 12.14. The extreme points in the disc $x^2 + y^2 \le 1$ are the points at the circle $x^2 + y^2 = 1$. Prove this. ∎

The image of a vertex under an affine map need ι.
Take the diamond S with 4 vertices in Figure 12.3 and p.
the x-axis, $[x,y]^T \mapsto x = [1,0][x,y]^T$. The image is an int
two vertices (endpoints), and two vertices of S go to the m
the interval.

Proposition 12.15. Let S be a convex set and f an affine transformation that takes distinct points of S to distinct points. Then
$F(x)$ is a vertex in $F(S)$ for any vertex x in S.

Proof. Suppose that x is a vertex in S and $f(x) = (f(y) + f(z))/2$
for some $f(y)$ and $f(z)$ in $f(S)$. Then $x = (y+y)/2$ hence $x = y = z$.
So $f(x) = f(y) = f(z)$. Thus, $f(x)$ is a vertex. ∎

For example, if we have a linear program P with all variables
required to be ≥ 0, then we can write it in a standard tableau so P is
affinely equivalent to the tableau problem P'. Proposition 12.15 says
that there is a 1-1-correspondence between vertices in the feasible
region for P and the vertices in the feasible region for P'. Now we
relate the latter vertices with feasible tableaux as follows.

Theorem 12.16. Consider a linear program given by a standard
row tableau. Then a point in the feasible region is a vertex if and
only if it is the basic solution for a feasible tableau.

Proof. Consider an arbitrary feasible tableau

$$\begin{array}{cc} x & 1 \end{array}$$
$$\begin{bmatrix} A & b \\ c & d \end{bmatrix} \begin{array}{l} = u \\ = z \to \min, \quad x \geq 0, \quad u \geq 0 \end{array} \tag{12.17}$$

and the corresponding basic solution $x = 0, u = Ax^t + b = b$. The
feasibility means that $b \geq 0$. Let us try to write

$$[0,b] = ([y, Ay^T + b] + [z, Az^T + b])/2$$

with feasible solutions $[y, Ay^T + b], [z, Az^T + b]$. Since $y, z \geq 0$, we
conclude that $y = z = 0$; hence $[y, Ay^T + b] = [z, Az^T + b] = [0,b]$.
Thus, $[0,b]$ is a vertex.

Conversely, let $[x', Ax'^T + b]$ be a vertex in the feasible region S.
(If S empty, there is nothing to prove.) We start with any feasible
tableau (12.17) and by degenerate pivot steps put as many variables,
which take the zero value at the vertex, as possible on the top.

Suppose now that in the tableau (12.17) we cannot switch any
variable on top taking a nonzero value with any variable at the right

margin taking the zero value. If all variables on the top take the zero value, then $[x', Ax'^T + b] = [0, b]$ is the basic solution, and we are done. Assume now that a variable on the top, x_0, takes a nonzero value x_0', and we are going to get a contradiction.

Permuting columns, we assume that this is the first variable on the top. Permuting rows, we assume that the zeros in the b-part, if any, are on the top of this part. Our tableau looks like

$$\begin{matrix} x_0 & y & 1 \end{matrix}$$
$$\begin{bmatrix} A' & * & 0 \\ * & * & + \\ * & * & d \end{bmatrix} \begin{matrix} = v \\ = w \\ = z \to \min, \quad x \geq 0, \quad u \geq 0 \end{matrix} \qquad (12.18)$$

where $+$ stands for positive entries. Since we cannot switch x_0 with any variable by a pivot step in v, then either there is no A', v in the tableau or $A' = 0$. Now we change the vertex a little bit replacing x_0 by $x_0 - \varepsilon$ and $x_0 + \varepsilon$ and keeping the same values y' for the other variables y on the top. For sufficiently small $\varepsilon > 0$, we obtain two distinct feasible solutions

$$[[x_0' - \varepsilon, y'], A[x_0' + \varepsilon, y']^T + b]$$

and

$$[[x_0' + \varepsilon, y'], A[x_0' + \varepsilon, y']^T + b],$$

and our vertex $[[x_0', y'], A[x_0', y']^T + b]$ is the half-sum of these two points, which is a contradiction. ∎

As a corollary of Theorem 12.16, we obtain that the basic solution of an optimal tableau is a vertex in the feasible region. Thus, if a linear program has an optimal solution, then one of the vertices in the feasible region is optimal. This statement is called the *corner principle*. A corner here refers to a vertex (extreme point).

Another consequence of Theorem 12.16 is as follows:

Corollary. The feasible region for any linear program has only finitely many vertices.

Proof. Let x be the column of n variables in our program P and m the number of inequalities in P that are not sign restrictions. If all n variables in our program P are required to be ≥ 0, then P is affinely equivalent to a linear program given by a standard tableau, so we can apply Theorem 12.6 and Proposition 12.5 and conclude that the number of vertices is at most

$$\frac{(m+n)!}{m! n!} = \binom{m+n}{m} = \binom{m+n}{n},$$

where n is the number of variables in P and $n + m$ is the number of inequalities in P (recall that to put P into a standard tableau P' we introduce a slack variable for each inequality which is not a sign constraint). In general we will prove that the number of vertices for P is at most $\binom{m+n}{m}$. Suppose we find a finite set of vertices V for P with $N > \binom{m+2n}{m}$ vertices (we still do not know whether P has only finitely many vertices). Then we can shift all variables, i.e., make transformation $x \mapsto y = x + s$ (the affine transformation going back is also a shift, $y \mapsto x = x - s$) where s is a constant column such that $y \geq 0$ for all vertices in the selected set V. Now we add at most n of sign restrictions on variables in the column y to obtain a new problem P'' with n variables $y \geq 0$. The problem P'' has still the same number m of other inequalities. Moreover the vertices in V correspond to vertices for P''. (Some vertices in P could be now cut off.) Since there are too many of them, we obtain a contradiction. We used a fact about vertices that is Exercise 18 in the end of this section. ∎

By Theorem 12.16, in Phase 2 of the simplex method, a pivot step corresponds to going from a vertex in the feasible region to another vertex. Every vertex can be described by the corresponding basic solution or by the system of linear equations: all nonbasic variables = 0. A degenerate pivot step does not change the basic solution but any pivot step changes the basis (the set of variables at the right margin) and the set of equations describing the solution. The change is not big: Only one basic variable is replaced and only one equation is replaced.

It is not difficult to show that the two vertices connected by a nondegenerate pivot step are distinct and *adjacent* in the following sense.

Definition 12.19 Two distinct vertices x, y in a convex set S are called *adjacent* if S stays convex after deleting the line segment connecting x, y. ∎

Indeed, suppose that x and y are the basic solutions of two feasible tableaux T and T' connected by a pivot step, and suppose that their convex combination $z = ax + (1 - a)y$ can be written as $(u + v)/2$ with feasible solutions u, v. We have to prove that then both u and v are convex combinations of x and y. Our pivot step switches the variables, say, x_1 and x_2. Then the other variables x' on top (nonbasic variables) take zero values at both vertices.

We consider the tableau T:

$$
\begin{array}{ccc}
x_1 & x' & 1
\end{array}
$$
$$
\begin{bmatrix}
\alpha^* & * & \beta \\
* & * & * \\
* & * & *
\end{bmatrix}
\begin{array}{l}
= x_2 \\
= x'' \\
\rightarrow \min,
\end{array}
$$

where for notational convenience, we pretend that x_1 is the first variable on the top, x' denotes the rest of the variables on the top, x_2 is the first variable at the right margin, and x'' is the rest of the decision variables on the side.

Since the variables in x' vanish at both x, y they also vanish at z, hence they vanish at u, v. So all five points x, y, z, u, v are determined by values of the variable x_1. The values of x_1 for x and y are 0 and $-\beta/\alpha$ respectively. These values and the values between (recall that the pivot entry α is negative in Phase 2 and β is positive) give feasible solutions, and other values give infeasible solutions. Thus, both u and v belong to the interval connecting x and y; this interval is the image of the interval $0 \leq x_1 \leq -\beta/\alpha$ under an affine transformation.

Thus, a degenerate pivot step leaves us at the same vertex while a nondegenerate pivot takes us to a adjacent vertex. In both cases, we change the set of linear equations giving this vertex by one equation. The equations in the question are

the nonbasic variables = 0.

We are still discussing the linear program given by the standard tableau (12.17), that is, the linear program

minimize $z = cx^T + d$, subject to $x \geq 0$, $Ax^T = u \geq 0$.

Let n be the number of variables on the top and m the number of variables at the right margin. The inequalities $x \geq 0$ represent n half-spaces, in the n-dimensional space R^n, and the constraints $Ax^T = u \geq 0$ give m more half-spaces. So the feasible region in the x-space R^n is given as the intersection of $m + n$ half-spaces.

Sometimes it is more convenient to think about the feasible region S as embedded in (x, u)-space R^{n+m}. In this case, it is given by $m + n$ inequalities and m equations, $Ax^T = u$.

Since the tableau is feasible, the basic solution $x = 0, u = b$ is in S. So S is a nonempty set. It is a convex "polyhedron" or "polytope" (bounded or unbounded; see Exercise 13 for more on bounded sets) of nonnegative dimension.

In a generic case (when b has no zero entries), the dimension of S is exactly the number n and there are exactly n hyperplanes (*active* constraints) passing through the basic solution; namely, we set the variables on top to be 0.

In Phase 2 of the simplex method, we start at a vertex, namely, the basic solution for the initial tableau. Each pivot step takes us to an adjacent vertex (see Definition 12.19) with the same or a better value of the objective function. In the nongenerate case, we strictly improve the value of the objective function at each pivot step. In the degenerate case, we actually stay at the same vertex, by replacing one of the active constraints by another one.

Getting a bad column means that we found an "edge" going to infinity, such that we drive the objective function to $-\infty$ as we follow the edge. This can happen only if S is unbounded.

As a corollary of Definition 12.8, we obtain that the basic solution of an optimal tableau is a vertex in the feasible region. Thus, if a linear program has an optimal solution, then one of the vertices in the feasible region is optimal. This statement is sometimes called the corner principle. A corner here refers to a vertex (extreme point).

Example 12.20. Here we consider a two-dimensional example with pictures. Let the linear program

$$x + 8y \rightarrow \max,\ y - 2x \leq 2,\ x + y \leq 5, 2x + y \leq 7,\ x \leq 3, x \geq 0, y \geq 0$$

be given. We are going to solve it by the simplex method. Since the constraints are in canonical form, we need only to multiply the objective function $f = x + 8y$ by -1 and introduce slack variables a, b, c, d to obtain a standard row tableau:

$$
\begin{bmatrix}
\begin{array}{rrr}
x & y & 1 \\
2 & -1 & 2 \\
-1 & -1 & 5 \\
-2 & -1 & 7 \\
-1 & 0 & 3 \\
-1 & -8 & 0
\end{array}
\end{bmatrix}
\begin{array}{l}
\\
= a \\
= b \\
= c \\
= d \\
= -f \quad \rightarrow \min .
\end{array}
$$

The tableau is feasible, so we pass to Phase 2. There are two minuses in the last row, and we can choose either to be in the pivot column. Let us choose y to be included into the basis. Then the maximal ratio is -2, so u_1 goes out of the basis. After the pivot step we obtain the tableau

$$
\begin{array}{ccc}
x & a & 1 \\
\end{array}
$$

$$
\left[\begin{array}{ccc}
2 & -1 & 2 \\
-3 & 1 & 3 \\
-4 & 1 & 5 \\
-1 & 0 & 3 \\
-17 & 8 & -16
\end{array}\right]
\begin{array}{l}
= y \\
= b \\
= c \\
= d \\
= -f \quad \rightarrow \min.
\end{array}
$$

In Figure 12.21 our initial tableau is represented by its basic solution, the vertex $[0, 0]$. The second tableau is represented by the vertex $[0, 2]$. Thus, the pivot step is a move from a vertex along the edge $x = 0$ to an adjacent vertex $[0, 2]$.

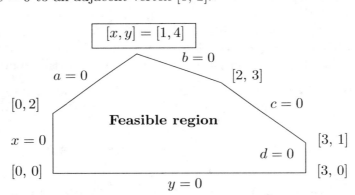

Figure 12.21. The feasible region for Example 12.20

Now we have to choose the x-column as the pivot column and the b-row as the pivot row. The pivot entry is -3. The pivot step gives an optimal tableau:

$$
\begin{array}{ccc}
b & a & 1 \\
\end{array}
$$

$$
\left[\begin{array}{ccc}
* & * & 4 \\
* & * & 1 \\
-4 & 1 & 3 \\
-1 & 0 & 2 \\
17/3 & 7/3 & -33
\end{array}\right]
\begin{array}{l}
= y \\
= x \\
= c \\
= d \\
= -f \quad \rightarrow \min.
\end{array}
$$

Since the tableau is optimal, we do not need the entries marked by $*$. So min $= -33$, max $= 33$ at $x = 1, y = 4$. In Figure 12.21 we moved from the vertex $[0, 2]$ along the edge $a = 0$ to adjacent vertex $[1, 4]$.

What would happen if we chose the other pivot column in the initial tableau? Then we would follow the path [0, 0], [3, 0], [3, 1], [2, 3], [1, 4] and reach the optimal solution in 4 pivot steps instead of 2 pivot steps.   ■

In conclusion, here is a quotation by Dantzig: "The tremendous power of the simplex method is a constant surprise to me."

Exercises

1–2. Prove that the diamond and the interval in Figure 12.3 are convex.

3. Show that the feasible region for the constraint $x^4 + y^4 \leq 1$ is convex. Can this region be the feasible region for a linear program ?

4–6. Show that the three sets in Figure 12.4 are not convex.

7. Show that the feasible region for the constraint $|x| \geq 1$ is not convex. Conclude that this region cannot be the feasible region for a linear program.

8. Give the disc $x^2 + y^2 \leq 1$ by an infinite system of linear constraints.

9. Can the open interval $0 < x < 1$ be given by a system of linear constraints?

10. Find the vertices in the five-dimensional simplex $x_1 + x_2 + x_3 + x_4 + x_5 + x_6 = 1$, all $x_* \geq 0$.

11. Consider a feasible tableau with two columns (one variable on top) and m rows ($m - 1$ basic variables on side). What is maximal possible number of pivot steps in the simplex method? *Hint* : consider first small values of m.

12. Consider a feasible tableau with three columns (two variables on top) and m rows ($m - 1$ basic variables on the side). What is maximal possible number of pivot steps in Phase 2? What if you can guess a better choice for the first pivot entry? *Hints*: Consider first small values of m and assume that all pivot steps are nondegenerate; use the graphical method.

13. Let S be a set of vectors $[x_1, \ldots, x_n]$ with n real entries. For example, S could be the feasible set of a linear program with the decision variables x_i. Show that the following five properties of S are equivalent:

(a) Every affine function on S is bounded.
(b) The function x_i is bounded on S for $i = 1, \ldots, n$.
(c) The function $x_1^2 + \cdots + x_n^2$ is bounded on S.
(d) The function $|x_1| + \cdots + |x_n|$ is bounded on S.
(e) The function $\max_{1 \leq i \leq n} |x_i|$ is bounded on S.

Remark. The set S is called *bounded* if one of these conditions (and hence all of them) holds. In some publications, the boundedness is included into the concept of polyhedron or polytope. A bounded polyhedron is the convex hull of its vertices (i.e., consists of all convex combinations of its vertices). Conversely, the convex hull S of any finite set is a bounded polyhedron, so S can be given by a finite set of linear constraint. Recall that a function $f(x)$ on a set X is *bounded* if there is a number C such that $f(x) \leq C$ for all x in S.

14. Let a convex set S be given by a system of linear constraints. Prove that a point x is a vertex if and only if no other point has the same tight constraints.

15. Suppose we minimize a linear form over a convex set and there is an exactly one optimal solution x. Prove that x is a vertex in S. *Hint*: Write $x = (y + z)/2$ and consider

$$\min = f(x) = (f(y) + f(z))/2.$$

16. Given any linear program and its feasible region S, show that for each vertex x in S there is a linear form f such that x is the unique optimal solution when we minimize f over S. Construct a convex set S and a vertex x for which there is no such f.

17. Let x be an extreme point in a convex set S. Prove that there is a nonconstant linear form f such that x is an optimal solution for maximization of f over S. In other words, there is a tangent hyperplane to S touching the point x. *Hint*: To do this problem in general, you have to be at a Ph.D. level in mathematics. Try to solve this problem for some simple convex sets you know, like convex polygons in plane.

18. Let x be a vertex in a convex set S, and S' be another convex set containing x. Then x is a vertex in the intersection $S \cap S'$. For example, $x = 0$, S is the interval $0 \leq x \leq 1$, and S' is the ray $x \leq 1/2$.

Chapter 5

Duality

§13. Dual Problems

Let us recall the canonical form of a linear programming problem:

$$cx + d \to \min, Ax \le b, x \ge 0.$$

Here x is a column of variables (unknowns) and the data consist of a row c, a number d, a matrix A, and a column b. In more detail, we want to

minimize the objective function

$$f(x) = c_1 x_1 + c_2 x_2 + \cdots + c_m x_m$$

subject to the linear constraints

$$a_{11} x_1 + a_{12} x_2 + \cdots + a_{1m} x_m \le b_1$$
$$a_{21} x_1 + a_{22} x_2 + \cdots + a_{2m} x_m \le b_2$$

$$\cdots$$

$$a_{n1} x_1 + a_{n2} x_2 + \cdots + a_{nm} x_m \le b_n$$

and the nonnegativity constraints

$$x_i \ge 0, \ i = 1, 2, \ldots, m.$$

The first n linear constraints can be written as linear equalities by introducing the slack variables, $u_j \ge 0$, where $j = 1, 2, \ldots, n$:

$$u_j = b_j - a_{j1} x_1 - a_{j2} x_2 - \ldots - a_{jm} x_m, \ j = 1, \ldots, n$$

(i.e., $u = b - Ax \ge 0$ in matrix form).

Thus, in matrix notation, we can write the linear programming problem as follows:

$$\text{minimize } cx$$
$$\text{subject to } b - Ax = u$$
$$x \geq 0, \ u \geq 0$$

or, even shorter,

$$cx \to \min, b - Ax = u \geq 0, x \geq 0,$$

where c is the $1 \times m$ row matrix $c = [c_1 \ c_2 \ \ldots \ c_m]$, x is the $m \times 1$ column matrix

$$x = \begin{bmatrix} x_1 \\ x_2 \\ \vdots \\ x_m \end{bmatrix},$$

b is the $n \times 1$ column matrix

$$b = \begin{bmatrix} b_1 \\ b_2 \\ \vdots \\ b_n \end{bmatrix},$$

u is the $n \times 1$ column matrix

$$u = \begin{bmatrix} u_1 \\ u_2 \\ \vdots \\ u_n \end{bmatrix},$$

and A is the $n \times m$ matrix

$$A = \begin{bmatrix} a_{11} & a_{12} & \cdots & a_{1m} \\ a_{21} & a_{22} & \cdots & a_{2m} \\ \vdots & \vdots & \ddots & \vdots \\ a_{n1} & a_{n2} & \cdots & a_{nm} \end{bmatrix}.$$

We can write our linear program in standard row tableau:

$$\begin{array}{cc} x^T & 1 \\ \begin{bmatrix} A & b \\ c & d \end{bmatrix} & \begin{array}{l} = u \\ = f \to \min, \quad x \geq 0, \quad u \geq 0. \end{array} \end{array}$$

We also can write the same problem in a column tableau:

$$\begin{matrix} -x \\ 1 \end{matrix} \begin{bmatrix} -A^T & c^T \\ b^T & -d \end{bmatrix} \quad x \geq 0, \ u \geq 0$$

$$\begin{matrix} \| & \| \\ u^T & -f \end{matrix} \quad \rightarrow \max$$

We call such a column tableau *standard*. Notice that the matrix

$$\begin{bmatrix} -A^T & c^T \\ b^T & -d \end{bmatrix}$$

is an arbitrary given matrix, and all variables are distinct. We replaced x by $-x$ to keep the same pivot rule (8.6):

$$\begin{matrix} -x \\ -y \end{matrix} \begin{bmatrix} \alpha^* & \beta \\ \gamma & \delta \end{bmatrix} \quad \mapsto \quad \begin{matrix} -u \\ -y \end{matrix} \begin{bmatrix} 1/\alpha & -\beta/\alpha \\ \gamma/\alpha & \delta - \beta\gamma/\alpha \end{bmatrix}$$

$$\begin{matrix} \| & \| \\ u & v \end{matrix} \qquad\qquad \begin{matrix} \| & \| \\ x & v \end{matrix}$$

So we pivot standard column tableaux using the same pivoting rule we used for standard row tableaux. A pivot entry is always a nonzero entry of the matrix. We do not choose it in the last row or the last column to keep our tableaux standard. We change signs of the variables that are moved. This is the only difference from (8.6). There is no difference in the matrix, only at the margins.

Definition. The *basic solution* associated to the standard column tableau

$$\begin{matrix} -y \\ 1 \end{matrix} \begin{bmatrix} A & b \\ c & d \end{bmatrix} \quad y \geq 0, v \geq 0$$

$$\begin{matrix} \| & \downarrow \\ v & \max \end{matrix}$$

is $y = 0$, $v = c$. ∎

This basic solution is feasible if and only if $c \geq 0$. In this case the column tableau is called (column) *feasible*. The corresponding value of the objective function is d.

Definition. A standard column tableau is called *optimal* if $b \geq 0$ and $c \geq 0$; that is, if it is both, row and column feasible. ∎

In this case the basic solution $y = 0$, $v = c$ is optimal, and $\min = d$.

Example 13.1. Let us write the diet problem (Example 2.1) in a standard column tableau:

$$
\begin{array}{c}
-a \\
-b \\
-c \\
-d \\
-e \\
1
\end{array}
\left[
\begin{array}{cccc}
-0.3 & -73 & -9.6 & 10 \\
-1.2 & -96 & -7 & 15 \\
-0.7 & -20253 & -19 & 5 \\
-3.5 & -890 & -57 & 60 \\
-5.5 & -279 & -22 & 8 \\
-50 & -4000 & -1000 & 0
\end{array}
\right]
$$
$$
\begin{array}{cccc}
= u_1 & = u_2 & = u_3 & = -C \quad \to \text{max}.
\end{array}
$$

Note that this tableau is row feasible, so the simplex method, Phase 2 can be applied. We apply it to an imaginary row problem, which later will be called the dual problem. In terms of the original (primal) diet problem, this is the dual simplex method.

In general, we can write any linear program in a standard column tableaux, and then use the simplex method. This is the dual simplex method. It may be more efficient than the simplex method that we discussed in Chapter 4.

Let a linear program be given using a standard row tableau:

$$
\begin{array}{cc}
x & 1
\end{array}
$$
$$
\left[
\begin{array}{cc}
A & b \\
c & d
\end{array}
\right]
\begin{array}{l}
= u \\
= z \to \text{min}, \quad x \geq 0, \quad u \geq 0
\end{array}
\tag{13.2}
$$

We will refer to this linear program as the *primal* or *row* problem. Note that in a standard row tableau we use the top and the right margins for the row of variables, x, and the column of variables, u, respectively. The *dual* or *column* problem of the primal problem given previosly is defined to be the following linear program, which we write in the standard column tableau form:

$$
\begin{array}{c}
-y \\
1
\end{array}
\left[
\begin{array}{cc}
A & b \\
c & d
\end{array}
\right]
\quad y \geq 0, \ v \geq 0
\tag{13.3}
$$
$$
\begin{array}{cc}
\| & \| \\
v & w \quad \to \text{max},
\end{array}
$$

where all variables in the matrices y and v are distinct and different from the variables in the matrices x and u and the objective variable w. The variables in the matrices y and v are called *dual* variables.

The standard column tableau uses the remaining m the left and the bottom margins) for the variables. The varia the row v and the objective variable w are multiplied by -1 in or to have a minimization dual problem rather than a maximization problem.

Both linear programming problems can be written in the same tableau, which now uses all the margins:

$$
\begin{array}{c}
\quad x \quad\ 1 \\
\begin{array}{c} -y \\ 1 \end{array}
\left[\begin{array}{cc} A & b \\ c & d \end{array} \right]
\begin{array}{l} = u \\ = z \to \min \end{array} \\
\ = v\ \ = w \quad \to\ \ \max
\end{array}
\qquad
\begin{array}{c}
x \geq 0, u \geq 0 \\
y \geq 0, v \geq 0.
\end{array}
\qquad (13.4)
$$

Note that every (primal) variable in the row (primal) problem is coupled with a (dual) variable in the column (dual) problem. For instance, the variables in the row matrix x are coupled with the variables in the row matrix v, and the variables in the column matrix u are coupled with the variables in the column matrix y. We will call this situation duality. A remarkable feature of the duality is that (13.2) and (13.3) are dual to each other. In other words, the dual problem of the primal problem (13.2) is problem (13.3) and vice versa (so we can say that the "dual of the dual is the primal"). This is because rewriting a standard row tableau as an equivalent column tableau and rewriting a standard column tableau as an equivalent row tableau both result in the same operation on the data matrix:

$$
\begin{bmatrix} A & b \\ c & d \end{bmatrix} \mapsto \begin{bmatrix} -A^T & c^T \\ b^T & -d \end{bmatrix},
$$

and this operation repeated gives back the original matrix [recall that $(e^T)^T = e$ and $-(-e) = e$ for any matrix e].

Dropping the basic variables u in the row problem in (13.4), we obtain the canonical form $cx^T + d \to \min$, $-Ax^T \leq b, x \geq 0$. The canonical form for the dual problem is $b^t y^t - d \to \min$, $A^T y^T \leq c^T, y \geq 0$. So it is very easy to write the dual for a problem in canonical form, bypassing slack variables and tableaux.

For a linear program in standard form $cx^T + d \to \min$, $-Ax^T = b, x \geq 0$, its dual can be obtained via canonical form. A simpler dual has variables corresponding to the linear equations, which are not required to be nonnegative. An alternative is to solve the system $Ax = b$ in the standard form, which allows us to get a smaller canonical form and the dual problem.

.ationship between the feasible values of a
, of its dual. First we need a definition.

1 an optimization problem, the values of the
he feasible region are called *feasible values*. ∎
gion is empty (i.e., the system of constraints is
here are no feasible values. Now we apply this
of linear programs written in (13.4).

Any ic̖ value for the row problem is $cx^T + d$, where the
row x is a feasible solution (i.e., $x \geq 0$ and $Ax^T + b = u \geq 0$.) Any
feasible value for the column problem is $-y^T A + d$, where the column
y is a feasible solution (i.e., $y \geq 0$ and $-y^T A + c = v \geq 0$). Putting
these equalities together, we obtain that

$$cx^T + d = (v + y^T A)x^T + d = vx^T + y^T(Ax^T) + d$$

$$= vx^T + y^T(u - b) + dvx^T + y^T u - y^T b + d \geq -y^T b + d,$$

because both numbers vx^T and $y^T u$ are nonnegative.

Thus, we obtain the inequality $cx^T + d \leq -y^T b + d$ between the
feasible values of our two problems and hence the following result.

Fact 13.6. Every feasible value for a linear minimization program is
greater than or equal to every feasible value of the dual maximization
problem. ∎

This fact does not violate the symmetry between the problem
and its dual because we multiply the objective functions by -1 when
we transpose the situation. Here is a shorter way to write this fact:
min $\geq$ max whenever both problems are feasible. In fact with our
conventions about max and min for unbounded and infeasible problems, the inequality min $\geq$ max holds in all cases. Namely, when
the row problem is unbounded, we obtain that the dual problem has
no feasible values (i.e., it is infeasible; hence max = min = $-\infty$).
When the column problem is unbounded, then the row problem is
infeasible; hence max = min = ∞)

It is also clear that the inequality becomes equality if and only
if $vx^T + y^T u = 0$ (i.e., whenever a variable in one of the programs
takes a nonzero value, the corresponding dual variable is zero). The
last condition is called *complementary slackness*. So we obtain the
following result, which is very useful to prove that a solution is optimal.

Fact 13.7. If we have a feasible solution for a linear program and a feasible solution for the dual problem and the feasible values are the same (i.e., the complementary slackness holds), then both solutions are optimal. ∎

The duality theorem (included in Theorem 13.8) is a deeper result which replaces the inequality min ≥ max by equality min = max in the case when both problems are feasible. In particular, it gives the converse of the last fact (see Theorem 13.9). The duality theorem also relates properties of a given linear program, such as feasibility and boundness, with that of its dual problem. The idea is to apply pivoting on both the primal linear program and its dual in order to obtain outcomes for both problems. According to Theorem 11.3 there are three possible outcomes for each of these problems. However, as we will see, the outcomes of both problems are closely related.

To see how this theorem applies, we write both problems in the same tableau [as in (13.4)] and apply the simplex method. Then, in a finite number of pivoting steps, the following three outcomes are possible.

Case 1: We obtain an optimal tableau ($b \geq 0$, $c \geq 0$);

Case 2: We obtain a tableau with a bad row;

Case 3: We obtain a row feasible tableau ($b \geq 0$) with a bad column.

In Case 1, the optimal tableau ($b \geq 0$, $c \geq 0$) gives an optimal solution for either problem. An optimal solution for the primal problem is given by $u = b$, $x = 0$; an optimal solution for the dual problem is given by $v = c$, $y = 0$. Moreover, the optimal value is $d = \min(z) = \max(w)$ for both z and w.

In Case 2, $\min(z) = -\infty$ and the column problem (13.3) has no feasible solutions. So $\max(w) = -\infty$.

In Case 3, the row problem has no feasible solutions [that is, $\min(z) = +\infty$]. To see what happens with the column problem, we transpose the tableau and apply the simplex method. When we perform this operation, the original row problem becomes a column problem and the original column problem is now the row problem. Note that we cannot obtain an optimal tableau since this would give an optimal solution for the original row problem which is now the column problem in this new tableau. So only two outcomes are now possible:

a tableau with a bad row

or

a feasible tableau with a bad column.

In the first case neither problem has any feasible solutions. In the second case the new row problem (that is, the original column problem) is unbounded. It can be shown that by extra pivoting we can obtain a tableau with a bad row and a bad column.

Thus, we obtain the main theorem of linear programming—the theorem on four alternatives—which includes the duality theorem.

Theorem 13.8. Given a pair (13.4) of dual linear programs, one and only one of the following four situations occurs:

(1) Both problems have optimal solutions with the same optimal value $[\min(z) = \max(w)]$.

(2) Both problems are infeasible.

(3) The row problem is infeasible and the column problem is unbounded $[\min(z) = \infty = \max(w)]$.

(4) The column problem is infeasible and the row problem is unbounded $[\min(z) = -\infty = \max(w)]$.

Proof. This theorem follows from the existence of a simplex method that always works (one that avoids cycling) and the previous discussion. ∎

Sometimes this is called the theorem in three alternatives, because (3) and (4) are symmetric. But usually the theorem on three alternatives refers to a less precise classification, namely, Cases 1–3. We prefer to stick to the name "theorem on four alternatives."

The duality theorem usually refers to the equality of the optimal values in the cases (1), (2), (4). But we will sometimes refer to Theorem 13.8 as the duality theorem for short.

The duality theorem is a deep fact with several interpretations and applications. Geometrically, the duality theorem means that certain convex sets can be separated from points outside them by hyperplanes. Traditionally, this is how the duality theorem is proved. But we chose another way of proving it: It follows from the existence of a simplex method that always works (one that avoids cycling). We will see another interpretation of the duality theorem in the theory of matrix games (cf. Chapter 7). For problems in economics, the dual problems and the duality theorem also have important economic interpretations (see examples in the next section).

As a corollary of the duality theorem, we obtain the following result called *the complementary slackness theorem*:

Theorem 13.9. Given a feasible solution for a linear problem and a feasible solution for the dual problem, they are both optimal if and only if the feasible values are the same (i.e., the complementary slackness condition holds). ∎

Problem 13.10. Check whether $x_1 = 1, x_2 = 2, x_3 = 3, x_4 = 4, x_5 = 0$ is an optimal solution for the linear program

$$x_1 - x_2 + x_3 - 2x_4 + 6x_5 \to \min,$$
$$x_1 + x_2 + x_3 + x_4 + x_5 \geq 10,$$
$$x_1 + 2x_2 + 3x_3 + 4x_4 + 5x_5 \geq 20,$$
$$x_1 - x_2 + x_3 - x_4 + x_5 \geq -2,$$
$$\text{all } x_i \geq 0.$$

Solution. First we check that the given solution x is feasible. We find that it is, and the first, the third, and the last constraints hold as equalities (such constraints are called *active* or *tight*).

Then we put the problem in a standard row tableau with slack variables at the right margin, and we write dual variables y_i at the left and bottom margin in the same tableau, where all $x_i, y_i \geq 0$:

	x_1	x_2	x_3	x_4	x_5	1	
$-y_6$	1	1	1	1	1	-10	$= x_6$
$-y_7$	1	2	3	4	5	-20	$= x_7$
$-y_8$	1	-1	1	-1	1	-2	$= x_8$
1	1	-1	1	-2	6	0	$= z \to \min$
	$= y_1$	$= y_2$	$= y_3$	$= y_4$	$= y_5$	$= w$	$\to$ max.

Note that $x_6 = x_8 = 0$ for our slack variables in x corresponding to the active constraints. On the other hand, $x_7 = 10 \neq 0$. Assume now that our x is optimal and consider an optimal solution y for the dual problem that exists by the duality theorem. By complementary slackness, $y_1 = y_2 = y_3 = y_4 = 0$ and $y_7 = 0$. Thus, our column problem becomes

	x_1	x_2	x_3	x_4	x_5	1
$-y_6$	1	1	1	1	1	-10
$-y_8$	1	-1	1	-1	1	-2
1	1	-1	1	-2	6	0
	$= 0$	$= 0$	$= 0$	$= 0$	$= y_5$	$= w$	$\to$ max.

The second column says that $y_8 - y_6 = 1$ while the fourth column says that $y_8 - y_6 = 2$, which gives a contradiction. So the given solution is not optimal.

Exercises

1–2. Write in a standard column tableau:

1. $5x - 6y + 2z \to \min$, $[x, y, z] \geq 0, y \leq 7, x + y \geq 3$.

2. Minimize
$$.39a + .11b + .18c + .21d + .35e + .44f + .25g + .25h + .23i + .24j$$
subject to
$$15a + 15b + 15c + 15d + 15e + 15f + 10g + 15h + 15i + 10j \geq 100,$$
$$25a + 25b + 25c + 6f + 10g + 25h + 10j \geq 100,$$
$$a + 25b + 25c + 25d + 30f + 25g + 25h + 25i + 25j \geq 100,$$
$$a + 25c + 25d + 30e + 25f + 25g + 25h + 25i + 25j \geq 100,$$
$$a + 25b + 25c + 25d + 30e + 25f + 25g + 25h + 25i + 25j \geq 100,$$
$$3a + 2b + 1c + 2d + 4e + 2f + g + 2h + 3i + 3j \geq 70,$$
$$[a, b, c, d, e, f, g, h, i, j] \geq 0. \qquad \blacksquare$$

3. Rewrite the linear programs in Exercise 1 and Exercise 2 in standard row tableaux.

4. Solve the linear problem in Problem 13.10.

5. Prove that for any linear program, the set of feasible values is a convex set on line.

6–10. Check whether the following solution is optimal for the problem given by the standard row tableau

$$
\begin{bmatrix}
\begin{array}{cccccc}
x_1 & x_2 & x_3 & x_4 & x_5 & 1 \\
7 & -2 & -6 & 6 & -1 & 15 \\
-1 & 1 & 1 & -2 & 2 & -4 \\
-1 & 0 & 1 & 0 & -1 & -2 \\
4 & 0 & -3 & 5 & 3 & 4
\end{array}
\end{bmatrix}
\begin{array}{l}
\\
= x_6 \\
= x_7 \\
= x_8 \\
= z \;\to\; \min.
\end{array}
$$

6. The basic solution.

7. all $x_i = 1$.

8. $x_1 = 4, x_3 = 7, x_5 = x_7 = 1$, all other $x_i = 0$.

9. $x_1 = 1, x_2 = 2, x_3 = 3$, all other $x_i = 0$.

10. $x_3 = 3, x_2 = x_4 = x_5 = 1$, all other $x_i = 0$.

§14. Sensitivity Analysis and Parametric Programming

In real-life problems we often do not know exact values or these values may change with time. For example, RDAs in Example 2.1 (diet problem) are of necessity only vague estimates and they depend on country and year. Sensitivity analysis is concerned with how changes in data affect the optimal value and the optimal solutions. Most often, this is about small changes, while large changes are studied in *parametric programming.* In this section we consider both. We have already considered problems where some values in the data were not given explicitly but we were asked to solve the problem for all possible values. These values are called *parameters.*

First we show how to find the changes in the optimal values caused by small changes in resource limits or required limits using optimal tableaux.

Suppose that in the diet problem (Example 2.1) we want to know the change in the minimal cost if

$$\text{the protein requirement is changed from 50 to 51;} \qquad (14.1)$$

$$\begin{array}{c} \text{all three requirements are changed:} \\ \text{from 50, 4000, 1000 to} \\ 50 + \varepsilon_1, 4000 + \varepsilon_2, 1000 + \varepsilon_3, \\ \text{respectivel,y with } |\varepsilon_i| \leq 1, \ i = 1, 2, 3; \end{array} \qquad (14.2)$$

$$\begin{array}{c} \text{all five prices are changed by } \varepsilon_a, \varepsilon_b, \varepsilon_c, \varepsilon_d, \varepsilon_e, \\ \text{respectively, with absolute value of every change } \leq 1 \end{array} \qquad (14.3)$$

$$\begin{array}{c} \text{all requirements are changed as in (14.2)} \\ \text{and all prices are changed as in (14.3)} \end{array} \qquad (14.4)$$

Now we answer these four questions. Note that the second question includes the first one, and the last question includes all four questions.

First we write the diet problem in a standard row tableau:

$$
\begin{array}{cccccc}
a & b & c & d & e & 1 \\
\end{array}
$$
$$
\begin{bmatrix}
0.3 & 1.2 & 0.7 & 3.5 & 5.5 & -50 \\
73 & 96 & 20253 & 890 & 279 & -4000 \\
9.6 & 7 & 19 & 57 & 22 & -1000 \\
8 & 10 & 15 & 5 & 60 & 0
\end{bmatrix}
\begin{array}{l}
= u_1 \\
= u_2 \\
= u_3 \\
= C \to \min
\end{array}
$$

(cf. Example 13.1).

It takes at least two pivot steps to obtain an optimal tableau:

$$
\begin{array}{cccccc}
a & b & u_1 & d & u_3 & 1 \\
\begin{bmatrix}
-0.519 & -0.136 & -0.2469 & -2.654 & 0.0617 & 49.38 \\
-10425 & -2710 & -4941 & -52951 & 1248 & 996931 \\
0.011 & -0.20 & 0.21 & -0.30 & -0.0079 & 2.806 \\
7.499 & 12.71 & 0.471 & 44.34 & 0.2458 & 269
\end{bmatrix}
\begin{array}{l}
= c \\
= u_2 \\
= e \\
= C
\end{array}
\end{array}
$$

with numbers given approximately; the exact value for the optimal value is $80,000/297 \approx 269.36$, not 269. Surplus variables u_1 and u_3 for protein and calcium are on the top, so the corresponding constraints are active (the constraints hold as equalities for the basic solution) while the constraint in vitamin A is not active (there is a surplus). So it is clear that a small change in the vitamin A requirement of 4000 would not affect the optimal solution.

One way to find what happens when 50 is replaced by another number is to replace 50 by a parameter, say, p, in the original tableau and pivot the modified tableau. The parameter would stay in the last column. However, there is another way to proceed that also allows us to get the modified optimal tableau without pivoting again. Instead of replacing 50 by p, we replace the restriction $u_1 \geq 0$ on the slack variable by the condition $u_1 \geq p - 50$. This makes our tableau nonstandard, but we do not do any additional pivoting.

To answer (14.1), our constraint on u_1 is $u_1 \geq 1$, so we cannot set it to 0 together with the other variables on the top to get an optimal solution. The best we can do is to set $u_1 = p - 1 = 1$, which gives the optimal value $\approx 0.47 + 269.36 = 269.83$. It is important to check the feasibility: The values for c and u_2 go down but stay positive. Thus, the answer to (14.1) is that the minimal cost increases by approximately 0.47 (the exact number is $140/297$).

Similarly, if we replace the requirement 50 by, say, $p = 48$, we relax the condition $u_1 \geq 0$ to $u_1 \geq -2$ which allows us to take $u_1 = -2$ and get improvement in C equal $280/297$ (now the minimal cost goes down). Again it is important to check the feasibility: the value for e goes down but stays positive.

Thus, we see that the minimal cost, $mc = mc(p)$, is an affine function of p when we replace the requirement 50 by a parameter p that stays close to 50 (so the optimal tableau for $p = 50$ produces feasible solutions when we change the basic value $u_1 = 0$ for the slack variable to the value $u_1 = p - 50$) and that the number $140/297 \approx 0.471$ is the slope of $mc(t)$. This number is known as the *shadow price* of protein in our diet problem. We can compare it with prices for protein from alternative sources to make decisions about our diet.

Moreover, we can tell that if we replace all required limits 50, 4000, 1000 by parameters t_1, t_2, t_3 respectively, then for $|t_1 - 50|$, $|t_2 - 4000|$, and $|t_3 - 1000|$ sufficiently close to 0, the minimal cost, mc $\approx 0.471t_1 + 0.240t_3 + 269.36$. So it is an affine function of t_1, t_2, t_3 and it is independent of t_2. It is not a surprise that it is independent of t_2 because the corresponding constraint is not active for the basic solution—we have a surplus of vitamin A. Therefore, the answer to (14.2) is $\approx 0.471\varepsilon_1 + 0.240\varepsilon_3$.

Also, we can write the last column of the (standard) optimal tableau of the modified problem:

$$
\begin{matrix}
& 1 \\
\begin{bmatrix}
49.38 - 0.2469\varepsilon_1 + 0.0617\varepsilon_3 \\
996931 - 4941\varepsilon_1 + 1248\varepsilon_3 + \varepsilon_2 \\
2.806 + 0.21\varepsilon_1 - 0.0079\varepsilon_3 \\
269 + 0.471\varepsilon_1 + 0.2458\varepsilon_3
\end{bmatrix}
&
\begin{matrix}
= c \\
= u_2 \\
= e \\
= C \rightarrow \min.
\end{matrix}
\end{matrix}
\qquad (14.5)
$$

The rest of the tableu stays the same, independent of ε_i. It is easy to compute now for which values of ε_i the tableau stays optimal. It is clear that the last column is positive when $|\varepsilon_i| \leq 1$.

Next we address the question of (14.3). Since the dual problem has the same optimal value and the coefficients of the objective function for the row problem are the constant terms in the constraints of the dual problem, we can use duality to get the answer. Also, we can argue directly that a small change in data would result in a small change in the optimal tableau, so the tableau stay optimal. Therefore, the answer is approximately $48.38\varepsilon_c + 2.806\varepsilon_e$ for ε_* sufficiently close to 0. But is the number 1 in (14.3) sufficiently close to 0?

To answer this question, it is time to write the last row of the modified tableau (approximately; the other rows do not depend on parameters):

$$
\begin{bmatrix}
7.499 - 0.519\varepsilon_c + 0.011\varepsilon_e + \varepsilon_a \\
12.71 - 0.136\varepsilon_c - 0.20\varepsilon_e + \varepsilon_b \\
0.471 - 0.2469\varepsilon_c + 0.21\varepsilon_e \\
44.34 - 2.654\varepsilon_c - 0.30\varepsilon_e + \varepsilon_d \\
0.2458 + 0.0617\varepsilon_c - 0.0079\varepsilon_e \\
269 + 49.38\varepsilon_c + 2.806\varepsilon_e
\end{bmatrix}^T
\cdot
\qquad (14.6)
$$

Now it is clear that this row is positive for $|\varepsilon_*| \leq 1$.

Finally, we have to face (14.4) when we modify both the requirements and prices. Then, in the modified optimal tableau, the first three entries of the last column are the same as in (14.5), the first five entries of the last row are the same as in (14.6), and the last entry in the last row or column is (approximately)

$$[\varepsilon_c, \varepsilon_e, 1] \begin{bmatrix} -0.2469 & 0.0617 & 49.38 \\ 0.21 & -0.0079 & 2.806 \\ 0.471 & 0.2458 & 269 \end{bmatrix} \begin{bmatrix} \varepsilon_1 \\ \varepsilon_3 \\ 1 \end{bmatrix}.$$

Dropping the constant term here, we get an approximate answer for (14.4).

The same trick works in general. Namely, suppose we have a linear program of the form $cx \to \min, Ax \geq t, x \geq 0$, and we ask how the optimal value depends on the requirements t when they are close to certain values $t = b$. Suppose that the program has an optimal solution for $t = b$ and that this optimal solution is the basic solution for an optimal tableau where all basic variables take positive values. (To put the problem in a standard row tableau, we introduce the surplus variables $u = Ax - b$.) Then the minimal value, $\mathrm{mv} = \mathrm{mv}(t)$, is an affine function $c_1(t - b) + \bar{d}$, where $\bar{d} = \mathrm{mv}(b)$ is the last entry in the last row in the optimal tableau and the row c_1 consists of entries of the last row of the optimal tableau corresponding to active constraints and zeros for the parameters corresponding to nonactive constraints. When the objective function is interpreted as the cost, the entries of the row c_1 are called the *shadow prices* corresponding to the active constraints. Note that small changes in the requirements in nonactive constraints do not change the optimal value. The corresponding shadow prices are zeros. When a basic slack variable takes the zero value, two different shadow prices for this constraint are possible: one corresponding to increasing requirement, and the other corresponding to decreasing requirement (see the discussion of parametric programming in §14).

Thus, the basic solution for the dual problem tells us how the optimal value depends on some changes in data. Similarly, the optimal tableau tells us how the optimal value depends on small changes in the objective function. Again the dependence is affine.

Remark. More generally, we can replace b and c by functions, not necessarily affine, $F(t)$ and $G(t)$ of parameters t. This makes the optimal value a function $\mathrm{mv}(t)$ of t (not necessarily defined for all t). Then the last row and column (if positive except the last entry) of the optimal tableau for the problem with $b = F(t_0)$ (if it exists) allows us to express the partial derivatives of $\mathrm{mv}(t)$ as a linear function of partial derivatives of $F(t)$ and $G(t)$ (if they exist).

Using the optimal tableau, it is also possible to find what happens with the optimal tableau under a small change of all entries in the initial tableau, not only those in the last row and column. For simplicity we assume that we have only one parameter t. So our setting is now as follows. We have a standard row tableau

$$\begin{array}{cc} x & 1 \end{array}$$
$$\left[\begin{array}{cc} A(t) & b(t) \\ c(t) & d(t) \end{array}\right] \begin{array}{l} = u \\ \rightarrow \min \end{array} \qquad (14.7)$$

depending on a parameter t such that all functions are differentiable at $t = t_0$ and an optimal tableau for $t = t_0$

$$\begin{array}{cc} \bar{x} & 1 \end{array}$$
$$\left[\begin{array}{cc} \bar{A} & \bar{b} \\ \bar{c} & \bar{d} \end{array}\right] \begin{array}{l} = \bar{u} \\ \rightarrow \min \end{array} \qquad (14.8)$$

has all entries in $\bar{b}$ and $\bar{c}$ positive. Then we show that the linear program has exactly one optimal solution $x = x(t)$ for every t sufficiently close to $t = t_0$, and we compute the derivative $z'(t_0)$ of the optimal value $z(t) = c(t)x(t)^T$ at $t = t_0$.

To simplify notations, we permute rows and columns in the tableaux to arrange that $x = [x_1, x_2]$, $u = \left[\begin{array}{c} u_1 \\ u_2 \end{array}\right]$, $\bar{x} = [u_1, x_2]$, $\bar{u} = \left[\begin{array}{c} x_1^T \\ u_2 \end{array}\right]$. The row x_2 consists of all variables that stay on the top, and it could be vacuous. The row x_1 consists of variables on top that go to the right margin at the optimal tableau, and it could be vacuous too, in which case the answer is trivial: $z'(t_0) = d'(t_0)$.

We rewrite the initial matrix (14.7) and the optimal tableau (14.8) accordingly:

$$\begin{array}{ccc} x_1 & x_2 & 1 \end{array}$$
$$\left[\begin{array}{ccc} \alpha(t) & \beta(t) & b_1(t) \\ \gamma(t) & \delta(t) & b_2(t) \\ c_1(t) & c_2(t) & d(t) \end{array}\right] \begin{array}{l} = u_1 \\ = u_2 \\ \rightarrow \min, \end{array} \qquad (14.9)$$

$$\begin{array}{ccc} u_1^T & x_2 & 1 \end{array}$$
$$\left[\begin{array}{ccc} * & * & \bar{b}_1 \\ * & * & * \\ \bar{c}_1 & * & \bar{d} \end{array}\right] \begin{array}{l} = x_1^T \\ = u_2 \\ \rightarrow \min. \end{array} \qquad (14.10)$$

The pivot steps, taking the tableau (14.9) with $t = t_0$ to (14.10), take the parametric tableau (14.9) to

$$
\begin{bmatrix}
 & x_1 & x_2 & 1 & \\
 & \alpha(t)^{-1} & -\alpha^{-1}\beta(t) & -\alpha(t)^{-1}b_1(t) & = u_1 \\
 & \gamma(t)\alpha(t)^{-1} & * & b_2(t) - \gamma(t)\alpha(t)^{-1}b_1(t) & = u_2 \\
 & c_1(t)\alpha(t)^{-1} & * & d(t) - c_1(t)\alpha(t)^{-1}b_1(t) & \rightarrow \min
\end{bmatrix}
$$

$$(14.11)$$

(see Remark 8.11). Since all functions in (14.9) are continuous at $t = t_0$, we have the following. For all t sufficiently close to t_0 : the matrix $\alpha(t)$ is invertible [because $\alpha(t_0)$ is invertible]; the last column in (14.10) without the last entry is strictly positive (because $\bar{b} > 0$); the last row in (14.11) without the last entry is strictly positive (because $\bar{c} > 0$). So the tableau (14.11) is optimal and its basic solution is the only optimal solution (for t close to t_0). The last entry in the last row is the optimal value. Its derivative at $t = t_0$ is

$$
z'(t_0) = d'(t) - c_1'(t_0)\bar{b}_1 - \bar{c}_1 b_1'(t_0) + \bar{c}_1\alpha'(t_0)\bar{b}_1.
$$

In the case when $A(t)$ is constant, the last term drops. When both $A(t)$ and $c(t)$ are constant and both $b(t)$ and $d(t)$ are affine or both $A(t)$ and $b(t)$ are constant and both $c(t)$ and $d(t)$ are affine, the optimal value is an affine function of t for t close to t_0. ∎

What happens when our optimal tableau has zero values for some basic variables (such tableaux are called *degenerate*)? In this case the slopes may depend on direction in the change of parameters.

Example 14.12. This example is adapted from Section 4-10 of *Linear programming and economic analysis*, by Dorfman, Samuelson and Solow (McGraw-Hill, 1958).

A chemical firm processes a certain raw material by the use of two major types of equipment, called stills and retorts. Four different production processes are available to the firm. If Process 1 is used to treat 100 tons of the raw material, it will utilize 7% of the weekly capacity of the stills and 3% of the weekly capacity of the retorts. The value of the product and the costs vary with the process used. If 100 tons are treated by Process 1, the net profit to the firm is $60. The firm plans to process 1500 tons of raw material weekly. Obviously, the company wants to *maximize* the profit it will accrue by processing 1500 tons of raw material weekly, using an *optimal* combination of all four processes. The following table gives the pertinent information for all four processes:

Production processes	(1)	(2)	(3)	(4)	Available
Raw material (tons/week)	100	100	100	100	1500
Still capacity (%)	7	5	3	2	100
Retort capacity (%)	3	5	10	15	100
Profit ($/week)	60	60	90	90	

We can write these data in row and column standard tableaux, where x_i is the level (intensity) of Process i ($x_1 = 1$ means that a hundred tons of the raw material is treated by Process 1), and y_j is a slack variable. Here is the standard column tableau form of this linear program

$$
\begin{array}{c}
-x_1 \\
-x_2 \\
-x_3 \\
-x_4 \\
1
\end{array}
\left[
\begin{array}{cccc}
100 & 7 & 3 & -60 \\
100 & 5 & 5 & -60 \\
100 & 3 & 10 & -90 \\
100 & 2 & 15 & -90 \\
1500 & 100 & 100 & 0
\end{array}
\right]
\quad
\begin{array}{l}
x_1, x_2 \geq 0, \\
x_3, x_4 \geq 0; \\
y_1, y_2, y_3 \geq 0
\end{array}
$$

$$\quad = y_1 \quad = y_2 \quad = y_3 \quad = f \qquad \rightarrow \max.$$

We also include the standard row tableau, on which we will pivot:

$$
\begin{array}{ccccc}
x_1 & x_2 & x_3 & x_4 & 1 \\
\end{array}
$$
$$
\left[
\begin{array}{ccccc}
-100 & -100 & -100 & -100 & 1500 \\
-7 & -5 & -3 & -2 & 100 \\
-3 & -5 & -10 & -15 & 100 \\
-60 & -60 & -90 & -90 & 0
\end{array}
\right]
\begin{array}{l}
= y_1 \\
= y_2 \\
= y_3 \\
= -f \rightarrow \min.
\end{array}
$$

Let us work with the row form. It is (row) feasible, since the first three entries in the last column, 1500, 100, and 100, are $\geq$ 0. Therefore, we go to Phase 2 of the simplex method. Applying pivoting twice, we obtain an optimal tableau:

$$
\begin{array}{ccccc}
y_1 & x_2 & y_3 & x_4 & 1 \\
\end{array}
$$
$$
\left[
\begin{array}{ccccc}
-1/70 & -5/7 & 1/7 & 5/7 & 50/7 \\
61/700 & 6/7 & -4/7 & -13/7 & 185/7 \\
3/700 & -2/7 & -1/7 & -12/7 & 55/7 \\
33/70 & 60/7 & 30/7 & 150/7 & -7950/7
\end{array}
\right]
\begin{array}{l}
= x_1 \\
= y_2 \\
= x_3 \\
= -f \rightarrow \min.
\end{array}
$$

So the (unique) answer is that maximal profit $f = \$7950/7$ per week (approximately \$1135.71 per week) at $x_1 = 50/7$, $x_2 = 0$, $x_3 = 55/7$, $x_4 = 0$. We used completely the raw material available ($y_1 = 0$) and the retorts ($y_3 = 0$), but the stills are underused.

Can we make more profit by buying and using more raw material? The answer depends on comparison of the price at which we can buy extra material and the *shadow price*, the change in the optimal value of the linear program when we replace 1500 by 1501 or 1499. So we consider the linear program with 1500 replaced by $1500 + \varepsilon$ as a function of the parameter ε. Here is the perturbed problem in a standard row tableau:

$$
\begin{array}{ccccc}
x_1 & x_2 & x_3 & x_4 & 1 \\
\left[\begin{array}{ccccc}
-100 & -100 & -100 & -100 & 1500 + \varepsilon \\
-7 & -5 & -3 & -2 & 100 \\
-3 & -5 & -10 & -15 & 100 \\
-60 & -60 & -90 & -90 & 0
\end{array}\right.
\end{array}
\begin{array}{l}
\left.\right] \\
\end{array}
\begin{array}{l}
= y_1 \\
= y_2 \\
= y_3 \\
= -f \to \min.
\end{array}
$$

Pivoting twice, we obtain an optimal tableau:

$$
\begin{array}{ccccc}
y_1 & x_2 & y_3 & x_4 & 1 \\
\left[\begin{array}{ccccc}
-1/70 & -5/7 & 1/7 & 5/7 & 50/7 + \varepsilon/70 \\
61/700 & 6/7 & -4/7 & -13/7 & 185/7 - 61\varepsilon/700 \\
3/700 & -2/7 & -1/7 & -12/7 & 55/7 - 3\varepsilon/700 \\
33/70 & 60/7 & 30/7 & 150/7 & -7950/7 - 33\varepsilon/70
\end{array}\right.
\end{array}
\begin{array}{l}
\left.\right] \\
\end{array}
\begin{array}{l}
= x_1 \\
= y_2 \\
= x_3 \\
\to \min.
\end{array}
$$

Here is how we obtain the answer (for small ε) *without* pivoting. We relax the condition $y_1 \geq 0$ replacing it by $y_1 \geq -e$. From the optimal tableau, we see that the optimal value for f in the perturbed (relaxed) problem is $7950/7 + 33\varepsilon/70$ for small positive ε.

Note that in the optimal tableau of the perturbed problem, the last column is a perturbation of the last column of the original problem; that is, the difference between these two columns is given by a column matrix multiplied by ε. Moreover, this column matrix is precisely the negative of the first column of the original problem.

The answer to our question lies in the shadow price (i.e., the coefficient $33/70$ of ε in the optimal value of the perturbed problem; i.e., the slope of the optimal value as a function of the parameter). If the price of the raw material is less than the shadow price, then buy more raw material and increase the profit. Otherwise, do not buy any more raw material.

An economic interpretation of the slope, depending on the situation, can be the shadow price (i.e., the maximal price we are willing to pay for a small additional amount of a resource) or the marginal cost (i.e., the additional cost needed to produce a small additional amount of a product while we are minimizing the total cost).

Let us now consider the previous problem with $1500 + \varepsilon$ instead of 1500, where ε ranges over all real numbers (before we were interested only in small ε). This is an example of *parametric linear programming*. Note that ε is not a variable under our control, but a *parameter*. The optimal value (the maximal profit in our problem) is a function $F(\varepsilon)$ of the parameter. When $\varepsilon \leq -1500$, there are no feasible solutions. We do need the raw material to get any profit, so $F(-1500) = 0$. We have computed above $F(0) = 7950/7$. Here is a plot of $F(\varepsilon)$:

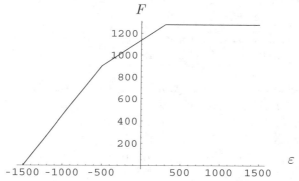

The graph of the function $F(\varepsilon)$ consists of line segments (such functions are often called piecewise linear; see the next definition). The function is nondecreasing because we are not required to use all resources available. The slope of the function is the marginal cost of the row material. The slope changes at the *break points*. The slope decreases (such functions are called concave, see Definition 14.14 below)—the *law of diminishing returns*.

The optimal value of any linear program as a function of parameters in data has similar properties. To describe those properties we give a few definitions.

Definition 14.13. A function $f(t)$ of k variables $t = [t_1, \ldots, t_k]^T$ defined on a convex set P is called piecewise affine if the set P is the union of finitely many convex subsets such that $f(t)$ is an affine function on every subset. ∎

The term *piecewise linear* is often used instead of *piecewise affine*. In most common case $k = 1$ the graph of a piecewise affine function is made from finitely many line segments. The convex sets on a line are easy to list: the empty set, points, intervals, rays, the whole line.

Remark. The following quote of Hermann Weyl (1885–1955) may help you to grasp the concept of piecewise linear function (The Mathematical Way of Thinking, an address given at the Bicentennial Conference at the University of Pennsylvania, 1940):

> Our federal income tax law defines the tax y to be paid in terms of the income x; it does so in a clumsy enough way by pasting several linear functions together, each valid in another interval or bracket of income. An archeologist who, five thousand years from now, shall unearth some of our income tax returns together with relics of engineering works and mathematical books, will probably date them a couple of centuries earlier, certainly before Galileo and Vieta.

Definition 14.14. A function $f(t)$ of k variables $t = [t_1, \ldots, t_k]^T$ defined on a convex set P is called convex if the set of points above its graph is convex. ∎

Sometimes the terms *concave upward* or *convex downward* are used instead of *convex*. We call a function f *concave* if the function $-f$ is convex (i.e., the set of points below the graph of f is convex).

There are several equivalent definitions of a convex function. For example, a function (defined on a convex set) is convex if and only if it is the maximum of a family of affine functions. In the case of a finite family, the maximum is a piecewise affine convex function.

Now we state a general result of parametric programming.

Theorem 14.15. Consider a linear program $cx \to \min$, $Ax \leq b, x \geq 0$ in canonical form. Assume that all entries of either the row c or the column b are affine functions of k parameters $t_1, \ldots, t_k$. Consider the set P of values $t = [t_1, \ldots, t_k]$ of the parameter for which the linear program has an optimal solution, and let $f(t)$ be the optimal value. Then P is a convex set, and $f(t)$ is the minimum of a finite set of affine functions on P. So $f(t)$ is a piecewise affine and concave function. If b is a nondecreasing function of t or c is a nonincreasing function of t, then $f(t)$ is a nondecreasing function of t.

Proof. If P is empty, then we have nothing to prove so let us assume that our program has an optimal solution for at least one value of t.

Using the duality theorem, we see that it suffices to consider the case when the parameter is in the objective function $c = c' + c''t$. Then the feasible region S is independent of t. To prove convexity of P, consider two numbers t', t'' in P and their convex combination $t = at' + (1-a)t''$. We have to show that t is in P, i.e., the function $c(t)x$ is bounded from below on S. Since $c(t)$ is an affine function of t, $c(t) = ac(t') + (1-a)c(t'')$. Since $c(t')x, c(t'')x \geq C$ for all x in S for a number C, we have $c(t)x = ac(t')x + (1-a)c(t'')x \geq aC + (1-a)C = C$ for all x in S; hence $c(t)x$ is bounded from below on S.

Each tableau for our LP has parameters only in the last row, and every entry in the last row is an affine function of parameters. This is true for the initial tableau, and pivot steps preserve this property. The total number of these tableaux is finite (it was bounded in §10).

Now we consider the subset of tableaux that are optimal for at least one value of (vector) parameter t (the value could be different for different tableaux) and the last entries in the last rows in these tableaux. For each t all these tableaux are feasible and at least one is optimal. So the optimal value $f(t)$ is the minimum over these last entries.

Finally, it is obvious that if all coefficients of the objective function increase or stay the same, then the minimal value $f(t)$ cannot improve. ∎

Objective functions with parameters appear in *goal programming* when we want to combine several objectives or goals into one objective function. A way to do this is to take a linear combination of several objectives we want to minimize with nonnegative coefficients. The choice of coefficients (the weights) could be controversial, but they could be considered as parameters. If the functions we combine are affine, then the resulting objective function is also affine, and its coefficients are affine functions of parameters. So Theorem 14.15 can be applied.

Problem 14.16. Solve $P = 3x + 4y \rightarrow \max, 2x + 2y \leq 200, x + 3y \leq t; x, y \geq 0$, t a given number.

Solution. This problem can be solved graphically (see §3). Here is an outline. When $t < 0$, the feasible region is empty, so the problem is infeasible. When $0 \leq t \leq 100$, the first constraint is redundant, the feasible region is a triangle, and $\max = 3t$ at $x = t, y = 0$. When $100 \leq t \leq 300$, the optimal solution is the intersection of the lines corresponding to the first two constraints: $2x + 2y = 200, x + 3y = t$; hence $x = 150 - t/2, y = t/2 - 50$, and $\max = 250 + t/2$. When $t \geq 300$, the second constraint is redundant, the feasible region is a triangle, and $\max = 400$ at $x = 0, y = 100$.

Now we will solve the problem by the simplex method. We introduce two slack variables $u = 100 - x - y \geq 0$ and $v = t - x - 3y \geq 0$ and write the problem in a standard row tableau:

$$
\begin{bmatrix}
x & y & 1 \\
-1 & -1 & 100 \\
-1 & -3 & t \\
-3 & -4 & 0
\end{bmatrix}
\begin{matrix}
\\
= u \\
= v \\
= -P \quad \rightarrow \min.
\end{matrix}
\tag{14.17}
$$

When $t < 0$, the v-row is bad, so the problem is infeasible. Assume now that $t > 0$. The tableau is feasible, so we go to Phase 2. We chose the y-column as the pivot column.

If $t \geq 300$, then we pivot on -1 and obtain

$$
\begin{bmatrix}
x & u & 1 \\
-1 & -1 & 100 \\
2 & 3 & t - 300 \\
1 & 4 & -400
\end{bmatrix}
\begin{matrix}
\\
= y \\
= v \\
= -P \quad \rightarrow \min.
\end{matrix}
\tag{14.18}
$$

which is an optimal tableau; hence max $= 400$ at $x = 0, y = 100$.

Assume now that $0 < t < 300$. Then according to the simplex method we pivot the tableau (14.17) on -3 and obtain

$$
\begin{bmatrix}
x & v & 1 \\
-2/3 & -1/3 & 100 - t/3 \\
-1/3 & -1/3 & t/3 \\
-5/3 & 4/3 & -4t/3
\end{bmatrix}
\begin{matrix}
\\
= u \\
= y \\
= -P \quad \rightarrow \min.
\end{matrix}
\tag{14.20}
$$

Now the x-column is the pivot column. We have to compare $(100 - t/3)/(-2/3)$ and $(t/3)/(-1/3)$ to choose a pivot entry. When $0 < t \leq 100$, we pivot on $-1/3$ and obtain

$$
\begin{bmatrix}
y & v & 1 \\
2 & 1 & 100 - t \\
-3 & -1 & t \\
5 & 3 & -3t
\end{bmatrix}
\begin{matrix}
\\
= u \\
= x \\
= -P \quad \rightarrow \min.
\end{matrix}
\tag{14.21}
$$

hence max $= 3t$ at $x = t, y = 0$.

When $100 \leq t \leq 300$, we pivot (14.20) on $-2/3$ and obtain

$$
\begin{bmatrix}
u & v & 1 \\
-3/2 & 1/2 & 150 - t/2 \\
1/2 & -1 & t/2 - 50 \\
5/2 & 1/2 & -t/2 - 250
\end{bmatrix}
\begin{matrix}
\\
= x \\
= y \\
= -P \quad \rightarrow \min;
\end{matrix}
\tag{14.22}
$$

hence max $= t/2 + 250$ at $x = 150 - t/2, y = t/2 - 50$.

Note that the slope is decreasing (the law of diminishing return):

interval	$0 \le t \le 100$	$100 \le t \le 300$	$300 \le t$
slope	3	1/2	0.

The last tableau is optimal when $t = 200$ (the case in a film on linear programming), and the slope is the last entry in the v-column. If we start to change the first limit 200 (instead of the second limit 200), then the slope would be $5/4$.

Exercises

1–4. Solve the following linear programs, where all the variables $a, b, c, d, e, f, g, h, i$, and j are required to be nonnegative. *Hint.* The row and column problem in a nonstandard tableau need not be dual to each other.

1.

		a	b	c	1		
	g	1	0	-1	-2	$= d$	
	h	2	-1	0	-3	$= c$	
	-1	0	2	1	0	$= w$	$\to$ min
		$= i$	$= j$	$= k$	$= u$	$\to$	min .

2.

		a	b	c	1		
	g	1	2	3	-1	$= d$	
	h	2	0	1	3	$= c$	
	-1	-1	1	0	0	$= w$	$\to$ min
		$= i$	$= j$	$= k$	$= u$	$\to$	min .

3

a	a	a	-1	
0	0	0	0	$= d$
0	0	0	0	$= e$
0	0	0	0	$= w \to$ min.

4.

a	b	-1	
1	0	$1 + \varepsilon$	$= c$
0	1	1	$= d$
1	2	0	$= w \to$ max,

where ε is a given number.

§15. More on Duality

The duality theorem has many facets and interpretations, and it was published in different forms by many authors including Fourier (1826), Gordan (1873), Minkowski (1896), and Farkas (1901).

In Theorem 6.16, we gave a version of duality for systems of linear equations. Here is another form of this: The system of linear equations $Ax = b$ has a solution if and only if $yb = 0$ for any row matrix y such that $yA = 0$. Here is how Fredholm (1903) stated this.

Exactly one of the following two systems is feasible:

$$\text{(a) } Ax = b; \quad \text{(b) } yA = 0, yb > 0.$$

We are going to state a version of duality for systems of linear inequalities. But first we show a way to deduce the dual problem from the primal problem. The main idea is to find a lower bound for the optimal value of the primal problem using the linear constraints of the primal problem. Namely, we combine linearly given constraints to obtain the lower bound. The dual program turns out to be the following: Maximize this lower bound.

Let us start with a simple example. Suppose that we want to solve the linear program:

$$\begin{cases} \text{Minimize} & x + y \\ \text{subject to} & x + y \geq 2. \end{cases}$$

It is clear that $\min(x + y) = 2$.

Here is a more complicated example:

$$\begin{cases} \text{Minimize} & 4x + 5y \\ \text{subject to} & x + 3y \geq 2, \\ & 2x - y \geq 3. \end{cases}$$

If we multiply the first constraint by 2 and add the result to the second constraint, then we obtain that $4x + 5y \geq 7$. This constraint, which is a linear combination of given constraints with positive coefficients, gives a low bound for the objective function: $\min(4x+5y) \geq 7$ (under our conditions). Moreover, it is easy to see in this example that this lower bound can be reached; that is, it is sharp. Thus,

$$\min(4x + 5y) = 7.$$

Let us try this approach to a bigger LP, given by the following standard row tableau:

$$
\begin{array}{ccccc}
x_1 & x_2 & x_3 & x_4 & 1 \\
\left[\begin{array}{ccccc}
1 & 2 & 3 & 4 & 5 \\
1 & -2 & 0 & 4 & -1 \\
0 & 2 & -3 & 1 & 0 \\
1 & 2 & 3 & -4 & 5
\end{array}\right]
\begin{array}{l}
= u_1 \geq 0 \\
= u_2 \geq 0 \\
= u_3 \geq 0 \\
= z \to \min,
\end{array}
\end{array}
\qquad \text{all} x_i \geq 0
$$

or, using matrices,

$$
\begin{array}{cc}
x & 1 \\
\left[\begin{array}{cc}
A & b \\
c & d
\end{array}\right]
\begin{array}{l}
= u \geq 0 \\
= z \to \min,
\end{array}
\end{array}
\qquad x \geq 0
$$

or, using matrix multiplication,

$$
Ax^T \geq -b, \; x \geq 0, \; cx^T + d \to \min.
$$

If y is a column matrix such that $y \geq 0$ and $y^T \cdot A \leq c$, then we obtain that $cx^T \geq u^T \cdot Ax^T \geq -y^T \cdot b$. Thus, we obtain a lower bound

$$
z = cx^T + d \geq -y^T \cdot b + d
$$

for the objective function z using a linear combination of given constraints $Ax^T \geq -b$ with nonnegative coefficients y.

Next we try to pick y as before such that the bound is as sharp as possible:

$$
-u^T \cdot b + d \to \max, \; y \geq 0, \; y^T \cdot A \leq c.
$$

This is the dual problem, which can be written as the column problem in the same standard tableau:

$$
\begin{array}{cc}
 & x \quad\;\; 1 \\
\begin{array}{c} -y \\ 1 \end{array}
\left[\begin{array}{cc}
A & b \\
c & d
\end{array}\right]
\begin{array}{l}
= u \\
= z \to \min
\end{array}
\quad
\begin{array}{l}
x \geq 0, u \geq 0 \\
y \geq 0, v \geq 0
\end{array} \\
\quad\;\; = v \;\; = w \;\; \to \max.
\end{array}
$$

In fact, passing to matrices made our computation easy, and they work for the LP given by an arbitrary standard row tableau.

Finding $y \geq 0$ such that $y^T \cdot A \leq c$ is equivalent to writing the linear function cx as a linear combination $y^T Ax + ux$ of the left-hand sides Ax, x of all given constraints $Ax \geq -b, x \geq 0$ with nonnegative coefficients y, u, where u is the row of basic variables for the dual (column) program.

So how is the duality theorem related to our situation? There is a possibility that our problem is unbounded. Then obviously we cannot get any bounds for z by any method, so the dual problem must be infeasible, which gives us a part of the theorem. There is a possibility that our problem is infeasible. Then the theorem says that the dual problem is either unbounded or infeasible. In terms of linear combinations, this means that either we can get arbitrary good bounds or no bound can be obtained as a linear combination. Finally, it may happen that our problem has an optimal solution. Then the theorem says that the optimal value is the best bound that can be obtained by linear combinations.

Now we ask ourselves for which numbers e the constraint $cx \geq d'$ follows from the given constraints $Ax \geq -b, x \geq 0$. In other words (see §4), does every feasible solution for the system satisfy the constraint as well? It is clear that the answer depends on the relation between $\min(cx)$ over the feasible region of the system and the number d'. If $\min(cx) < d'$ (including the case when $min = -\infty$, i.e., the program is unbounded), then the answer is no. If $\min(cx) \geq e$ (including the case when $\min = \infty$, i.e., the program is infeasible), then the answer is no.

Therefore, we can easily get our version of duality theorem from the following version of the duality theorem.

Theorem 15.1. Given any system of inequalities $A'x' \geq b'$ and another inequality $c'x' \geq d'$ that follows from the system, then either the inequality is a linear combination of the constraints in the system together with the constraint $0 \geq -1$ with nonnegative coefficients, or the system is infeasible and the constraint $0 \geq 1$ is a linear combination of the constraints of the system. ∎

We put primes into the condition of the theorem because the data are not the same as in the tableau. To apply the theorem, we take $A' = \begin{bmatrix} A \\ 1_n \end{bmatrix}$, n the number of variables in x, $x' = x$, $b' = -b$, and so on.

Theorem 15.1 can be obtained easily from our duality theorem using the standard trick $x = x' - x''; x', x'' \geq 0$ to write data in a standard row tableau.

Now we state a version of Theorem 6.16 for systems of inequalities. Suppose we are given a system of inequalities $Ax \geq b$ and another inequality $cx \geq d$ of the same type $\geq$. Is the latter constraint follows from the system? In other words (see §4), does every feasible solution for the system satisfy the constraint as well?

It is clear that the answer depends on the relation between $\min(cx)$ over the feasible region of the system and the number d. If $\min(cx) < d$ (including the case when $\min = -\infty$, i.e., the program is unbounded), then the answer is *no*. If $\min(cx) \geq d$ (including the case when $\min = \infty$, i.e., the program is infeasible), then the answer is *yes*.

Remark 15.2 The duality theorem implies that any primal-dual pair of linear programs

$$cx + d \to \min, \ Ax \geq -b, \ x \geq 0; \qquad -yb + d \to \max, \ yA \geq c, \ y \geq 0$$

can be written as a system of linear constraints:

$$Ax \geq -b, \ x \geq 0, \ cx + yb = 0, \ y \geq 0, \ yA \leq c.$$

Every feasible solution $[x^T, y]$ to this system of linear constraints gives optimal solutions x and y for both programs, and the optimal solutions x and u for the optimization problems give a solution $[x^T, y]$ for this system of constraints. In this sense linear programming is about solving systems of linear constraints. Another way to put it is that finding an optimal solution for an LP can be reduced to Phase 1 for another LP. ∎

When a linear program comes from a real-life situation, its dual also can be interpreted in real-life terms. We consider now Examples 2.1, 2.2, 2.3, 2.4 (in more general forms) and give economic interpretations of the dual problems.

Example 15.3. Consider the general diet problem (a generalization of Example 2.1):

$$Ax \geq b, x \geq 0, C = cx \to \min,$$

where m variables in x represent different foods and n constraints in $Ax \geq b$ represent ingredients. We want to satisfy given requirements b in ingredients using given foods at minimal cost C.

On the other hand, consider a warehouse that sells the ingredients at prices $y_1, \ldots, y_n \geq 0$. Its objective is to maximize the profit $P = yb$, matching the price for each food: $yA \leq c$.

We can write both problems in a standard tableau using slack variables $u = Ax - b \geq 0$ and $v = c - yA \geq 0$:

$$
\begin{array}{c}
\quad \quad x^T \quad \ 1 \\
\begin{array}{cc}
-y^T \\
1
\end{array}
\left[
\begin{array}{cc}
A & -b \\
c & 0
\end{array}
\right]
\begin{array}{l}
= u \\
= C \to \min \\
\end{array}
\\
\ \ = v \ \ = P \quad \to \quad \max.
\end{array}
\qquad
\begin{array}{l}
x \geq 0, u \geq 0 \\
y \geq 0, v \geq 0
\end{array}
$$

So these two problems are dual to each other. In particular, the simplex method solves both problems, and if both problems are feasible, then $\min(C) = \max(P)$. The shadow prices mentioned in §14 turn out to be the optimal prices for the ingredients in the dual problem, so they are called *dual prices* as well.

Another economic interpretation for the same mathematical problem is that the variables in x are intensities of different industrial processes, the constraints correspond to different products, with b being the federal order to be fulfilled (or demand to be satisfied); $C = cx$ is the total cost that you want to minimize using given processes and satisfying given production quotas. With this interpretation for the primal problem, here is an interpretation for the dual problem: A competitor, Ann, who lost the government contract, says that you van you to buy the products from her at her low prices $y \geq 0$, matching unit cost for every process you got (i.e., $yA \leq c$) and maximizing her profit yb.

With this interpretation, the optimal prices y are *marginal costs* for you (i.e., they answer to the question what is the additional cost to produce an additional unit of each product). In §14, we saw that the marginal costs decrease with increase in volume (in linear programming). ∎

Example 15.4. Consider the general mixing problem (a generalization of Example 2.1):

$$Ax = b, x \geq 0, C = cx \to \min,$$

where m variables in x represent different alloys and n constraints in $Ax \geq b$ represent elements. We want to satisfy given requirements b in elements using given alloys at minimal cost C.

On the other hand, consider a dealer who buys and sells the elements at prices $y_1, \ldots, y_n$. The positive price means that the dealer sells, and negative price means that the dealer buys. The dealer's objective is to maximize the profit $P = yb$, matching the price for each alloy: $yA \leq c$.

To write the problems in standard tableaux, we use the standard tricks and artificial variables:

$$u' = Ax - b \geq 0, \ u'' = -Ax + b \geq 0;$$
$$v = c - yA \geq 0; y = y' - y'', \ y' \geq 0, y'' \geq 0.$$

Now we manage to write both problems in the same standard tableau:

$$
\begin{array}{c}
\\
-y'^T \\
-y''^T \\
1
\end{array}
\begin{array}{c}
x^T \quad\quad 1 \\
\left[
\begin{array}{cc}
A & -b \\
-A & b \\
c & 0
\end{array}
\right]
\begin{array}{l}
= u' \\
= u'' \\
= C \ \rightarrow \min
\end{array}
\\
\quad\quad = v \ \ = P \quad \rightarrow \quad \max.
\end{array}
\qquad
\begin{array}{l}
x \geq 0, u \geq 0 \\
y', y'' \geq 0, v \geq 0
\end{array}
$$

Remark 15.5. Adding an arbitrary constant d to the linear objective C, we get an arbitrary LP in standard form, and replacing the last zero in the last row by d, we have this LP in a standard row tableau.

Example 15.6. Consider a generalization of the manufacturing problem in Example 2.3:

$$P = cx \rightarrow \max, Ax \leq b, x \geq 0,$$

where the variables in x are the amounts of products, P is the profit (or revenue) you want to maximize, constraints $Ax \leq b$ correspond to resources (e.g., labor of different types, clean water you use, pollutants you emit, scarce raw materials), and the given column b consists of amounts of resources you have. Then the dual problem

$$yb \rightarrow \min, yA \geq c, y \geq 0$$

admits the following interpretation. Your competitor, Bob offers to buy you out at the following terms: You go out of business, and he buys all resources you have at price $y \geq 0$, matching your profit for every product you may want to produce, and he wants to minimize his cost.

Again Bob's optimal prices are your resource shadow prices by the duality theorem. The shadow price for a resource shows the increase in your profit per unit increase in the quantity b_0 of the resource available or decrease in the profit when the limit b_o decreases by one unit. While changing b_0 we do not change the limits for the other resources and any other data for our program. There are only finitely many values of b_0 for which the downward and upward shadow prices are different. One of these values could be the borderline between the values of b_0 for which the corresponding constraint is binding or nonbinding (in the sense that dropping of this constraint does not change the optimal value).

In §14, we saw that the shadow price of a resource cannot increase when supply b_0 of this resource increases (the law of diminishing returns).

Example 15.7. This transportation problem is similar to Example 2.4, even though we have a different geographic setting. There are warehouses in Bedford and Scranton. They can supply 220 and 280 units, respectively. The retail stores are in State College, Altoona, and Harrisburg. These need 170, 120, 210 units, respectively. The shipping cost table is:

	State College	**Altoona**	**Harrisburg**
Bedford	77	39	105
Scranton	150	186	122

The constraints are

$$x_{ij} \geq 0, \quad i = 1, 2; \quad j = 1, 2, 3 \qquad (i)$$

$$\begin{cases} x_{11} + x_{12} + x_{13} \leq 220 \\ x_{21} + x_{22} + x_{23} \leq 280 \end{cases} \qquad (ii)$$

and

$$\begin{cases} x_{11} + x_{21} \geq 170 \\ x_{12} + x_{22} \geq 120 \\ x_{13} + x_{23} \geq 210. \end{cases} \qquad (iii)$$

Notice that in this LP the sum of units available in the warehouses is 500, which equals the amount of units needed by the retail stores; this was not the case in the previous example (the warehouses had 130 widgets available, whereas the retail stores needed a total of 100). By looking at the constraints in (ii), we obtain

$$x_{11} + x_{12} + x_{13} + x_{21} + x_{22} + x_{23} \leq 500. \qquad (iv)$$

On the other hand, (*iii*) yields

$$x_{11} + x_{12} + x_{13} + x_{21} + x_{22} + x_{23} \geq 500. \qquad (v)$$

Combining the inequalities (*iv*) and (*v*), we obtain the equality

$$x_{11} + x_{12} + x_{13} + x_{21} + x_{22} + x_{23} = 500.$$

This equality means that the total supply equals the total demand, and it is called the *balance condition*. This condition allows us, if we wish so, to replace all the inequality signs in (*ii*) and (*iii*) by equality signs. The balance condition forces the slack variables to be zero for all feasible solutions, so we obtain an equivalent problem if we replace all $\leq$ and $\geq$ in (*ii*) and (*iii*) by equality signs.

This small problem can be solved easily by the simplex method. In the next chapter we will see that the simplex method works particularly well for transportation problems. Our goal now is to give an economic interpretation for the dual problem. First we use Example 15.7 to introduce potentials. To put the problem in a standard row tableau, we introduce slack variables. Then we write the dual variables in the same tableau.

	x_{11}	x_{12}	x_{13}	x_{21}	x_{22}	x_{23}	1	
$-u_1$	-1	-1	-1	0	0	0	220	$= y_1$
$-u_2$	0	0	0	-1	-1	-1	280	$= y_2$
$-v_1$	1	0	0	1	0	0	-170	$= z_1$
$-v_2$	0	1	0	0	1	0	-120	$= z_2$
$-v_3$	0	0	1	0	0	1	-210	$= z_3$
1	77	39	105	150	186	105	0	$= C$
	$\|\|$	$\|\|$	$\|\|$	$\|\|$	$\|\|$	$\|\|$	$\downarrow$	
	w_{11}	w_{12}	w_{13}	w_{21}	w_{22}	w_{23}	max	

with $u_i, v_j, w_{ij}, x_{ij}, y_i, z_j \geq 0$.

The objective function to be maximized by the dual problem is

$$-220u_1 - 280u_2 + 170v_1 + 120v_2 + 210v_3 \qquad (15.8)$$

The control variables u_i, v_j of the dual problem are called *potentials*. While the potentials correspond to the constraints on each retail store and each warehouse (or to the corresponding slack variables), there are other variables w_{ij} in the dual problem that correspond to the decision variable x_{ij} of the primal problem.

They are the slack variables for the six dual constraints

$$w_{ij} = c_{ij} + u_i - v_j \geq 0 \; \forall \; i, j, \tag{15.9}$$

where $i = 1, 2$, $j = 1, 2, 3$, and c_{ij} are the entries of the cost matrix

$$c = \begin{bmatrix} 77 & 39 & 105 \\ 150 & 186 & 122 \end{bmatrix}.$$

So what is a possible meaning of the dual problem?

Imagine that you want to be a mover and suggest a simplified system of tariffs. Namely, you assign a "zone" $u_i \geq 0 \; i = 1, 2$ to each of the warehouses and a "zone" $v_j \geq 0 \; j = 1, 2$ to each of the retail stores. To beat competition, you want the constraints (15.9). Your profit is (15.8), and you want to maximize it.

Remark. 15.10. Observe that the problem is invariant under the change

$$u_i \rightarrow u_i + t, v_j \rightarrow v_j + t$$

using any fixed value of t for all i, j. This allows us to ignore the conditions $u, v \geq 0$ if we wish so.

Exercises

1–4. Find whether the last equation in the system is redundant (i.e., follows from the others).

1. $\begin{cases} x + 2y + 3z = 4 \\ 5x + 6y + 7z = 8 \\ 6x + 8y + 10z = 0 \end{cases}$
2. $\begin{cases} x + 2y + 3z = 4 \\ 5x + 6y + 7z = 8 \\ 7x + 10y + 13z = 16 \end{cases}$

3. $x = 6, y = 5, z = 0, 2x - 8y + 3z = 7$

4. $\begin{cases} x_1 + 2x_2 + 3x_3 + 4x_4 + 5x_5 = 6 \\ 6x_1 + 5x_2 + 4x_3 + 3x_4 + 2x_5 = 1 \\ x_1 - x_2 + x_3 - x_4 + x_5 = 0 \end{cases}$

5–8. Find whether the last constraint in the system is redundant (i.e., follows from the others).

5. $\begin{cases} x + 2y + 3z = 4 \\ 5x + 6y + 7z = 8 \\ 6x + 8y + 10z \geq 0 \end{cases}$
6. $\begin{cases} x + 2y + 3z \geq 4 \\ 5x + 6y + 7z \geq 8 \\ 7x + 10y + 13z \geq 16 \end{cases}$

7. $x = 6, y = 5, z = 0, 2x - 8y + 3z \leq 7$

8.
$$\begin{cases} x_1 + 2x_2 + 3x_3 + 4x_4 + 5x_5 = 6 \\ 6x_1 + 5x_2 + 4x_3 + 3x_4 + 2x_5 = 1 \\ x_1 - x_2 + x_3 + -x_4 + x_5 \geq 0 \end{cases}$$

9–14. Solve the linear program, where all $x_i \geq 0$. . *Hint*: Solve the dual problem by graphical method.

9.

x_1	x_2	x_3	x_4	x_5	x_6	x_7	x_8	x_9	1	
0	8	−5	6	7	8	3	5	4	−1	≥ 0
−1	2	−2	1	1	1	2	2	5	0	$\rightarrow$ min

10.

x_1	x_2	x_3	x_4	x_5	x_6	x_7	x_8	x_9	1	
6	8	5	6	7	8	3	5	4	−1	≥ 0
1	2	5	1	1	1	5	0	5	0	$\rightarrow$ min

11.

x_1	x_2	x_3	x_4	x_5	x_6	x_7	1	
3	4	1	1	1	2	3	−1	$= x_8$
4	3	1	0	0	2	4	−2	$= x_9$
5	5	1.4	5	6	3	6	0	$= z$ $\rightarrow$ min

12.

x_1	x_2	x_3	x_4	x_5	x_6	x_7	x_8	1	
6	4	1	10	5	2	8	7	−1	$= x_9$
19	9	10	1	9	10	6	7	−2	$= x_{10}$
20	10	10	10	10	11	10	10	0	$= z$ $\rightarrow$ min

13.

x_1	x_2	x_3	x_4	x_5	x_6	x_7	x_8	x_9	1	
6	8	5	6	7	8	3	5	4	−1	≥ 0
29	29	14	14	13	13	4	8	3	−2	≥ 0
31	32	15	15	15	15	5	20	5	0	$\rightarrow$ min

14.

x_1	x_2	x_3	x_4	x_5	x_6	x_7	x_8	x_9	1	
6	8	5	6	−7	8	3	5	4	−1	≥ 0
−9	0	−4	1	13	13	4	8	3	−2	≥ 0
−1	2	−5	15	15	15	5	0	5	0	$\rightarrow$ min

Chapter 6

Transportation Problems

§16. Phase 1

Have you ever had the opportunity to read magazines from years gone by? The advertisements in these magazines are fascinating. In one old magazine, blue jeans, touted as an essential piece of clothing apparel for agricultural workers, were advertised for $2.99 a pair. However, in small print, was the following phrase:

Prices may be higher west of the Rockies.

Why did the East Coast manufacturers of these blue jeans feel compelled to put that disclaimer in their advertisement? Obviously, they were faced with the daunting task, and equally daunting expense, of transporting their product across the Rocky Mountains to the consumers in the West who wanted blue jeans. Now, with manufacturing plants located throughout the United States, instead of primarily in the East, and with improved means of shipping, it might not be necessary to put such disclaimers in advertisements. Nonetheless, transportation costs continue to be an important consideration for business and industry.

In Example 2.4 of Chapter 1, we discussed a particular case of transportation problems and solved this problem by using a trial-and-error method. Since transportation problems provide particularly simple and nice examples of linear programs, we will revisit these examples, setting them up as linear programs and then solving them using the methods we have just learned. Historically speaking, these problems were investigated in detail many years before the general linear program was. Interestingly, the concept of duality, which we considered in Chapter 5, first appeared in transportation problems.

In this chapter we will see that Phase 1 of solving linear programs —finding a feasible solution—can be done by hand. Also, Phase 2—finding an optimal solution—is easier in transportation problems than for general linear programs.

To introduce graphs and tables, we consider again Example 15.7 and begin with a graphical representation (Fig. 16.1).

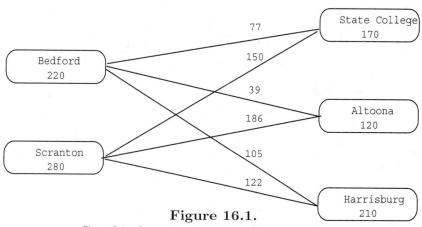

Figure 16.1.
Graphical representation of Example 15.7

Also, the numerical data can be written in a table:

	SC	A	H	Supply
B	77	39	105	220
S	150	186	122	280
Demand	170	120	210	

Table 16.2. Example 15.7 tabled

We left room in the table for values of the unknowns x_{ij} which are sometimes referred to as a *flow*. Any transportation problem can be treated similarly. Note that there exists no feasible solution unless the *balance condition* holds: The total supply (the sum of the numbers in the right margin) equals the total demand (the sum of the numbers across the bottom). Example 2.4 presented an open transportation problem, that is, a problem without the balance condition. That problem can be written as a closed (balanced) transportation problem, with the balance condition satisfied, by introducing a fictitious store, *dummy demand point,* where any surplus product goes at no cost. If we set the demand at the fictitious store to be exactly the total surplus supply, then the balance condition is restored.

A position in the table corresponds to an arrow in the graphical representation, and entering a number in the i, j position in the table corresponds to allocating stock from warehouse i to store j. Here is a way to find a feasible solution. First check the balance condition, which is satisfied in this problem so we can proceed. Pick a position in the table and write the maximum possible number, namely the available stock at the corresponding warehouse.

For example, in Example 15.7 we choose the first position in the first row and write there 170.

170			~~220~~ 50
			280
~~170~~	120	210	

We suppressed the objective function because we do not use it yet. Note that the first column of the table, corresponding to the store in State College, is done. We indicate this by crossing out the demand 170. The second entry in the first column must now be zero, because the demand is satisfied. We suppress entering this zero. We keep track of unallocated warehouse stock by entering a new number 50 instead of 220 at the right margin of the table.

Thus, we entered one number, crossed out one number, and adjusted another number. The first column is done (we found the flow in it), and we proceed with a smaller, 2-by-3 table.

We pick a second position in the table and again write the maximum possible number $50 = \min(50, 120)$:

170	50		~~220~~ ~~50~~
			280
~~170~~	~~120~~ 70	210	

Now both the first row and the first column are done. We repeat this procedure for remaining 1-by-2 matrix, as long as any choice remains:

170	50		~~220~~ 50
	70		~~280~~ 210
~~170~~ 0	~~120~~ ~~70~~	210	

At the last step the allocation will have been determined by the previous choices:

170	50		~~220~~ 50
	70	210	~~280~~ ~~210~~
~~170~~	~~120~~ ~~70~~	~~210~~	

The corresponding total cost is

$$77 \cdot 170 + 39 \cdot 50 + 186 \cdot 70 + 122 \cdot 210.$$

Here is the resulting feasible solution, with two zeros suppressed

170	50		220
	70	210	280
170	120	210	

This method of finding a feasible solution works for any transportation problem under the balance condition. Now it is time to state an arbitrary transportation problem.

The General Transportation Problem.

A manufacturer has m warehouses and n retail stores. The warehouse #i has a_i units of product available and the store #j needs b_j units.

It is assumed that the following *balance condition* holds:

$$\sum_{i=1}^{m} a_i = \sum_{j=1}^{n} b_j$$

For instance, in Example 15.2 the balance condition is satisfied. This is not the case in Example 2.4, as we observed. The cost of shipping one unit from warehouse #i to store #j is denoted by c_{ij} and the number of units shipped from warehouse #i to store #j is denoted by x_{ij}. The linear program can be stated as follows:

$$\begin{cases} \text{minimize} & C(x_{11}, \ldots, x_{mn}) = \sum_{i=1}^{m} \sum_{j=1}^{n} c_{ij} x_{ij} \\ \text{subject to} & \sum_{i=1}^{n} x_{ij} = b_j, \quad j = 1, \ldots, m \\ & \sum_{j=1}^{m} x_{ij} = a_i, \quad i = 1, \ldots, n \\ & x_{ij} \geq 0, \; i = 1, \ldots, n, \; j = 1, \ldots, m, \end{cases}$$

where C is the total cost function to be minimized, the constraints

$$\sum_{i=1}^{n} x_{ij} = b_j, \quad j = 1, \ldots, m$$

represent the number of units that store #j receives from warehouse #i, and the remaining constraints represent the number of units that warehouse #i can supply to store #j, in addition to the usual non-negativity constraints.

In this chapter we will show how the simplex method solves this problem. For now, here are some questions you might want to consider:

(a) We have been looking at ways to minimize shipping costs. However, does it ever make sense to maximize shipping costs rather than minimize them? Consider this situation: You are working for an interstate mover and get paid for shipping. In this situation, would you try to minimize or maximize the shipping cost?

(b) Is the general transportation problem *general* enough? Is Example 2.4 a particular case of the general transportation problem? In other words, can the constraints in Example 2.4 be written as equations rather than inequalities? Note that the total supply in that example is 130, which is larger than the total demand of 100. So if we just replace the inequalities by equations, then we would not have any feasible solutions.

Remark 16.3. Historically, transportation problems were the first linear programs that were explicitly stated, studied theoretically, solved, and used in industry.   ∎

Now we observe that the method of finding a feasible solution works for any transportation problem. We select any position and write there the maximal possible entry – namely, the minimum of the number in the right margin and the number in the bottom margin. Then we cross out one of these two numbers, which equals the written entry, and adjust the other one. At the last step, when a 1×1 table is left, we cross out both numbers. Thus, the total number of steps is $m + n - 1$, where m is the number of rows and n is the number of columns. Our method gives a feasible solution with at most $m+n-1$ nonzero values for mn variables.

To proceed with finding an optimal solution, it is convenient to make this number of selected positions exactly $m+n-1$ by crossing out only one margin entry (and making the other 0) in the case when both are the same (unless this is the last step,) and allowing the value 0 to be entered at a selected position if either of the numbers at the margin is 0. By this method, we cross out one row or one column at each step, except the last. We should not cross out the last remaining row or column unless it is the last step, step $m + n - 1$.

The $m + n - 1$ positions we obtain this way correspond to the basic variables (those at the right margin) in a feasible standard row tableau. We write in the table the values of the basic variables and we do not write the zero values for nonbasic variables to leave room for the values of dual variables.

Thus, a feasible solution can be found easily by hand for any problem small enough so the data can be written down by hand. But in fact, what we are doing is Phase 1 of the simplex method. We use smaller tables rather than standard tableaux. Initially, all variables x_{ij} are on the top, with slack variables at the right margin. The selected positions (i, j) correspond to variables x_{ij} which we pivot to the right margin. The pivot entries are 1 or -1.

Example 16.4. Find a basic feasible solution for the transportation problem

			2
			3
2	2	1	

We did not give an objective function, because we do not need it to find a feasible solution. We use the *northwest* method. According to this method, we choose the position in the upper-left corner. Note that we cross out 2 under the first row and adjust 2 at the right margin to 0:

2			~~2~~ 0
			3
~~2~~	2	1	

Now the first column is done, and the northwest method tells us to choose the second position in the first row, and the maximal number we can put there is 0:

2	0		~~2~~ ~~0~~
			3
~~2~~	2	1	

Now both the first row and first column are done. The next position we choose is in the second row and the second column:

2	0		~~2~~ ~~0~~
	2		~~3~~ 1
~~2~~	~~2~~	1	

Now only the last row and column remain. We write the last entry 1 and cross out the last two numbers at the margin:

2	0		~~2~~ ~~0~~
	2	1	~~3~~ ~~1~~
~~2~~	~~2~~	~~1~~	

In this example, $m = 2, n = 3$, and we wrote $m + n - 1 = 4$ numbers in the table, one of them being 0. ∎

In graphical terms, the chosen positions are $m + n - 1$ edges (arcs) that connect all vertices (nodes), forming a graph (network) called a *tree*. This tree has exactly $m + n - 1$ arrows (or edges), is connected, and has no loops (cycles).

As mentioned previously, if the problem is expressed as a standard row tableau, the chosen positions correspond to the basic variables at the right margin of a row feasible tableau. The other positions correspond to the nonbasic variables across the top of a row feasible tableau. The method just described in fact implements Phase 1 of the simplex method. It is not necessary to work with the entire tableau for transportation problems, because the A block of the tableau always consists of 0s, 1s, and -1s. The b block of the tableau consists of the values in the chosen positions of the table. Construction of the c block will be discussed later.

We always obtain a feasible solution if the balance condition holds, and if it does not we declare the problem infeasible and stop. This is why we check the balance condition before beginning to choose entries. If the supply and demand data consist of integers, then the feasible solution determined by this method is likewise integral. It is always a vertex of the feasible region. The proof is left as an exercise.

There are many ways for choosing the table positions in our method. The northwest method ignores the objective function. But other methods for choosing positions may result in a better feasible solution, depending on the objective function. For example, in the minimum cost method, we go for a position with minimal cost.

To proceed with the second stage of the simplex method, we must recover the c-part of the tableau for the problem from the b-part, that is, from the table entries chosen to construct a feasible solution. The method for doing this is best described in terms of *potentials*. We introduced the potentials in Example 15.7.

In the notation of the general transportation problem, the dual problem is written

$$\text{maximize} - \sum_{i=1}^{m} a_i u_i + \sum_{j=1}^{n} b_j v_j$$

subject to

$$w_{ij} = c_{ij} + u_i - v_j \geq 0 \; \forall \; i, j$$

$$u \geq 0, v \geq 0.$$

In Example 15.7,

$$[a_1, a_2] = [220, 280], [b_1, b_2, b_3] = [170, 120, 210], [c_{i,j}] = \begin{bmatrix} 1 & 2 & 3 \\ 1 & 2 & 2 \end{bmatrix}.$$

While the potentials correspond to the constraints on each retail store and each warehouse (or to the corresponding slack variables), there are other variables w_{ij} in the dual problem that correspond to the decision variable x_{ij} of the primal problem. They are the slack variables for the dual problem written in terms of potentials. We will call the variables w_{ij} *discrepancies*.

At start, all x_{ij} are at the right margin of a standard tableau (see Example 15.7) and all w_{ij} are at the left margin. After a few pivot steps, the variables x_{ij} corresponding to the selected positions (i, j) are on the top of a feasible tableau, and the corresponding w_{ij} are at the bottom margin. Their current values are in the last row of the tableau, its "c-part."

To proceed with the second stage of the simplex method, we have to recover the c-part of a row feasible tableau from the b-part. We know that if x_{ij} is a basic variable for the row problem—that is, it appears in the right margin of the tableau—then the corresponding dual variable w_{ij} appears in the left margin and w_{ij} takes the value 0 in the basic solution of the column problem (involving the potentials).

Thus, $w_{ij} = c_{ij} + u_i - v_j = 0$ for all chosen positions (i, j). It turns out that these equations determine potentials uniquely up to an additive constant. This follows from the fact that the chosen positions determine a tree in the graphical representation (Figure 16.5).

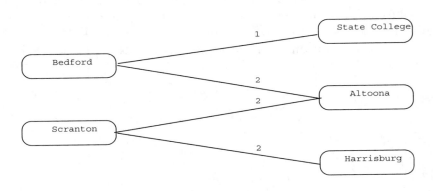

Figure 16.5. Tree
for the feasible solution of Example 15.7

We continue now with Example 15.7. We marked with a * the positions that were chosen as we found a feasible solution (Table 16.6).

77 *	39 *	105	220
150	186 *	122 *	280
170	120	210	

Table 16.6. Basic positions
for the feasible solution of Example 15.2

The edges correspond to the selected positions, and the cost per unit is written on them. Because we seek a solution up to an additive constant, to solve for the potentials we can start by setting $u_1 = 0$. Then the equation $c_{11} = v_1 - u_1$ forces $v_1 = 77$. Similarly, the equation $c_{12} = v_2 - u_1$ forces $v_2 = 39$. Using this value of v_2 in the equation for c_{22} forces $u_2 = -147$, which, when substituted into the equation for c_{23}, forces $v_3 = -25$. Thus we have found all the potentials based on the choice of $u_1 = 0$. All other solutions to the system of equations just solved are clearly obtained by adding the same constant to all potentials. We represent the potentials in Figure 16.7 and Table 16.8.

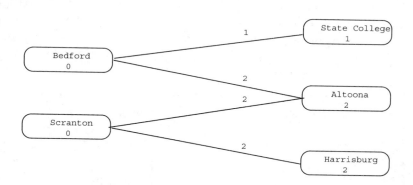

Figure 16.7. Tree Representation
of the potentials in Example 15.7

	77	39	−25	
0	77	39	105	
	77 *	39 *	105	220
−147	150	186	122	
	150	186 *	122 *	280
		170	120	210

Table 16.8. Potentials
for the feasible solution of Example 15.7

Now we can compute the values $w_{ij} = c_{ij} + u_i - v_j$ for the other (nonbasic) positions. We put them in Table 16.8 in parentheses (Table 16.9). Because one of w_{ij} in this example is negative, our tableau is not optimal. We could show the basic solution for the primal problem by replacing each * representing a chosen position with the value assigned to x_{ij} at the time of that choice.

	77	39	−25	
0	77	39	105	
	77 *	39 *	105 (130)	220
−147	150	186	122	
	150 (−74)	186 *	122 *	280
		170	120	210

Table 16.9. The basic solution
for the dual problem to Example 15.7

As mentioned previosly, the computed values of w_{ij} correspond to the entries of the last row of the tableau which we have in mind without writing it down, except for the last entry in the tableau which is the current value of the objective function. Now we are ready for a pivot step.

Exercises

1–3. Find a feasible solution for each transportation problem.

1.

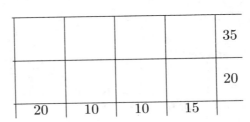

				35
				20
20	10	10	15	

2.

35	0	0	0	~~35~~ 0
5	4	0	0	~~9~~ ~~4~~ 0
0	6	5	0	~~11~~ ~~5~~ 0
0	6	5	15	~~20~~ ~~5~~ 0
40	10	10	15	

40 ~~5~~ 0 10 ~~6~~ 0 10 ~~5~~ 0 15 0

3.

				35
				90
				111 102 11
				20 10
140	91	10	19	

162
−91
———
11

140
−35
———
105

140 91 10 19
~~129~~ ~~39~~ 0 ~~90~~ 0
256

260

Demand > supply
⇓
infeasible

4–7. Find a feasible solution and the corresponding total transportation cost.

4.

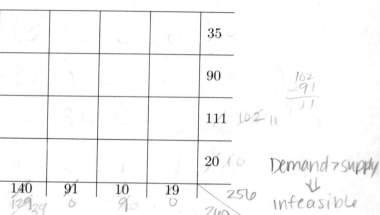

1	2	3	1	2	3	1	2	3	
									10
0	3	2	0	1	2	1	2	1	
									20
2	1	0	1	2	1	2	1	1	
									12
1	1	1	2	2	2	2	2	1	
									8
2	3	4	5	1	2	3	14	16	

5.

1	2	3	1	2	3	1	2	3	15
0	3	2	0	1	2	1	2	1	15
2	1	0	1	2	1	2	1	1	12
1	1	1	2	2	2	2	2	1	8
2	8	4	15	11	2	3	4	1	

6.

1	2	3	1	2	3	1	2	3	15
0	3	2	0	1	2	1	2	1	5
2	1	0	1	2	1	2	1	1	12
1	1	1	2	2	2	2	2	1	18
12	8	14	5	1	2	3	4	1	

7. Example 2.4.

55	30	40	50	40	0	50
35	30	100	45	60	0	30
40	60	95	35	30	0	50
25	10	20	30	15	30	

8. Prove that any feasible solution derived by our method is a vertex of the feasible region.

§17. Phase 2

Suppose that after computing potentials, some w_{ij} is negative. Then according to the simplex method, Phase 2, we have to exclude w_{ij} from the basis and include another variable instead. In terms of the row problem, we must include x_{ij} into the basis instead of another variable. When we do this, the objective function should improve or at least keep the same value. Graphically, this change of basis corresponds to adding a new edge to the tree and removing one edge from the resulting loop to form a new tree. We will use Example 2.4 to illustrate this process. We begin by using the northwest method to find a basic feasible solution (Table 17.1).

55	30	40	50	40	0	
25	10	15				50
35	30	100	45	60	0	
		5	25			30
40	60	95	35	30	0	
			5	15	30	50
25	10	20	30	15	30	

Table 17.1. First feasible solution of Example 2.4

The corresponding transportation cost is $25 \cdot 55 + 10 \cdot 30 + 15 \cdot 40 + 5 \cdot 100 + 25 \cdot 45 + 5 \cdot 35 + 15 \cdot 30 + 30 \cdot 0 = 1375 + 300 + 600 + 500 + 1125 + 175 + 450 + 0 = 4525$. We will see that ignoring the objective function will result in this case in many pivot steps on the way to an optimal table. Usually, it pays to spend some time at Phase 1 of solving a transportation problem so we need to spend less time at Phase 2. As before, we proceed to find the potentials (Table 17.2).

	55	30	40	-15	-20	-50	
0	55	30	40	50	40	0	
	25	10	15				50
-60	35	30	100	45	60	0	
			5	25			30
-50	40	60	95	35	30	0	
				5	15	30	50
	25	10	20	30	15	30	

Table 17.2. Potentials for Table 17.1

We took the first potential 0 for the first row arbitrarily. The transportation cost can be also computed using potentials instead of the flow: $55\cdot 25+30\cdot 10+40\cdot 20-15\cdot 30-20\cdot 15-50\cdot 30-(0\cdot 50-60\cdot 30-50\cdot 30) = 4525$. As in the previous example, the feasible solution determines a tree in the graphical representation (Figure 17.3).

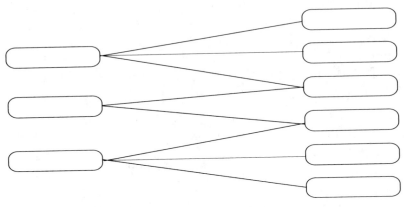

Figure 17.3. Tree for Table 17.1

Next we compute the values for the dual slack variables (discrepancies) w_{ij}. In tableau form, each nonbasic w_{ij} appears at the left margin with a basic x_{ij} at the right margin, and its value at the basic solution is 0. The basic w_{ij} corresponds to nonbasic x_{ij} and appears at the bottom of the tableau, so their values belong to the last row. Discrepancies w_{ij} are determined by the selected positions in Table 17.1. Recall that the discrepancies are zeros at the selected positions and are not written there and are written in parentheses elsewhere. In other words, the discrepancies, like the flow, are determined by the tree (see Figure 17.3).

	55	30	40	-15	-20	-50	
0	55	30	40	50	40	0	
	25	10	15				50
				(65)	(60)	(50)	
-60	35	30	100	45	60	0	
	(-80)	(-60)	5	25	(20)	(−10)	30
-50	40	60	95	35	30	0	
	(-65)	(-20)	(5)	5	15	30	50
	25	10	20	30	15	30	

Table 17.4. Discrepancies w_{ij} for Table 17.2

Some of the basic w_{ij} are negative, so the solution is not optimal. We must pick a negative basic w_{ij} and switch it with a nonbasic $w_{i'j'}$, which is accomplished in the table by switching a nonbasic x_{ij} with a basic $x_{i'j'}$. We pick $w_{21} = -80$ and decide to switch x_{21} with a basic variable. In graphical terms, we are adding an edge from warehouse 2 to store 1, which destroys the tree structure of the graph by creating the loop $(2,1), (2,3), (1,3), (1,1)$ (Figure 17.5).

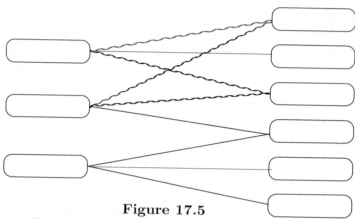

Figure 17.5
Cycle arising from adding x_{21} to the basis

We start a flow of $\varepsilon \geq 0$ on the new chosen edge, from warehouse 2 to store 1; that is, we set $x_{21} = \varepsilon$. We must adjust the flow around the loop to satisfy the balance constraint. Therefore we subtract ε from x_{23} so as not to exceed the supply at warehouse 2. Then we add ε to x_{13} to meet the demand of store 3. On the last edge of the loop we subtract ε from x_{11}, which avoids exceeding the supply of warehouse 1 and the demand at store 1 (Table 17.6).

		55	30	40	−15	−20	−50	
0	55	30	40	50	40	0		
	25 − ε	10	15 + ε	(65)	(60)	(50)	50
−60	35	30	100	45	60	0		
	(−80)ε	(−60)	5 − ε	25	(20)	(−10)	30
−50	40	60	95	35	30	0		
	(−65)	(−20)	(5)	5	15	30	50
	25	10	20	30	15	30		

Table 17.6.
Adjusting by ε the flow around the loop

The change in the transportation cost is $(35 - 100 + 40 - 55)\varepsilon = -80\varepsilon$. Note that the coefficient -80 is exactly the value of w_{ij}! To get the best improvement in the objective function, we give ε the maximum possible value—namely 5. This determines the x_{ij} to deselect, or, equivalently, the edge to remove from the graph to restore it to a tree. We move x_{21} into the basis and x_{23} out of the basis to arrive at a new basic solution (Table 17.7).

55	30	40	50	40	0	
20	10	20				50
35	30	100	45	60	0	
5			25			30
40	60	95	35	30	0	
			5	15	30	50
25	10	20	30	15	30	

Table 17.7.
Second feasible solution to Example 2.4

The transportation cost is now $20 \cdot 55 + 10 \cdot 30 + 20 \cdot 40 + 5 \cdot 35 + 25 \cdot 45 + 5 \cdot 35 + 15 \cdot 30 + 30 \cdot 0 = 4125 = 4525 - 80 \cdot 5$. We proceed to find the new values for the dual variables (Table 17.8).

	55	30	40	65	60	30	
0	55	30	40	50	40	0	
	20	10	20	(-15)	(-20)	(-30)	50
20	35	30	100	45	60	0	
	5	(20)	(80)	25	(20)	(-10)	30
30	40	60	95	35	30	0	
	(15)	(60)	(85)	5	15	30	50
	25	10	20	30	15	30	

Table 17.8.
Potentials and discrepancies w_{ij} for Table 17.7

We have executed one pivot step, replacing x_{23} with x_{21} as a basic variable, resulting in a new tree (Figure 17.9).

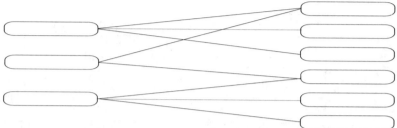

Figure 17.9. Tree resulting from removal of x_{23} from the basis

Some w_{ij} are still negative, however, so the table is not optimal. We have three negative entries in the table from which to choose. Choosing $w_{14} = -15$ corresponds to adding an edge to the graph from warehouse 1 to store 4, creating the loop

$$(1,4), (1,1), (2,1), (2,4).$$

Choosing $w_{15} = -20$ corresponds to adding an edge to the graph from warehouse 1 to store 5, creating the loop

$$(1,5), (1,1), (2,1), (2,4), (3,4), (3,5) \quad \text{(Figure17.10)}.$$

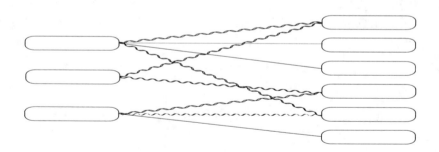

Figure 17.10. Loop resulting from adding x_{15} to the basis

To illustrate the process of finding a new basic feasible solution with a longer loop, we proceed with the latter choice. We start a flow of ε on the edge from warehouse 1 to store 5; that is, we set $x_{ij} = \varepsilon$. Then we subtract ε from x_{11} because we cannot exceed the supply of warehouse 1. Then we add ε to x_{21} to meet the demand at store 1.

We continue in this manner around the loop, subtracting ε from x_{24}, adding ε to x_{34}, and, finally, subtracting ε from x_{35} (see Table 17.11).

	55	30	40	65	60	30	
0	55 / $20-\varepsilon$	30 / 10	40 / 20	50 / (-15)	40 / (-20) ε	0 / (-30)	50
20	35 / $5+\varepsilon$	30 / (20)	100 / (80)	45 / $25-\varepsilon$	60 / (20)	0 / (-10)	30
30	40 / (15)	60 / (60)	95 / (85)	35 / $5+\varepsilon$	30 / $15-\varepsilon$	0 / 30	50
	25	10	20	30	15	30	

Table 17.11.
Adjusting by ε the flow around the new loop

The change in the value of the objective function is
$$\varepsilon(c_{15} - c_{35} + c_{34} - c_{23} + c_{21} - c_{11})$$
$$= \varepsilon(40 - 30 + 35 - 45 + 35 - 55) = -20\varepsilon = w_{15}e.$$
In this case, the maximum value that ε can take is 15, limited by the demand at store 5. Adjusting the x_{ij} as prescribed by the flow with $\varepsilon = 15$ results in a new feasible solution that decreases the objective function to $4125 - 20 \cdot 15 = 3825$ (Table 17.12). In the new solution, x_{15} replaces x_{35}, restoring our graphical representation to a tree (Figure 17.13).

55	30	40	50	40	0	
5	10	20		15		50
35	30	100	45	60	0	
20			10			30
40	60	95	35	30	0	
			20		30	50
25	10	20	30	15	30	

Table 17.12. Third feasible solution to Example 2.4

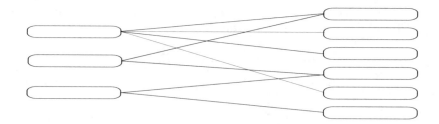

Figure 17.13. Tree resulting
from deleting x_{35} from the basis

We proceed as before to find the new potentials, the new w_{ij}, a
negative $w_{14} = -15$, creating the loop $(1,4), (1,1), (2,1), (2,4)$, and
the adjusted flow around the loop (Table 17.14).

	55	30	40	65	40	30	
0	55 $5 - \varepsilon$ ■	30 10	40 20	50 (-15) ε ■	40 15	0 (-30)	50
20	35 $20 + \varepsilon$ ■	30 (20)	100 (80)	45 $10 - \varepsilon$ ■	60 (40)	0 (-10)	30
30	40 (15)	60 (60)	95 (85)	35 20	30 (20)	0 30	50
	25	10	20	30	15	30	

Table 17.14. Dual variables and adjusted flow for Table 17.13

We set $\varepsilon = 5$, the maximum as constrained by the supply at
warehouse 1, to obtain a new feasible solution which reduces the
objective function by $w_{14}\varepsilon = 75$ (Table 17.15).

55	30	40	50	40	0	
55	30 10	40 20	50 5	40 15	0	50
35 25	30	100	45 5	60	0	30
40	60	95	35 20	30	0 30	50
25	10	20	30	15	30	

Table 17.15. Fourth feasible solution to Example 2.4

We compute the potentials, the new w_{ij}, a new negative $w_{16} = -15$ and the new flow of ε around the loop

$$(1,6), (1,4), (3,4), (3,6)$$

(Table 17.16).

	40	30	40	50	40	15	
0	55 (15)	30 10	40 20	50 $5-\varepsilon$	40 15	0 (-15) ε	50
5	35 25	30 (5)	100 (65)	45 5	60 (25)	0 (-10)	30
15	40 (15)	60 (45)	95 (70)	35 $20+\varepsilon$	30 (5)	0 $30-\varepsilon$	50
	25	10	20	30	15	30	

Table 17.16. Dual variables and adjusted flow for Table 17.15

Again ε is constrained to 5 by the supply at warehouse 1, and setting $\varepsilon = 5$ gives a new feasible solution that improves the objective function by 75 (Table 17.17).

55	30 10	40 20	50	40 15	0 5	50
35 25	30	100	45 5	60	0	30
40	60	95	35 25	30	0 25	50
25	10	20	30	15	30	

Table 17.17. Fifth feasible solution to Problem 2.4

We compute again the new potentials and the new w_{ij}, find a negative $w_{35} = -10$, create the loop $(3,5),(3,6),(1,6),(1,5)$, and set $x_{35} = \varepsilon$, $x_{36} = 25 - \varepsilon$, $x_{16} = 5 + \varepsilon$, and $x_{15} = 15 - \varepsilon$ (Table 17.18).

	25	30	40	35	40	0	
0	55 (30)	30 10	40 20	50 (15)	40 $15-\varepsilon$	0 $5+\varepsilon$	50
-10	35 25	30 (-10)	100 (50)	45 5	60 (10)	0 (-10)	30
0	40 (15)	60 (30)	95 (55)	35 25	30 (-10) ε	0 $25-\varepsilon$	50
	25	10	20	30	15	30	

Table 17.18. Dual variables and adjusted flow for Table 17.17

Setting ε to the maximum value of 15, constrained by the demand at store 5, yields a new feasible solution that reduces the transportation cost by 150 (Table 17.19).

55	30 10	40 20	50	40	0 20	50
35 25	30	100	45 5	60	0	30
40	60	95	35 25	30 15	0 10	50
25	10	20	30	15	30	

Table 17.19. Sixth feasible solution to Example 2.4

Then we compute the new potentials u_i and v_j and the new dual slack variables w_{ij}. This time we choose position $(2,6)$ where $w_{26} = -10$. We adjust the flow around the loop by setting $x_{26} = \varepsilon$, $x_{24} = 5 - \varepsilon$, $x_{34} = 25 + \varepsilon$, and $x_{36} = 10 - \varepsilon$.

Here is the new table:

	25	30	40	35	30	0	
0	55 (30)	30 10	40 20	50 (15)	40 (10)	0 20	50
−10	35 25	30 (−10)	100 (50)	45 $5-\varepsilon$	60 (20)	0 (−10) ε	30
0	40 (15)	60 (30)	95 (55)	35 $25+\varepsilon$	30 15	0 $10-\varepsilon$	50
	25	10	20	30	15	30	

Table 17.20. Dual variables and adjusted flow for Table 17.19

We set ε to the maximum of 5, constrained by the supply at store 2, to get a feasible solution that improves the objective function by 50 (Table 17.21).

55	30 10	40 20	50	40	0 20	50
35 25	30	100	45	60	0 5	30
40	60	95	35 30	30 15	0 5	50
25	10	20	30	15	30	

Table 17.21 Seventh feasible solution to Example 2.4

We compute again the potentials and the dual slack variables—that is, the c-part of the corresponding tableau (Table 17.22).

	35	30	40	35	30	0	
0	55 (20)	30 10	40 20	50 (15)	40 (10)	0 20	50
0	35 25	30 (0)	100 (60)	45 (10)	60 (30)	0 5	30
0	40 (5)	60 (30)	95 (55)	35 30	30 15	0 5	50
	25	10	20	30	15	30	

Table 17.22. Potentials and w_{ij} for Table 17.21

Finally all the basic w_{ij}, which correspond to the c-part of a standard tableau, are nonnegative. So the seventh feasible solution is optimal. In fact this is the solution found by an educated guess in Example 2.4, although educated guesses would be difficult in larger problems. The transportation cost is $10 \cdot 30 + 20 \cdot 40 + 20 \cdot 0 + 25 \cdot 35 + 5 \cdot 0 + 30 \cdot 35 + 15 \cdot 30 + 5 \cdot 0 = 3475$. Figure 17.23 shows a graphical representation of our optimal solution.

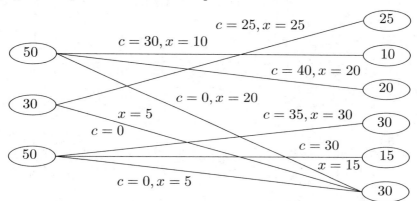

Figure 17.23. Optimal solution corresponding to Table 17.22

Since $w_{22} = 0$, x_{22} can be increased without hurting the objective function. Other optimal solutions may arise if this x_{ij} can be increased without violating any constraints. We can investigate this possibility by considering adding an edge to our graph from warehouse 2 to store 2, creating the loop $(2,2), (2,6), (1,6), (1,2)$. It is evident that the per-unit cost of shipping from warehouse 1 to store 2 equals the cost from warehouse 2 to store 2, and the cost of leaving units in stock—that is, shipping to store 6 is free for all warehouses. Thus we conclude that supplying some of the demand for store 2 from warehouse 2 instead of warehouse 1 and leaving fewer units in stock at warehouse 2 and more in stock at warehouse 1 would not change the transportation cost. We can only adjust the allocation in this way up to 5 units, because warehouse 2 only has 5 units to spare. All optimal solutions are obtained in this way, because all other w_{ij} are strictly positive.

By putting some extra numbers in our graph, we can represent the full solution (Figure 17.24). Each edge in the graph is marked by the number x_{ij} of units to be shipped from warehouse i to store j. Remember that "store 6" is ficticious.

So for each i, x_{i6} is actually the number of units that remain in stock at warehouse i. We can mark each warehouse with total supply and each store with total demand, which at store 6 equals the total excess supply across all warehouses. We can include all the optimal solutions by adding the edge from warehouse 2 to store 2 and making x_{12} and x_{22} functions of ε (Figure 17.24).

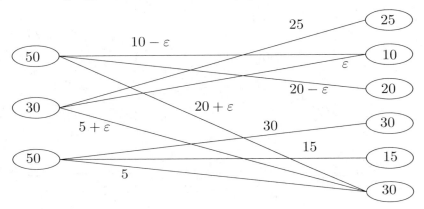

Figure 17.24. All optimal solutions, $0 \le \varepsilon \le 5$

The possibility of cycling in transportation problems is very remote, and Bland's method prevents it. If the given supply and demand data are integral, then the optimal solution is also integral. Moreover, the size of integers in feasible solutions does not exceed the size of integers in the supply and demand.

It is possible to use old potentials to compute new ones. This decreases the amount of computations but increases the complexity of the method and creates new opportunities for errors.

Remark 17.25.   The dual simplex method for transportation problems is known as the Hungarian method. It is based on the observation that adding a constant to a row or a column of the cost matrix does not change the optimality region because it changes the objective function by a constant. If we manage to obtain a cost matrix with all entries ≥ 0 and enough zero entries to place the whole flow at the positions with zero costs, then we obtain an optimal solution (with zero adjusted total cost). The dual method is particularly advisable for transportation problems that come from job assignment problems (see the next section) where degeneracy slows down the primal method. In those problems, $m = n$ and $a_i = b_j = 1$ for all i, j.

Here we outline the Hungarian method for these problems. The method works with nonnegative matrices having a zero in every row and column. To get an inital $n \times n$ matrix, we subtract the least element of each row from that row of the cost matrix $[c_{i,j}]$. Then we do likewise for each column.

1. Given any matrix C with nonnegative entries, we draw the minimum number t of lines through the rows and columns to cover all zeros in C. If $t = n$, then it is very easy to find a basic feasible solution as in §16 selecting positions with zero costs and crossing out these n lines one after another. Otherwise, we proceed to Step 2.

2. Now $t < n$. We compute the minimum m of all uncovered entries. Then we subtract m from each uncovered entry and add it to each twice-covered entry (i.e., covered by both a horizontal and a vertical line). Return to Step 1.

Exercises.

1–2. Compute the feasible solution and the corresponding dual basic solution to the transportation problem, where the stars (*) mark the basic variables for the row problem. Put both basic solutions into the same table. Are those solutions optimal?

1.

1 *	2 *	3	200
1	2 *	2 *	300
175	125	200	

2.

1	2 *	3 *	200
1 *	2 *	2	300
175	125	200	

3–5. Solve the transportation problems in Exercises 4–6 of §16.

6–8. Solve the transportation problem in Example 2.4 with

6. The supply 30 replaced by $30 + t$ where t is a parameter and $|t| \leq 1$

7. The supply 30 replaced by a parameter t

8. The unit cost $c_{23} = 100$ is replaced by a parameter t

Hint: Use Table 17.22 to start.

§18. Job Assignment Problem

We introduced this class of problems in Chapter 1 (see Example 2.5 and Exercises 6–9 in §2.) In general, in terms of variables and constraints, the problem can be stated as follows. We set

$$x_{ij} = 0 \text{ or } 1 \tag{18.1}$$

depending on whether the person i is assigned to do the job j. The total time then is then

$$\sum_i \sum_j c_{ij} x_{ij} \text{(to be minimized)}. \tag{18.2}$$

The condition that every person i is assigned to exactly one job is

$$\sum_j x_{ij} = 1 \text{ for all } i. \tag{18.3}$$

The condition that exactly one person is assigned to every job j is

$$\sum_i x_{ij} = 1 \text{ for all } j. \tag{18.4}$$

Thus, the problem is stated as an optimization problem (18.1–18.4.) The objective function (18.2) is linear, as are the constraints (18.3) and (18.4). However constraints (18.1) are not linear.

In view of (18.3) and (18.4), constraints (18.1) can be replaced by the following constraints: $x_{ij} \geq 0, x_{ij}$ are integers. This is an example of an integer programming problem.

If we drop the condition that x_{ij} are integers (keeping the conditions $x_{ij} \geq 0$), then mathematically our problem looks like a particular case of the transportation problem with all given supplies and demands equal to 1 and the balance condition meaning that the number of persons equals the number of jobs.

Let us forget about the integrability condition and solve the resulting linear problem by the simplex method. So we regard each worker as a "warehouse" supplying work and each job as a "store" with a demand for work. We set both the supply of each worker and the demand of each job equal to 1. Total time becomes the total transportation cost.

Because integral data guarantee an integral solution, our optimal solution has only integral values for all variables. So we obtain an optimal solution for the job assignment problem.

Thus, we have reduced any job assignment problem to a transportation problem. A variation of the assignment problem involves maximizing the sum S of ratings instead of minimizing the total cost (instead of time) of completing a set of jobs. Using a standard trick, we can reduce maximization to minimization. The resulting transportation problem we may have negative coefficients c_{ij} which we did not see in previous examples. However, the simplex method does work the same way no matter whether coefficients are positive, negative, or zeros, or any mixture of those. So this version can be also solved as previously.

Problem 18.5. Solve the job assignment problem

2	2	2	1	2
2	3	1	2	4
2	0	1	1	1
2	3	4	3	3
3	2	1	2	1

where the table data are interpreted as cost per person per job and the goal is to minimize total cost.

Solution. Unless you like pivoting, it pays to spend more time on Phase 1 rather than use the northwest method. We observe the minimal cost per person, that is., the minimal cost in each row: 1, 1, 0, 2, 1. So the minimal total cost is at least 5 (i.e., min $\geq$ 5.) Let us try to get this cost by choosing a position with minimal cost in each row. This works nicely and uniquely until we come to the last row, where there are two positions with cost 1. But one of them collides with the selected position in the second row, so we select the last position in the last row. Therefore, we get an optimal solution and it is unique:

Note that there are 5! = 120 feasible solutions, and trying all of them would take more time than our common-sense approach.

Problem 18.6. Solve the job assignment problem with data as in Problem 18.5, but now the table data are interpreted as the rating of each worker on each job and the goal is to maximize efficiency.

Solution. Again before indulging in pivoting we try to use common sense. Let us try first to choose the best job for each worker (this worked so well in the previous problem). Here is the pattern of maximal entries in rows:

We get the max $\leq 2 + 4 + 2 + 4 + 3 = 15$, but there is no way to do job 4 if we choose the best number in each row. Let us then try to choose a maximal entry in each column:

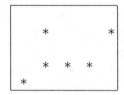

Now the first and third workers have nothing to do, and our upper bound for the maximum is max $\leq 3 + 3 + 4 + 3 + 4 = 17$, which is weaker than the previous bound.

In fact since choosing the maximal number in each row does not work, we conclude that max < 15. Since the maximum is an integer, we obtain that max ≤ 14. But now looking again at the data and the pattern of maximal entries in each row, we easily see a feasible solution with total rating 14:

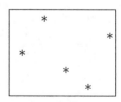

where the workers 1–4 take the best jobs for each and the worker 5 takes the second best job 4.

Problem 18.7. Minimize total grief in the matching problem.

1	1	2	1
2	3	2	3
4	3	3	4
4	4	5	6

Solution. Taking a minimal entry

in each column does not produce any feasible solutions, and we get the bound min > 5. Taking a minimal entry

*	*		*
*		*	
	*	*	
*	*	*	*

in each column does produce four optimal solutions. However, for a change, we ignore the optimal solution and use the simplex method. The point of this exercise is that although common sense and special tricks give shortcuts for some problems, the simplex method works for every linear program.

So we treat this problem as a transportation problem. As an initial basic solution we take the one found by a student in 2000:

$^1 1$	1	$^2 0$	1	$+$ 0
2	3	2	$^3 1$	$+$
4	$^3 1$	3	4	$+$
4	$^4 0$	$^5 1$	$^6 0$	$+$ 0
$+$	$+$	$+$	$+$	
	0		0	

Note that the total cost at this solution is $1 + 3 + 3 + 5 = 12$ which is better than $1+2+3+6 = 13$ given by the northwest method.

Then we find potentials and w_{ij}:

	5	5	6	7
4	$^1 1$	$^1(0)$	$^2 0$	$^1(-2)$
4	$^2(2)$	$^3(3)$	$^2(1)$	$^3 1$
2	$^4(1)$	$^3 1$	$^3(-1)$	$^4(-1)$
1	$^4(0)$	$^4 0$	$^5 1$	$^6 0$

The first potential 5 on top was chosen arbitrarily. Now there are three negative w_{ij}. We select the position (3, 3) with negative $w_{33} = -1$ (there were two other possible choices). This leads to a loop of length 6. (The choice of $w_{14} = -2$ would lead to a loop of length 4 and a degenerate pivot step.) Here is an adjustment of the flow along the loop:

$^1 1$	$^1(0)$	$^2 0$	$^1(-2)$
$^2(2)$	$^3(3)$	$^2(1)$	$^3 1$
$^4(1)$	$^3 1 - \varepsilon$	$^3 \varepsilon$	$^4(-1)$
$^4(0)$	$^4 0 + \varepsilon$	$^5 1 - \varepsilon$	$^6 0$

We take $\varepsilon = 1$ and deselect the position (4, 3) [another possible choice was the position (3, 2)]. The objective function improved by 1. Here are our new basic feasible solution, new potentials, and new adjustment along of a loop of length 6:

	5	6	6	8
4	$^1 1$	$^1(-1)$	$^2 0 - \varepsilon$	$^1(-3)\ \varepsilon$
6	$^2(3)$	$^3(3)$	$^2(2)$	$^3 1$
2	$^4(2)$	$^3 0 - \varepsilon$	$^3 1 + \varepsilon$	$^4(-1)$
1	$^4(1)$	$^4 1 + \varepsilon$	$^5(1)$	$^6 0 - \varepsilon$

We have selected the position (1, 4) as a new basic position (there were two other choices). We must take $\varepsilon = 0$, so it is going to be a degenerate pivot step that does not change the feasible solution (but it changes the basis, i.e., the set of basic variables). Now we deselect the position (4, 4) [the other possible choices were (1, 3) and (3, 2)], compute again the new potentials and w_{ij}, and find a negative w_{ij} (once we find one we do not need to compute other w_{ij}) and the corresponding loop.

Here is our next table:

	5	6	6	5
4	$^1 1$	$^1(-1)$	$^2 0 - \varepsilon$	$^1 0 + \varepsilon$
3	$^2(0)$	$^3(0)$	$^2(-7)\,\varepsilon$	$^3 1 - \varepsilon$
3	$^4(2)$	$^3 0$	$^3 1$	$^4(2)$
2	$^4(1)$	$^4 1$	$^5(1)$	$^6(3)$

We have selected the position $(2, 3)$. Again $+\varepsilon = 0$. We deselect the position $(1, 3)$ (no other choice this time). Next we compute new potentials and w_{ij}:

	5	5	5	5
4	$^1 1$	$^1(0)$	$^2(1)$	$^1 0$
3	$^2(0)$	$^3(1)$	$^2 0$	$^3 1$
2	$^4(1)$	$^3 0$	$^3 1$	$^4(1)$
1	$^4(0)$	$^4 1$	$^5(1)$	$^6(2)$

Now all $w_{ij} \geq 0$, so the table is optimal. The optimal assignment it gives is

The optimal value is $\min = 11$. Note that we had the optimal solution already after the first pivot step, but the last table proves that the solution is optimal.

Remark 18.8. A job assignment problem with n workers and jobs has $n!$ feasible solutions. After it is written as a transportation problem, these solutions become the vertices of the feasible region which is a simplex of dimension $n! - 1$ (in particular, an interval when $n = 2$). For each vertex, there are 2^{n-1} choices for the basis. There are $(n^2)!$ ways to place the basic variables at the left margin of the standard tableau, and $(n^2)!$ ways to place the nonbasic variables at the top margin. So we can imagine $[n^2)!]^2 \cdot 2^{n-1} \cdot n!$ standard tableaux for the problem. Fortunately, the simplex method does not need to go through all these tableaux. In fact, the simplex method for transportation problems proved to be very efficient in practice.

Exercises

1–4. Solve the job assignment problem where the table data are interpreted as cost per job per person and the goal is to minimize total cost.

1.

$$
\begin{array}{|ccccc|}
\hline
1 & 2 & 3 & 1 & 2 \\
3 & 1 & 2 & 1 & 0 \\
0 & 1 & 3 & 2 & 1 \\
2 & 3 & 4 & 3 & 3 \\
1 & 3 & 3 & 1 & 2 \\
\hline
\end{array}
$$

.

2.

$$
\begin{array}{|cccccc|}
\hline
4 & 2 & 2 & 2 & 1 & 2 \\
2 & 3 & 1 & 2 & 4 & 4 \\
2 & 0 & 1 & 1 & 1 & 4 \\
2 & 3 & 4 & 3 & 4 & 3 \\
3 & 2 & 1 & 2 & 4 & 1 \\
1 & 2 & 1 & 2 & 0 & 1 \\
\hline
\end{array}
$$

.

3.

$$
\begin{array}{|ccccccc|}
\hline
2 & 1 & 4 & 3 & 4 & 3 & 1 \\
4 & 2 & 2 & 2 & 1 & 5 & 2 \\
2 & 3 & 5 & 1 & 2 & 4 & 4 \\
5 & 2 & 0 & 1 & 1 & 1 & 4 \\
2 & 3 & 4 & 3 & 4 & 3 & 5 \\
5 & 3 & 2 & 1 & 2 & 4 & 1 \\
1 & 2 & 1 & 2 & 0 & 1 & 5 \\
\hline
\end{array}
$$

.

4.

$$
\begin{array}{|ccccccccc|}
\hline
0 & 2 & 2 & 4 & 0 & 1 & 5 & 1 & 4 \\
2 & 3 & 1 & 3 & 4 & 3 & 2 & 4 & 1 \\
2 & 1 & 4 & 3 & 4 & 3 & 1 & 2 & 4 \\
4 & 2 & 4 & 2 & 2 & 2 & 1 & 5 & 2 \\
2 & 3 & 5 & 1 & 2 & 4 & 2 & 4 & 4 \\
5 & 2 & 2 & 4 & 0 & 1 & 1 & 1 & 4 \\
2 & 3 & 4 & 3 & 4 & 3 & 2 & 4 & 5 \\
5 & 2 & 4 & 3 & 2 & 1 & 2 & 4 & 1 \\
1 & 2 & 1 & 2 & 0 & 1 & 2 & 4 & 5 \\
\hline
\end{array}
$$

.

5–8. Solve the job assignment problem 1–4 with the table data interpreted as the rating of each worker on each job and the goal is to maximize efficiency.

Chapter 7

Matrix Games

§19. What Are Matrix Games?

You have probably heard of game theory, a topic of interest right now. We will barely scratch the surface of this topic, but we hope that the glimpse we provide here will whet your appetite for more study.

In Webster's dictionary there are several definitions for the word *game*. A common usage of the word is

1a (1): activity engaged in for diversion or amusement: PLAY

However, when we talk about games, we have in mind something more akin to

3c: a situation that involves contest, rivalry, or struggle esp: one in which opposing interests given specific information are allowed a choice of moves with the object of maximizing their wins and minimizing their losses.

Games can be grouped into several categories according to how many players are involved. In a one-person game, the player's objective is just to maximize his, her, or its payoff. However, when there is more than one player involved, they may have different objectives (payoffs) to maximize. Surprisingly enough, linear programming has something to do with a certain class of games called matrix games. In fact, there is a deep connection between these two topics. In particular, matrix games can be solved by the simplex method, and linear programs can be reduced to matrix games.

In this chapter we will study matrix games, which are two-person, zero-sum games. The term *zero-sum* means that what one player wins the other player loses. For instance, if you and your opponent place a wager of $1000 on the outcome of your chess game, then the $1000 that the victor collects must come out of the loser's pocket, resulting in a zero-sum payoff. Not every two-person game is a zero-sum game.

For example, imagine that a husband and wife are contemplating a move to Buffalo, New York, where he can find a job paying $60,000 a year but her salary would be only $20,000 a year. However, if they moved to Akron, Ohio, she could work for $60,000 a year but his salary would be only $20,000 a year. By staying put, they both face unemployment. What should they do? The answer to that question is beyond the scope of this book, but it is clear that the payoff in this game does not add up to zero!

Example 19.1. This small matrix game is known as Heads and Tails or Matching Pennies. We will call the players *He* and *She*. He chooses: heads (H) or tails (T). Independently, she chooses: H or T. If they choose the same, he pays her a penny. Otherwise, she pays hin a penny. Here is his payoff in cents:

$$
\begin{array}{cc}
 & \begin{array}{cc} She \\ H \qquad\quad T \end{array} \\
He \ \begin{array}{c} H \\ T \end{array} & \left[\begin{array}{cc} -1 & 1 \\ 1 & -1 \end{array} \right].
\end{array}
$$

Example 19.2. Another game is Rock, Scissors, Paper. In this game two players simultaneously choose *Rock*, *Scissors*, or *Paper*, usually by a show of hand signals on the count of three, and the payoff function is defined by the rules *Rock breaks Scissors, Scissors cuts Paper, Paper covers Rock*, and every strategy ties against itself. Valuing a win at 1, a tie at 0, and a loss at -1, we can represent the game with the following matrix, where, for both players, strategy 1 is *Rock*, strategy 2 is *Scissors*, and strategy 3 is *Paper*:

$$
A = \left[\begin{array}{ccc} 0 & 1 & -1 \\ -1 & 0 & 1 \\ 1 & -1 & 0 \end{array} \right].
$$

The payoff is given for the player who chooses a row. ∎

In general, a matrix game is given by a matrix (*paoff matrix*).. One player (the row player) chooses a row, and the other player chooses a column. The corresponding entry in the matrix represents what the column player pays to the row player (the payoff of the row player). The players could be humans, teams, computers, animals. Games like chess, football, and blackjack can be thought as (very large) matrix games.

Example 19.3. The following matrix gives a matrix game with four choices for the row player and five choices for the column player.

$$A = \begin{bmatrix} 5 & 0 & 6 & 1 & -2 \\ 2 & 1 & 2 & 1 & 2 \\ -9 & 0 & 5 & 2 & -9 \\ -9 & -8 & 0 & 4 & 2 \end{bmatrix}.$$

■

Equilibrium

Let $A = [a_{i,j}]$ be a matrix with m rows and n columns. Recall that the corresponding matrix game is defined as follows. Independently, he chooses a row i, and she chooses a column j. As a result, she pays him $a_{i,j}$ (in US dollars).

The matrix A is his payoff matrix. Her payoff matrix is $-A$. The rows of A correspond to his (pure) strategies (or moves). The columns of A correspond to her choices.

Definition 19.4. A pair (i,j) of strategies is called an equilibrium or a saddle point if neither player can gain by changing his or her strategy. ■

In other words, $a_{i,j}$ is both largest in its column and the smallest in its row.

As an example, consider Example 19.3. We mark the maximal entries in every column by *. Then we mark the minimal entries in each row by ′. The positions marked by both * and ′ are exactly the saddle points:

$$A = \begin{bmatrix} 5^* & 0 & 6^* & 1 & -2' \\ 2 & 1^{*\prime} & 2 & 1' & 2^* \\ -9' & 0 & 5 & 2 & -9' \\ -9' & -8 & 0 & 4^* & 2^* \end{bmatrix}.$$

In this example, the position $(i,j) = (2,2)$ is the only saddle point. The corresponding payoff is 1.

When we have a saddle point, it can be declared a solution for the game, because no player can do better by an unilateral change. It seems reasonable for both players to sit there. (If it would not be a zero-sum game, they might try making joint decisions and sharing payoffs.) We will see later in this section that the payoff at each saddle point is the same. It is called *the value of the game*.

However, in Examples 19.1 and 19.2, there are no saddle points. To solve matrix games without saddle points we have to extend the notion of strategy by introducing *mixed strategies*.

Consider again the general matrix game, given by a matrix $A = [a_{i,j}]$. A prudent way to choose a strategy for him is as follows. He assumes that she is a perfect player (she has unlimited intelligence and memory) and (maybe) can read his mind. So he reasons: If I choose the first strategy, then she would pick the smallest payoff for me, namely, $\min_j a_{1,j}$; if I choose the second strategy, she would pay me $\min_j a_{2,j}$, and so on. So the best choice is a strategy that gives me at least $\max_i \min_j a_{i,j}$. This is his "gain floor" or the *worst-case payoff*.

Similarly, if she believes that he is omniscient, she would choose a strategy that makes her loss at most $\min_j \max_i (a_{i,j})$ (her "loss ceiling").

Now, a fact of life is that

$$\max_i \min_j \ a_{i,j} \le \min_j \max_i \ a_{i,j} \tag{19.5}$$

always (for any matrix A). This is because his worst-case payoff cannot be better than his best-case payoff. We will prove (19.5) without using any game-theoretical interpretation of the matrix A.

Furthermore, $\max_i \min_j a_{i,j} = \min_j \max_i a_{i,j}$ if and only if there is a saddle point. Moreover, the payoff at any saddle point is exactly $\max_i \min_j a_{i,j} = \min_j \max_i a_{i,j}$, the values of game.

Proof of the inequality (19.5). Let $b = \max\min$ and $c = \min\max$. Let d be in the same row as b and in the same column as c. Since b is minimal in its row, $b \le d$. Since c is maximal in its column, $d \le c$. Thus, $b \le d \le c$, hence $b \le c$.

Proof of the implication

$$\max\min = \min\max \Rightarrow \text{there is an equilibrium.} \tag{19.6}$$

Let b, c, d be as before. Assume that $b = c$. Then $d = b = c$. So d is minimal in its row and maximal in its column. Thus, its position (the corresponding strategies) is an equilibrium.

Finally, if (i, j) and (i', j') are two saddle points, then (i, j') and (i', j) are also saddle points. This justifies the following definition: Strategies involved in equilibria are called *optimal strategies*. Thus, a saddle point is a pair (his optimal strategy, her optimal strategy).

Proof of the implication

$$\text{there is an equilibrium} \Rightarrow \max\min = \min\max. \tag{19.7}$$

Let d be the payoff at a saddle point. Since d is maximal in its column, $d \le \min\max$. Since d is is minimal in its row, $d \le \max\min$.

So max min $\geq d \geq$ min max. On the other hand, by (19.5), max min $\leq$ min max. So max min $= d =$ min max.

Problem 19.8. Find max min and min max for

$$A = \begin{bmatrix} 5 & 0 & 6 & 1 & -2 \\ 2 & 1 & 2 & 1 & 2 \\ -9 & 2 & 5 & 2 & 0 \\ -9 & -8 & 0 & 4 & 2 \end{bmatrix}$$

and check whether there is a saddle point.

Solution. First we compute the maximum in each column: $\max(A)$ $= [5, 2, 6, 4, 2]$, hence min max $= \min[5, 2, 6, 4, 2] = 2$. Then we compute the minimum in each row: $\min(A) = [-2, 1, -9, -9]^T$, hence max min $= \max[-2, 1, -9, -9] = 1$. Since $2 \neq 1$, there is no saddle point.

Mixed Strategies

We consider again a matrix game with matrix $A = [a_{i,j}]$ of size $m \times n$.

Definition. 19.9. A mixed strategy for a player is a probability distribution on the set of his or her pure strategies. ∎

We will write a mixed strategy for him as a column

$$p = [p_1, \ldots, p_m]^T,$$

where $p_i \geq 0$ for all i and $p_1 + \cdots + p_m = 1$. His original strategies, which we will call now his *pure strategies,* can be identified with his mixed strategies whose entries are 0 or 1, then a mixed strategy is the same as a convex combination of pure strategies. So his pure strategies are the vertices in the set of his mixed strategies, and her pure strategies are the vertices in the set of her mixed strategies.

We will write a mixed strategy for her as a row $q = [q_1, \ldots, q_n]$, where $q_j \geq 0$ for all j and $q_1 + \cdots + q_n = 1$.

Once both players have chosen their mixed strategies p and q, the corresponding payoff for him (mathematical expectation) is

$$p^T A q^T = \sum_{i=1}^{m} \sum_{j=1}^{n} p_i a_{i,j} q_j.$$

Her payoff is $p^T A q^T$.

In Example 19.1, it is natural for him to use the mixed strategy $[1/2, 1/2]^T$. That is, he makes his decision by tossing a (fair) coin.

Then his payoff is 0 no matter what (pure or mixed) strategy she uses. Similarly (the game is symmetric), she may use the mixed strategy [1/2, 1/2] to pay 0 no matter what he is doing. Thus, this pair of mixed strategies is an equilibrium (in mixed strategies).

In general, it is natural for him to use a mixed strategy (called an optimal strategy), which gives him at least

$$\text{his value} = \max_{q \in Q} \min_{p \in P} p^T A q^T,$$

where P is the set of all his mixed strategies and Q is the set of all her mixed strategies (in fact, Q here can be replaced by the set of her pure strategies).

Similarly, a reasonable choice for her would be a mixed strategy (called an optimal strategy) which guarantees her at least

$$\text{her value} = \min_{p \in P} \max_{q \in Q} p^T A q^T.$$

The Minimax Theorem. For any matrix A,

$$\text{his value } = \text{ her value.} \qquad \blacksquare$$

In other words, there is an equilibrium in mixed strategies.

To solve a matrix game is to find an optimal strategy for him and her (that is, an equilibrium) and the value of the game (that is, his value = her value). In the next section we will relate solution of matrix games with solution of linear programs. We will see that the minimax theorem in game theory is essentially the duality theorem in linear programming.

Example 19.10. *Two-finger morra* (P. Morris).
Players A and B both simultaneously hold up one or two fingers and shout out a number that is a guess as to the total number of fingers held up (thus, it would be 2, 3, or 4). If either player gets the total right and the other does not, that player receives $1 from the other. Otherwise, no money changes hands. Player A has six conceivable strategies for playing the game: (1, 2), (1, 3), (1, 4), (2,2), (2, 3), (2, 4), where (1, 2) means "hold up 1 finger and shout 2," and so on. Now (1, 4), (2, 2) are obviously stupid, so we ignore them. Player B can also choose from the same set of six strategies.

The situation can be summarized in a payoff matrix:

$$
\begin{array}{c|cccc}
A \setminus B & (1,2) & (1,3) & (2,3) & (2,4) \\
\hline
(1,2) & 0 & 1 & -1 & 0 \\
(1,3) & -1 & 0 & 0 & 1 \\
(2,3) & 1 & 1 & -1 & -1 \\
(2,4) & 0 & -1 & 1 & 0
\end{array}
$$

In this matrix, a positive number indicates a gain for A and a negative number is a loss for A. The matrix makes it clear that, if A and B are playing a long series of these games, neither player should stick exclusively to one of these strategies. For example, if A is dumb enough to play only $(2, 3)$, then B, unless he is even dumber, will play $(2, 4)$. The right way to play is to mix up these (or, maybe, some of these) "pure" strategies in a random way. A mixed strategy is a way of describing how the pure strategies are mixed together. For example, "play each of your four strategies at random so that each is used $1/4$ of the time" is a mixed strategy.

Problem 19.11. Check that in $([0, 1/2, 1/2, 0]^T, [0, 1/2, 1/2, 0])$ is an equilibrium—that is, $[0, 1/2, 1/2, 0]^T$ is an A's optimal strategy and $[0, 1/2, 1/2, 0]$ is a B's optimal strategy. Compute the value of the game.

Solution. We augment the payoff matrix A by

$$p = [0, 1/2, 1/2, 0]^T \text{ and } q = [0, 1/2, 1/2, 0]) :$$

$$
\begin{array}{c|ccccc}
 & (1,2) & (1,3) & (2,3) & (2,4) & q \\
\hline
(1,2) & 0 & 1 & -1 & 0 & 0 \\
(1,3) & -1 & 0 & 0 & 1 & 0 \\
(2,3) & 1 & 1 & -1 & -1 & 0 \\
(2,4) & 0 & -1 & 1 & 0 & 0 \\
p & 0 & 0 & 0 & 0 & 0^{*'}
\end{array}.
$$

Now it is clear that the position (p, q) is a saddle point, so the corresponding payoff 0 is the value of the game. ∎

This game is symmetric because its payoff matrix $A = -A^T$ is skew symmetric. It is clear that the value of any symmetric game is 0 and that s is an optimal strategy for a player if and only if s^T is optimal for the other player. But it is not always so easy to find an optimal strategy.

The main goal of this chapter is to relate matrix games with linear programming. We conclude this section with a few remarks about game theory in general.

The theory of games invokes mathematics to model decision-making situations. The beginnings of the theory are usually traced back to Cournot (1889), who put forth a model of how two firms would choose prices in a duopoly—that is, if they were the only two firms competing in the market. Another early contribution came from Zermelo (1913), who studied the game of chess and proved that either White can always win, Black can always win, or both players can always guarantee a draw. Zermelo did not find which of these three possibilities is actually true, let alone exactly how White or Black should play, or the game would not continue to present such a challenge to human and computer players. Zermelo's result was based on properties of the mathematical structure of the game and therefore applies to tic-tac-toe as well, which also has those properties. For this game, it is well known that both players can guarantee a draw.

Whether decisions are made for profit, for fun, or for many other reasons, such as winning a war or winning an election, game theory analyzes mathematical models of the decision-making process and the outcomes that result from the different possible combinations of decisions.

In game theory, the decision makers are called *players*, decisions are called *moves*, and each opportunity for a player to make a move is called a *position*. Moves are what change the position in the game, either deterministically or with some element of chance. Positions in which more than one player move simultaneously can be modeled by a set of positions, each of which presents a move to just one player, without affecting the strategic possibilities of the game. Therefore, we can assume, without loss of generality, that exactly one player moves at each nonterminal position.

A *terminal position* is one in which no player can move, and at each terminal position an *outcome* is specified. In real life, an outcome could be that one player wins and one player loses, or that each player wins or loses some amount of money, or a less clearly defined circumstance such as some players are happy and some players are sad. In game theory outcomes are usually modeled as numerical payoffs that represent *valuations* or *preferences* of the players on the outcomes. A preference is a relative valuation.

For a given player, a set of moves for every position in which that player makes a move is called a *pure strategy* for the player. A rule that specifies probabilities for what moves will be made by a player at every position is called a *behavioral strategy*. A result due to Kuhn (1950) is that if a player can remember, at any position, all the moves he or she made at previous positions, then any behavioral strategy for that player can be represented by a *mixed strategy*, which specifies a single probability distribution over a player's pure strategies rather than a separate probability distribution at each position. The outcome of a game can be computed as a function of the players' pure strategies. If mixed strategies are considered, only the expected outcome can be computed, because, in any one play of a game, only pure strategy outcomes can actually result. In either case, the *payoff function* of the game is the rule assigning a numerical *payoff* to each player as a function of the strategy followed by each player.

It is often convenient to represent a game by a set of strategies for each player and a payoff function rather than as a set of positions and moves. Because players can choose their strategies before the game actually begins, when games are reduced to strategy sets and a payoff function, it is assumed that the players choose simultaneously and thus without any knowledge of each others' choice. This is called the *strategic*, or *normal form*, of a game.

Games can be grouped into several categories according to how many players are involved, properties of the information available to the players, whether or not chance is involved, and properties of the payoff function. When there is more than one player involved, the outcomes that are good for one player may not be good for another. In general, a player cannot know what outcome will result from his or her strategy without knowing what the other players will do. The problem of how players will predict each others' strategies is a fascinating problem in game theory. In most game-theoretic analyses the assumption is made that all players try to move so as to maximize their numerical payoffs, which is a realistic assumption for many real life decision-making scenarios, including parlor games like chess and competition games like Cournot's duopoly game. When and why this assumption is valid is a topic of current research in game theory (Bolton, 1998; Kurland and Byrne, 2000; Byrne and Kurland, 2000). Because our focus is on linear programming, and not psychology and behavior, we will stick to this assumption in the present context.

The games we will analyze with linear programming are called *two-person zero-sum* games and include the examples of chess and tic-tac-toe mentioned previosly, but not Cournot's duopoly game. Zero-sum means that the sum of the payoffs to all players always equals zero. In a zero-sum game, what helps one player always hurts at least one other player. In a two-person zero-sum game, the players' interests are diametrically opposed, and these games are sometimes called *strictly competitive* because there is no opportunity for mutual gain through cooperation. Any two-player game that ends either by one player winning or by the two players drawing—that is, tying—can be modeled as a zero-sum game, neglecting such concerns as how important winning is to each player. If the set of pure strategies available to each player is finite, linear programming can be used to find strategies for each player that guarantee the players the best payoffs that can be guaranteed. We will make this precise shortly.

Note that not all games have a finite move space. For example, in *games of timing*, a move is a choice of a time from a continuous interval, such as the classic "duel with pistols," which cost the extraordinary French mathematician Galois his life at age 20 in 1832.

Exercises

1–10. Find max min and min max and check whether there is a saddle point for the matrix games with the following matrices. If there is no saddle point, try to find as good mixed strategies for both players as you can.

1.
$$A = \begin{bmatrix} 0 & 0 & 6 & 1 & -2 \\ -1 & 2 & 5 & 2 & 0 \\ -4 & -8 & 0 & 4 & 2 \end{bmatrix}.$$

2.
$$A = \begin{bmatrix} 0 & 0 & 6 & 1 & -2 \\ 2 & 1 & 2 & 1 & 2 \\ -1 & 2 & 5 & 2 & 0 \\ -4 & -8 & 0 & 4 & 2 \end{bmatrix}.$$

3.
$$A = \begin{bmatrix} 4 & 5 & 0 & -1 & 1 & -2 \\ 0 & 2 & 1 & 2 & 1 & -2 \\ 1 & -1 & 2 & 5 & 2 & 0 \\ -4 & -9 & -8 & 0 & 4 & 2 \end{bmatrix}.$$

4.

$$A = \begin{bmatrix} 4 & -4 & 5 & 0 & 0 & -1 & 1 & -2 \\ 5 & 0 & 2 & 1 & 2 & 1 & 0 & -2 \\ -4 & 0 & -2 & 1 & 5 & 1 & 6 & 2 \\ 1 & -1 & 2 & 5 & 3 & 7 & 2 & 0 \\ -4 & -9 & -8 & 0 & 4 & 2 & 0 & -3 \end{bmatrix}.$$

5.

$$A = \begin{bmatrix} 4 & -4 & 3 & 0 & 0 & 0 & -1 & 1 & -2 \\ -1 & 0 & 2 & 1 & -2 & -2 & 1 & 0 & -2 \\ -4 & 0 & -2 & -2 & 1 & -1 & 1 & 6 & 2 \\ 1 & 2 & 2 & 5 & 3 & 3 & 7 & 2 & 0 \\ -4 & -9 & -8 & 0 & 4 & 2 & 2 & 0 & 3 \end{bmatrix}.$$

6.

$$A = \begin{bmatrix} 4 & -1 & 4 & -1 & 5 & 0 & 0 & -1 & 1 & -2 \\ -1 & 0 & 2 & 1 & 2 & 1 & 2 & 1 & 0 & -2 \\ -4 & 0 & -2 & 1 & 0 & 1 & 6 & 2 & 6 & -2 \\ 1 & -1 & 2 & 5 & 3 & 5 & 3 & 1 & 2 & 0 \\ -4 & -9 & -8 & 0 & 4 & 0 & 4 & 2 & -6 & -3 \end{bmatrix}.$$

7.

$$A = \begin{bmatrix} 7 & -1 & 3 & -1 & 5 & 0 & 0 & -1 & 1 & -2 \\ -1 & 0 & 0 & 1 & 2 & 1 & 2 & 1 & 0 & -2 \\ -4 & 0 & -2 & 1 & 0 & 1 & 1 & 2 & 6 & -2 \\ 0 & 0 & -2 & 1 & 0 & 1 & 1 & 2 & 0 & -2 \\ 1 & -1 & 2 & 0 & 3 & 5 & 3 & 1 & 2 & 0 \\ -4 & -1 & -8 & 0 & 4 & 0 & 4 & -2 & -6 & 0 \end{bmatrix}.$$

8.

$$A = \begin{bmatrix} 8 & -1 & 4 & -1 & 5 & 0 & 0 & -1 & 1 & -2 \\ -1 & 0 & 2 & 1 & 2 & 1 & 2 & 1 & 0 & -2 \\ -4 & 0 & -2 & 1 & 0 & 1 & 6 & 2 & 6 & -2 \\ -4 & 8 & -2 & 1 & 0 & 1 & 8 & 2 & 6 & -2 \\ 1 & -1 & 2 & 5 & 3 & 5 & 3 & 1 & 2 & 0 \\ -4 & -9 & -8 & 0 & 4 & 0 & 4 & 0 & -6 & 3 \end{bmatrix}.$$

9.

$$A = \begin{bmatrix} 9 & 0 & -4 & 0 & 2 & 1 & 0 & 1 & 9 & 1 & 2 & 6 & -2 \\ -4 & 3 & -4 & 0 & -2 & 1 & 9 & 1 & 9 & 1 & 2 & 6 & 0 \\ 0 & 0 & 2 & 0 & -2 & 1 & 0 & 1 & 9 & 6 & 2 & -6 & -2 \\ -4 & 0 & -4 & 0 & -2 & 1 & 0 & 1 & 9 & 6 & 2 & 6 & 3 \\ 0 & 0 & 0 & 0 & -2 & 1 & 5 & -2 & 9 & 6 & 2 & 6 & 0 \\ 4 & 0 & -4 & 0 & -2 & 1 & 0 & 1 & 9 & 6 & -2 & 6 & -2 \\ -4 & 0 & 9 & 0 & -2 & 3 & 0 & 1 & 9 & 6 & 2 & 0 & -2 \end{bmatrix}.$$

10.

$$
A = \begin{bmatrix}
0 & -1 & 0 & 2 & 1 & 2 & 1 & 2 & 1 & 0 & -2 \\
-1 & 1 & 0 & -2 & 1 & 0 & 1 & 6 & 2 & 6 & -2 \\
-4 & 0 & 0 & -2 & 1 & 0 & 1 & 6 & 2 & 6 & -2 \\
0 & -4 & 0 & -2 & 1 & 1 & 1 & 6 & 2 & 6 & -2 \\
0 & -4 & 0 & -2 & 1 & 0 & 1 & 6 & 2 & 6 & -2 \\
2 & 0 & 0 & -2 & 1 & 0 & 1 & 1 & 2 & 6 & -2 \\
-4 & 0 & -2 & 1 & 0 & 1 & 0 & -1 & 2 & 6 & -2 \\
1 & -1 & 2 & 5 & 3 & 1 & 0 & 3 & 1 & 2 & 0 \\
-4 & -5 & -8 & 0 & 4 & 0 & 4 & 20 & & -6 & -3
\end{bmatrix}.
$$

11. Prove that if (i, j) and (i', j') are two saddle points, then (i, j') and (i', j) are also saddle points.

12. *Colonel Blotto* (P. Morris).

Army $\vec{A}$ has three divisions and is attacking a town that is defended by the four divisions of army B. There are two roads into the town and the armies must divide their forces between the two roads. If the attacker outnumbers the defender on a particular road, then the attacker wins the defender's divisions and captures the town. The town counts as being worth two divisions. If the defender outnumbers the attacker on a particular road, then the attacking army loses its divisions. The strategies for A are (3, 0), (2, 1), (1, 2), and (0, 3), where, for example (2, 1) means: two divisions on road #1 and one division on road #2. Army B has five strategies: (4, 0), (3, 1), (2, 2), (1, 3), and (0, 4). The payoff matrix is

$A \backslash B$	$(4,0)$	$(3,1)$	$(2,2)$	$(1,3)$	$(0,4)$
$(3,0)$	-3	0	4	3	2
$(2,1)$	0	-2	-1	2	1
$(1,2)$	1	2	-1	-2	0
(0.3)	2	3	4	0	-3

From A's point of view, strategies (2, 1), (1,2) seem the most sound since the potential loss is only two divisions. Choosing these involves giving up the largest potential gain, however. It is also interesting that if B knows A will choose either (2, 1) or (1, 2), he will choose (2, 2) and will always win.

Check that the value of the game is $1/4$, A's optimal strategy is

$$[9/68, 23/68, 28/68, 8/68]^T,$$

and B's optimal strategy is

$$[5/12, 0, 2/12, 0, 5/12].$$

§20. Matrix Games and Linear Programming

Solving Matrix Games by Linear Programming

Now that you have had practice formulating and solving linear programs, it is a real challenge to reduce matrix games to linear programming. In normal form, a two-person zero-sum game in which he has m pure strategies and she has n pure strategies can be represented by an $m \times n$ matrix $A = [a_{ij}]$, where a_{ij} is his payoff when he uses strategy i and she uses strategy j. We will often use the words *row* and *column* in place of *strategy* (e.g., he plays row i and she plays column j). In a two-person zero-sum game we do not need to specify her payoff because the condition that the payoffs add up to zero tells us that her payoff is the negative of his payoff. Thus, for zero-sum games it is customary to simply refer to the payoff, meaning his payoff. Therefore, we say that he wants to maximize the payoff and she wants to minimize the payoff.

We will allow mixed strategies and attempt to find the best mixed strategy for each player. As probability distributions on pure strategies, we will assume that his strategy is independent from hers. We can represent a mixed strategy for him as $p = [p_1, \ldots, p_m]^T$ and a mixed strategy for her as $q = [q_1, \ldots, q_n)$. If he uses p and she uses q, basic probability theory tells us that because the distributions are independent, his expected payoff is $p^T A q^T$. The product $p^T A$ is a linear combination of the rows of A in which the $i\,^{rmth}$ row is, in accordance with the mixed strategy p, weighted by the probability p_i that he plays that row. Thus, the $j\,^{rmth}$ column—that is, an entry, of the product $p^T A$—is the expected payoff if she plays column j. Weighting each column j of $p^T A$ by q_j and summing thus yields the total expected payoff. Therefore, he wants to maximize $p^T A q^T$ and she wants to minimize $p^T A q^T$.

In Example 19.2, Rock, Scissors, Paper, if he uses the mixed strategy $p = [1/2, 1/2, 0]^T$ (*Rock* half the time and *Scissors* half the time), then she would expect to win $1/2$ from him by playing $q = [1, 0, 0]$ (*Rock* all the time). Player 1 would expect to win $1/6$ by playing $q = [1/3, 2/3, 0]$. This is readily seen by computing $p^T A = [-1/2, 1/2, 0]$. If she uses the mixed strategy $q = [1/4, 1/4, 1/2]$ (*Rock* one-quarter of the time, *Scissors* one-quarter of the time, and *Paper* one-half of the time), then he would expect to win $1/4$ from her by playing $p = [1, 0, 0]$ (*Scissors* all the time). He would expect to win $1/12$ from her by playing $p = [1/3, 2/3, 0]^T$. This is seen by computing $A q^T = [-1/4, 1/4, 0]^T$.

The question at this point is how he can maximize $p^T A q^T$ without controlling q. Similarly, how can she minimize $p^T A q^T$ without controlling p? The problems stated as linear programs before involved only one decision maker, not two! This is where the strictly competitive nature of two-person zero-sum games provides a key to the solution by providing grounds for each player to predict the behavior of his or her opponent. Coupled with the assumption that each player wants to optimize his or her personal payoff, each player in a two-person zero-sum game can conclude that no matter what he or she does, the other player will try to make the opponent's score as bad as possible. Thus, the payoff they will expect from a given strategy is the worst payoff they could possibly get using that strategy. This trick (measuring the quality of each strategy by the payoff in th worst case) takes the other player's choice out of the payoff function! So he measures the quality $f(p)$ of his strategy p as min $p^T A$, where minimum is taken over all entries of the row $p^T A$. It is the same as $\min_q p^T A q^T$, where minimum is taken over all her mixed strategies q.

So his problem becomes

$$\text{maximize } f(p) = \min p^T A$$

and her problem becomes

$$\text{minimize } g(q) = \max A q^T.$$

We emphasize that not just any vectors p and q are allowed, only strategies (i.e., probability distributions). Looking back at the Rock, Scissors, Paper game, we see that

$$f([1/2, 1/2, 0]^T) = -1/2 \text{ and } g([1/4, 1/4, 1/2]) = 1/4.$$

In other words, he can guarantee that the payoff will be at least $-1/2$ by using $p = [1/2, 1/2, 0]^T$. It is this guaranteed minimum, or lower bound, that he wants to maximize. She, by using $q = [1/4, 1/4, 1/2]$, can guarantee that the payoff is at most $1/4$. It is this guaranteed maximum, or upper bound, that she wants to minimize. It is clear that any lower bound on the payoff enforced by him is a lower bound on the upper bound to the payoff that she desires. Likewise, any upper bound on the payoff enforced by her is an upper bound on the lower bound sought by him. That is, $g(q) \geq f(p)$. In the Rock, Scissors, Paper example, if he uses the mixed strategy $p^T = [1/3, 1/3, 1/3]$, then $p^T A = [0, 0, 0]$ so the expected outcome is a draw, no matter what she does [i.e., $f(p) = 0$].

Likewise, if she uses the strategy $q = [1/3, 1/3, 1/3]$, then $Aq^T = [0, 0, 0]^T$ so the expected outcome is a draw, no matter what he does, i.e., $g(q) = 0$. Thus, $f([1/3, 1/3, 1/3]^T) = 0 = g([1/3, 1/3, 1/3])$. Because he has a strategy that guarantees that the payoff is at least zero, and she has a strategy that guarantees that the payoff is at most zero, it is clear that neither player can do better. The strategies $p^T = q = [1/3, 1/3, 1/3]$ are therefore optimal for both players in this game, in the sense that each player's strategy guarantees the best payoff that can be hoped for by this player because of the opponent's ability to bound the payoff.

By this trick, we have resolved one complication—namely, the dependence of each player's objective function on the other player's strategy—but there is still another complication. The new objective functions for the players are not linear. Before we discuss this complication, two other points are worth addressing. First, we can view *any* $m \times n$ matrix $A = [a_{ij}]$ as representing a two-person zero-sum game in which he has m pure strategies, she has n pure strategies, and the payoff when he plays the ith row and she plays the jth column is a_{ij}. The question of whether the matrix represents any real-world game or not is a modeling question that is independent of whether the mathematical structure can be solved as a game. Two-person zero-sum games are often called *matrix games* because of this identification—that is, bijection—between all two-person zero-sum games with real-valued payoff functions and finitely many pure strategies with all finite matrices of real numbers.

The second point is that in the Rock, Scissors, Paper example, $-1 = f(p) < g(q) = 1$ for all *pure* strategies p and q. This is not always the case (cf. Example 19.3).

Historical Remarks

1. A game in which all players know the exact position of the game at all times is called a game of *perfect information*. Chess, tic-tac-toe, and blackjack are games of perfect information, but *Heads and Tails* and Rock, Scissors, Paper are not. The extensive form of a game is a graphical structure called a *tree*, in which positions are represented by nodes and moves are represented by directed edges linking each nonterminal nodes to other nodes. A game is said to be finite if its extensive form, so defined, is finite. Zermelo (1913) proved that *any finite game of perfect information has an equilibrium in pure strategies.* Such an equilibrium can be found by pulling payoffs from the terminal positions to all positions (dynamical programming).

.e point in a matrix is a pair of optimal pure strategies
players in the corresponding game, so Zermelo's theorem
; the normal form of any finite two-person zero-sum game
of perfect information has a saddle point.

2. Nash (1949) proved the existence of an *equilibrium* in mixed
strategies for any game with finitely many players and finitely many
pure strategies for each player. An equilibrium, now widely known
as the Nash equilibrium, is a set of strategies, one for each player,
such that each player's strategy is a best response, given that other
players play their strategies. In other words, no player can improve
his or her payoff by *unilaterally* changing strategies. The class of
games Nash discussed could be characterized as having *finite normal form*. (In theory, we could have a game with infinite extensive
form but nonetheless finite normal form, by defining payoffs for infinite branches of the extensive form whose terminal nodes might
never actually be reached.) Nash proved that every finite normal
form game has an equilibrium in mixed strategies, allowing for pure
strategies as a special case of mixed strategies.

Saddle points in matrix games are in fact equilibria in pure
strategies, and the optimal mixed strategies we seek for matrix games
without saddle points are also Nash equilibria. Nash used a rather
powerful fixed-point theorem from topology for his proof. Once we
see how to solve for optimal strategies using linear programming, we
will be able to deduce his theorem, for the special case of two-person
zero-sum games, from the duality theorem on the four alternatives
(that is, the four possible final forms of all linear programs). ∎

Let us return now to the problem of finding optimal mixed
strategies for the matrix game given by an arbitrary $m \times n$ matrix A. We succeeded earlier in expressing each player's objective as
a function of only a player's own strategy and not the opponent's
strategy. However, as minimum and maximum functions, $f(p)$ and
$g(q)$ are not linear! To cope with this complication we can use another trick. Many tricks are often required in practice to express a
problem as a linear program, or as a differential equation, or as any
other mathematical form. Those students pursuing careers in fields
where they will be asked to apply mathematical methods to solve
problems should strive to internalize the tricks presented in this book
and in the long run become adept at thinking up such tricks. In this
case the trick we will use will be to replace the nonlinear objective
$f(p)$ with an additional variable λ for the row player; we will replace
$g(q)$ with an additional variable μ for the column player.

For our trick to work we must ensure that maximizing λ will be equivalent to maximizing $f(p)$ and that minimizing μ will be equivalent to minimizing $g(q)$. We shall accomplish this by adding constraints relating $f(p)$ to λ and relating $g(q)$ to μ. Recall that $f(p) = \min\{p^T A q^T | q$ is a strategy for her$\}$. The condition that "q is a mixed strategy for her" can be expressed mathematically by the following $m + 1$ linear constraints:

$$q_j \geq 0 \text{ for all } j, \text{ where } q = [q_1, \ldots, q_n]$$

$$\text{and } \sum_{j=1}^{n} q_i = 1.$$

These conditions restrict the vector q to probability distributions on the finite set of her pure strategies. We can use these conditions on q to help make λ represent $f(p)$. Multiplication by q is the discrete version of integration against the probability distribution q. The result is a linear combination of the components of $p^T A$, and because of the constraints on q, it is a *convex combination*. The result thus lies in the *convex hull* of the components of $p^T A$. In common parlance we are taking a *weighted average* of the components of $p^T A$. Intuitively, the idea is that an average, even a weighted average, of n numbers is bounded above by the maximum of those numbers and bounded below by the minimum.

Now we are ready to formulate the problem of finding an optimal strategy for him, or for her, as a linear program. His linear program problem is

$$\lambda \to \max, \ p^T A - [\lambda, \ldots, \lambda] \geq 0, \ \sum_{i=1}^{m} p_i = 1, p \geq 0$$

and her linear program is

$$\mu \to \min, \ A q^T - [\mu, \ldots, \mu]^T \leq 0, \ \sum_{j=1}^{n} q_j = 1, q \geq 0.$$

To put these in a standard tableau, we must manipulate the expressions into the desired form, as in Chapter 2. We will write his problem in a column tableau. The variable we added—namely λ—cannot always be constrained to nonnegative values. There are different ways to handle this. The standard trick is to change variables and set $\lambda = \lambda' - \lambda''$, where λ' and λ'' can now be constrained to be non-negative (see Chapter 2). Similarly, $\mu = \mu' - \mu''$. Let I denote the row of m ones and let J denote the column of n ones.

Now we can write both linear programs in the following standard tableau:

$$
\begin{array}{c}
\quad\ q \quad\ \mu' \quad\ \mu'' \quad 1 \\
\begin{array}{c} -p \\ -\lambda' \\ -\lambda'' \\ 1 \end{array}
\left[
\begin{array}{cccc}
-A & I^T & -I^T & 0 \\
J^T & 0 & 0 & -1 \\
-J^T & 0 & 0 & 1 \\
0 & 1 & -1 & 0
\end{array}
\right]
\begin{array}{l}
= * \geq 0 \\
= * \geq 0 \\
= * \geq 0 \\
= \mu \to \min
\end{array}
\end{array}
$$

$$
\quad \| \qquad \| \qquad \| \qquad \|
$$

$$
* \qquad * \qquad * \qquad \lambda \to \quad \max.
$$

Note that her problem is the row problem, and his problem is the column problem. Their problems are dual to each other. Since both problems have feasible solutions (take, for example, $p = I^T/m, q = J^T/n$), the duality theorem says that $\min(f) = \max(g)$. That is, his value = her value. Thus, the minimax theorem follows from the duality theorem.

When the value $v = \min(f) = \max(g)$ of the game is > 0 (which can be arranged by adding a constant v_0 to all entries of the matrix A to make them positive; do not forget to subtract v_0 back after the modified problem solved to obtain the value of the original problem), we can assume that $\lambda > 0$ in his problem and set $x = p/\lambda$. Then his problem takes the form

$$\text{minimize } 1/\lambda = Ix \text{ subject to } x \geq 0, x^T A \geq J^T.$$

Her problem becomes

$$1/\mu = yI \to \max, Ay^T \leq I, y \geq 0,$$

where $y = q/\mu$. We can write both problems in the following standard tableau:

$$
\begin{array}{c}
\quad\quad y \quad\ 1 \\
\begin{array}{c} -x \\ 1 \end{array}
\left[
\begin{array}{cc}
-A & I^T \\
-J^T & 0
\end{array}
\right]
\begin{array}{l}
= * \geq 0 \\
= -1/\mu \to \min
\end{array}
\end{array}
$$

$$
\quad \| \qquad \|
$$

$$
* \quad -1/\lambda \to \quad \max.
$$

After this problem is solved, we can recover the value $v = 1/Ix$ of the game, his optimal mixed strategy $p = xv$, and her optimal strategy $q = yv$.

This trick saves two rows and two columns in the tableau makes sense if we are going to solve a matrix game by hand. A additional bonus, the tableau is row feasible so we can bypass Pha 1 of simplex method.

Example 20.1. Solve the matrix game

$$\begin{bmatrix} 1 & 3 & 5 \\ 4 & 0 & 1 \\ 1 & 2 & 0 \end{bmatrix}.$$

This matrix game can be solved easily using domination (which allows to drop the last row and the last column) and the graphical method (see §21). Here we show how to apply the simplex method. Before this, we find that the first row gives at least 1 to the row player, and the second column gives at most 3 to the row player. So the value of game is between 1 and 3. There are no saddle points. To find an equilibrium, we have to use mixed strategies, $p = [p_1, p_2, p_3,]^T, q = [q_1, q_2, q_3]$. Since the value of game is positive, we can save two rows and two columns in our standard tableau:

	q_1/μ	q_2/μ	q_3/μ	1	
$-p_1/\lambda$	-1	-3	-5	1	$= v_1$
$-p_2/\lambda$	-4	0	-1	1	$= v_2$
$-p_3/\lambda$	-1	-2	0	1	$= v_3$
1	-1	-1	-1	0	$= -1/\mu \quad \to \min$
	$= u_1$	$= u_2$	$= u_3$	$= -1/\lambda$	$\to \max.$

Pivoting first at the $(v_1, q_1/\mu)$-position and then at the $(v_2, q_2/\mu)$-position, we obtain an optimal tableau. The optimal solutions are $q_1/\mu = q_2/\mu = 1/4, q_3/\mu = 0, -1/\mu = -1/2 = \min$ and $p_1/\lambda = 1/3, p_2/\lambda = 1/6, p_3/\lambda = 0, -1/\lambda = -1/2 = \max$. So the optimal strategies are $[p_1, p_2, p_3,] = [2/3, 1/3, 0], q = [q_1, q_2, q_3] = [1/2, 1/2, 0]$ and the value of game is 2.

Example 20.2. Solve the matrix game with the payoff matrix

$$A = \begin{bmatrix} 0 & 2 & -1 & 0 \\ -2 & 0 & 4 & -3 \\ 1 & -4 & 0 & 2 \\ 0 & 3 & -2 & 0 \end{bmatrix}.$$

$-A^T$ is skew symmetric, so the game is sym-
is 0. Unfortunately, it does not help much in
rategy. We know that an optimal strategy for a
xed strategy q such that $Aq^T \leq 0$ (i.e., $qA \geq 0$).
ıny saddle points or domination. To save rows
ft all matrix entries by 1 to make the value of
game positive (____ely, 1) and write a standard row tableau:

$$
\begin{array}{c}
\begin{array}{ccccc} q_1 & q_2 & q_3 & q_4 & 1 \end{array} \\
\left[
\begin{array}{ccccc}
-1 & -3 & 0 & -1 & 1 \\
1 & -1 & -5 & 2 & 1 \\
-2 & 3 & -1 & -3 & 1 \\
-1 & -4 & 1 & -1 & 1 \\
-1 & -1 & -1 & -1 & 0
\end{array}
\right]
\begin{array}{l}
= v_1 \\
= v_2 \\
= v_3 \\
= v_4 \\
\to \min
\end{array}
\end{array}
$$

By three pivot steps we can switch q_1, q_2, q_3 with v_1, v_2, v_3 and obtain
the optimal tableau:

$$
\begin{array}{c}
\begin{array}{ccccc} v_1 & v_2 & v_3 & q_4 & 1 \end{array} \\
\left[
\begin{array}{ccccc}
* & * & * & * & 4/7 \\
* & * & * & * & 1/7 \\
* & * & * & * & 2/7 \\
* & * & * & * & * \\
* & * & * & * & -1
\end{array}
\right]
\begin{array}{l}
= q_1 \\
= q_2 \\
= q_3 \\
= v_4 \\
\to \min.
\end{array}
\end{array}
$$

So an optimal strategy for the column player is $[4/7,\ 1/7,\ 2/7,\ 0]$,
and an optimal strategy for the row player is $[4/7, 1/7, 2/7, 0]^T$.

Reduction of Any Linear Program to a Matrix Game
Now we want to reduce an arbitrary pair (13.4) of dual linear programs

$$
\begin{array}{c}
\begin{array}{cc} x & 1 \end{array} \\
\begin{array}{c} -y \\ 1 \end{array}
\left[
\begin{array}{cc}
A & b \\
c & d
\end{array}
\right]
\begin{array}{l}
= u \\
= z \to \min
\end{array}
\qquad
\begin{array}{l}
x \geq 0, u \geq 0 \\
y \geq 0, v \geq 0.
\end{array} \\
\begin{array}{cc} = v & = w \end{array} \qquad \to \quad \max
\end{array}
$$

to a matrix game. At first glance it seems impossible, because the
linear programs might not be feasible, while every matrix game has
optimal solutions.

However, we go ahead and consider the following matrix:

$$M = \begin{bmatrix} 0 & -A & -b \\ A^T & 0 & -c^T \\ b^T & c & 0 \end{bmatrix}.$$

Its size is $(m+n+1) \times (m+n+1)$, where $m \times n$ is the size of the matrix A. Since $M^T = -M$, the corresponding matrix game is symmetric. No player has an advantage. So the value of the game is 0. Indeed, suppose that he can win a number $\varepsilon > 0$ (no matter what she does); i.e., $p^T M \geq \varepsilon > 0$ for a column $p \geq 0$. Then she can use the same strategy $q = p^T$ against him and she wins at least the same number ε no matter what he does. When both use this strategy, his payoff $p^T A q^T$ should be both positive and negative, which leads to a contradiction.

Suppose $x = \bar{x}, u = \bar{u}$ is an optimal solution for the row problem and $y = \bar{y}, v = \bar{v}$ is an optimal solution for the column problem. Set e to be 1 plus the sum of all entries in $\bar{x}$ and $\bar{y}$.

Then $\bar{p} = [\bar{y}^T, \bar{x}, 1]^T / e$ is a mixed strategy for him. It gives the following payoff:

$$\bar{p}^T M = [\bar{y}^T, \bar{x}, 1] M / e$$

$$= [\bar{x} A^T + b^T, -\bar{y}^T A + c, -\bar{y}^T b - \bar{x} c^T) / e = [\bar{u}^T, \bar{v}, 0]/e \geq 0$$

(we have used that $z = w$ for optimal solutions). Thus, $\bar{p}$ is an optimal mixed strategy for him. Note that its last entry $1/e$ of the column $\bar{p}$ is not 0.

Conversely, given any optimal strategy $p = \bar{p}$ for him (that is, $\bar{p}$ is an optimal strategy for her) with a nonzero last entry, say $1/e$, we can write $\bar{p}^T = [\bar{y}^T, \bar{x}, 1]/e$ with nonnegative rows $\bar{y}^T, \bar{x}$ of appropriate sizes. Since $\bar{p}$ is optimal, $\bar{p}^T M \geq 0$. So

$$\bar{x} A^T + b^T \geq 0, -\bar{y}^T A + c \geq 0, -\bar{y}^T b - \bar{x} c^T \geq 0.$$

This shows that $x = \bar{x}, u = A\bar{x}^T + b$ is an optimal solution for the preceding row problem and $y = \bar{y}, v = c - \bar{y}^T A$ is an optimal solution for the column programs [recall again that $\min(z) = \max(w)$).]

Thus, there is a 1-1 correspondence between the optimal strategies of the game with nonzero last entries and the optimal solutions of the two linear programs. If there is no optimal strategy with a nonzero last entry, then there are no optimal solutions for the linear programs.

Exercises

1–4. Solve the matrix games:

1.
$$\begin{bmatrix} 1 & 3 & 5 \\ 4 & 0 & -1 \\ -1 & 2 & 0 \end{bmatrix}.$$

2.
$$\begin{bmatrix} 1 & -3 & 4 \\ 1 & 0 & 1 \\ 2 & 2 & 0 \end{bmatrix}.$$

3.
$$\begin{bmatrix} 1 & -3 & 5 & 1 & -3 & 5 \\ 1 & 0 & 1 & 1 & 0 & 1 \\ 1 & 2 & 0 & 1 & 2 & 0 \end{bmatrix}.$$

4.
$$\begin{bmatrix} 1 & -3 & 5 & 1 & -3 & 5 \\ 1 & 0 & 1 & 1 & 0 & 1 \\ 1 & 0 & 1 & 1 & 0 & 1 \\ 1 & 0 & 1 & 1 & 0 & 1 \\ 1 & 2 & 0 & 1 & 2 & 0 \end{bmatrix}.$$

5–10. Solve the matrix games in Exercises 1–6 of §19.

11–13. Given a linear program with all $x_i \geq 0$, write down a matrix game equivalent to that program and its dual.

11.
$$x_1 + 2x_2 + x_3 + x_4 + x_5 + 3x_6 \to \min,$$
$$3x_1 + x_2 + x_3 + 2x_4 + x_5 + x_6 \geq 5,$$
$$3x_1 + x_2 + x_3 + 2x_4 + x_5 + x_6 \geq 4,$$
$$3x_1 + x_2 + x_3 + 2x_4 + x_5 - x_6 = 3.$$

12.
$$x_1 + 2x_2 + 4x_3 + x_4 + x_5 + x_6 \to \max,$$
$$3x_1 + x_2 + x_3 + 2x_4 + x_5 - 2x_6 + x_7 \geq 5,$$
$$3x_1 + x_2 + x_3 + 2x_4 + x_5 + x_6 - x_7 \geq 6,$$
$$3x_1 + x_2 + x_3 + 2x_4 + x_5 - x_6 \leq 7.$$

13.
$$x_1 + 2x_2 + x_3 + x_4 + x_5 + 3x_6 + x_7 + x_8 \to \min,$$
$$3x_1 + x_2 + x_3 + 2x_4 + x_5 + x_6 + x_7 - 3x_8 \geq 1,$$
$$3x_1 + x_2 + x_3 + 2x_4 - x_5 + x_6 + x_7 + 3x_8 \geq 5,$$
$$3x_1 + x_2 + x_3 + 2x_4 + x_5 - x_6 + x_7 + x_8 \geq 1.$$

§21. Other Methods

A matrix game where a player has only one strategy can be solved very easily: The other player chooses the maximal payoff against the only strategy of the opponent. In fact, in this case the matrix game degenerates into a one-person game. An example of this situation is a casino game, the blackjack, since the strategy of the dealer is fixed.

We do not need the simplex method to solve a matrix game where a player has only two pure strategies (i.e., the payoff matrix has two rows or two columns). The graphical method solves such a game easily, as the next two examples show. Sometimes we can reduce the size of the payoff matrix by crossing out redundant strategies, which allows us to solve rather large games easily. Another simple idea that works sometimes is to check for saddle points. Either you find one and the game is solved, or you find an upper and lower bound for the value of game.

Domination and the Graphical Method

Example 21.1. Consider the matrix game with the following matrix:

$$A = \begin{bmatrix} -1 & 6 & 1 & 6 \\ 9 & 2 & 3 & 1 \\ 8 & -1 & 3 & 0 \end{bmatrix}.$$

If we do not have a computer at hand, how could we solve this game ? Note that his second strategy $r2$ gives a greater or equal payoff than his third strategy $r3$ in all cases. In other words,

the second row $\geq$ the third row.

We say that the strategy r2 *dominates* r3.
Given any mixed strategy

$$p = [p_1, p_2, p_3]^T = p_1 r1 + p_2 r2 + p_3 r3$$

for him, he would not loose if he replaces r3 in p by the second strategy. In particular, there is an optimal strategy for him which does not use r3 (the corresponding entry is 0), and we can drop the third row without changing the value of the game. Now we have a smaller game to solve:

$$A = \begin{bmatrix} -1 & 6 & 1 & 6 \\ 9 & 2 & 3 & 1 \end{bmatrix}.$$

Note next that her fourth strategy c4 dominates her second one c2:

the fourth column $\leq$ the second column.

So we drop the second column without changing the value of the game and obtain a smaller matrix game:

$$A = \begin{bmatrix} -1 & 1 & 6 \\ 9 & 3 & 1 \end{bmatrix}.$$

Now we do not see any domination between rows or columns. Since he has only two pure strategies (r1, r2), we can solve this matrix game graphically. Namely, we plot his mixed strategies as the unit interval on the horizontal axis. Her strategies c1, c3, c4 are represented by linear functions on the interval. The minimum of these functions is a convex (upward) function. Its maximal value is $v = $ minimax.

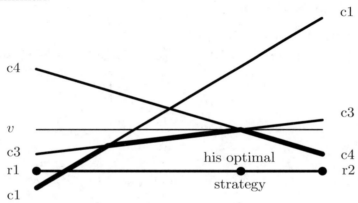

So his optimal strategy is $(1 - p_2)$r1+ p_2r2, where

$$1 \cdot (1 - p_2) + 3p_2 = 6(1 - p_2) + 1 \cdot p_2 = v,$$

hence $7p_2 = 5$. Thus, his optimal strategy is $(2/7)$r1 + $(5/7)$r2, and the value of the game is $v = 17/7$.

Her optimal strategy q_3c3 + $(1 - q_3)$c4 is the combination of c3 and c4 that gives the constant function 17/7 in our figure:

$$1 \cdot q_3 + 6(1 - q_3) = 17/7 = 3q_3 + 1 \cdot (1 - q_3),$$

hence $q_3 = 5/7$. So her optimal strategy is $(5/7)$c3+ $(2/7)$c4.

We give the final answer in terms of the original large matrix:

The value of the game is 17/7.

His optimal strategy is $p = (2/7)r1 + (5/7)r2 = [2/7, 5/7, 0]^T$.

Her optimal strategy is $q = (5/7)c3 + (2/7)c4 = [0, 0, 5/7, 2/7]$.

Remember that he is the row player, she is the column player, and the payoff is given for him.

Example 21.2. Suppose we have to solve the matrix game with the matrix

$$\begin{bmatrix} -1 & 2 & 2 & 0 \\ 2 & 0 & -1 & 2 \\ 1 & 1 & 1 & 1 \\ 0 & 2 & 1.5 & 0 \\ 0 & 0 & 1 & 2 \end{bmatrix}.$$

First we mark by * maximal entries in every column and by ' minimal entries in each row:

$$\begin{bmatrix} -1' & 2^* & 2^* & 0 \\ 2^* & 0 & -1' & 2^* \\ 1' & 1' & 1' & 1' \\ 0' & 2^* & 1.5 & 0' \\ 0' & 0' & 1 & 2^* \end{bmatrix}.$$

Since no position is marked twice, there are no saddle points. We have

$$\max \min = 1 \le \text{the value of game} \le \min \max = 2.$$

We call the row player "he," his pure strategies r1–r5, the column player "she," and her pure strategies c1–c4:

$$\begin{array}{c} & \begin{array}{cccc} \text{c1} & \text{c2} & \text{c3} & \text{c4} \end{array} \\ \begin{array}{c} \text{r1} \\ \text{r2} \\ \text{r3} \\ \text{r4} \\ \text{r5} \end{array} & \begin{bmatrix} -1 & 2 & 2 & 0 \\ 2 & 0 & -1 & 2 \\ 1 & 1 & 1 & 1 \\ 0 & 2 & 1.5 & 0 \\ 0 & 0 & 1 & 2 \end{bmatrix}. \end{array}$$

By domination we cross out c4 (compare it with c1), then r5 (compare it with r3 after c4 is gone), and c2 (compare it with r3 after r5 is gone). We are left with the following 4×2 matrix:

$$\begin{array}{c} & \begin{array}{cc} \text{c1} & \text{c3} \end{array} \\ \begin{array}{c} \text{r1} \\ \text{r2} \\ \text{r3} \\ \text{r4} \end{array} & \begin{bmatrix} -1 & 2 \\ 2 & -1 \\ 1 & 1 \\ 0 & 1.5 \end{bmatrix}. \end{array}$$

Now we make a figure where her mixed strategies are represented by the horizontal line segment connecting c1 and c2 and his strategies are represented by functions on this segment, the corresponding payoffs.

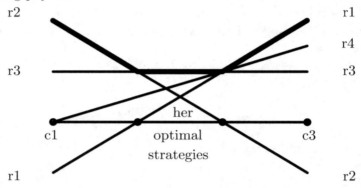

We take maximum over her choices first because she computes her worst-case payoff. Then we minimize this max, which is a piecewise linear function. It is clear that the min is 1 and it is achieved at every point between $[2/3, 1/3]$ (the left point at the figure) and $[1/3, 2/3]$ (the right end). His optimal strategy is r3. It is represented by a horizontal line. Now we write the final answer in terms of the original 5×4 matrix: The value of the game is 1, his optimal strategy is r3 (which is $[0, 0, 1, 0, 0]^T$ as a mixed strategy), and her optimal strategy is $(1/3)$c1 $+(2/3)$c3 $= [1/3, 0, 2/3, 0]$. All her optimal strategies are

$$a\text{c1} + (1 - a)\text{c3 with } 1/3 \le a \le 2/3.$$

To double check the answer, we augment the payoff matrix with her optimal strategies $q = [1/3, 0, 2/3, 0]$, $q' = [2/3, 0, 1/3, 0]$ and verify that we get equilibria:

	c1	c2	c3	c4	q	q'
r1	-1	2	2	0	1	0
r2	2	0	-1	2	2/3	1/3
r3	1	1	1	1	$1^{*\prime}$	$1^{*\prime}$
r4	0	2	1.5	0	2/3	1/3
r5	0	0	1	2	0	0

■

Now we point out some other tricks, scaling, shifts, and symmetry, which could be useful. If we multiply every entry of a matrix A by the same positive number t, then the optimal strategies stay the same and the new value is the old value times t.

If we add the same number t to every entry of a matrix A, then the optimal strategies stay the same and the new value is the old value plus t.

If we permute rows or columns, the value stays the same.

For example, the value of the matrix game with the payoff matrix

$$\begin{bmatrix} 3 & 2 & 0 \\ 2 & 1 & 3 \\ 1 & 4 & 2 \end{bmatrix}$$

is 2. To see this, add -2 to all entries and permute the first two columns to obtain a skew-symmetric matrix.

Fictitious Play (Brown's Method)

The players start with some strategies $p^{(0)}, q^{(0)}$. Then he finds an optimal response $p^{(1)}$ to her strategy $q^{(0)}$, and she finds an optimal response $q^{(1)}$ to his strategy $p^{(0)}$. Then he finds an optimal response $p^{(2)}$ to her strategy $(q^{(0)} + q^{(1)})/2$, and she finds an optimal response $q^{(1)}$ to his strategy $(p^{(0)} + p^{(1)})/2$.

At step t, he finds an optimal response $p^{(t)}$ to her strategy

$$\bar{q}^{(t-1)} = (q^{(0)} + \cdots + q^{(t-1)})/t,$$

and she finds an optimal response $q^{(t)}$ to his strategy

$$\bar{p}^{(t-1)} = (p^{(0)} + \cdots + p^{(t-1)})/t.$$

Note that her cumulative strategy $\bar{q}^{(t)}$ can be computed using only its previous value $q^{(t-1)}$, the current optimal response $q^{(t)}$, and time t:

$$\bar{q}^{(t)} = \bar{q}^{(t-1)}(1 - 1/t) + q^{(t)}/t.$$

The same is true for his cumulative strategies. To make the optimal response unique, we can mix all optimal responses with equal weights.

It turns out (proved by J. Robinson) that for an arbitrary matrix A, this method (devised by Brown) works. Namely, both average payoffs, $\min(\bar{p}^{(t)T}A)$ and $\max(A\bar{q}^{(t)T})$, converge to the value of the game as $t \to \infty$. Every limit point of the sequence $\bar{p}^{(t)}$ is an optimal strategy for him. Every limit point of the sequence $\bar{q}^{(t)}$ is an optimal strategy for her.

Since every linear program can be reduced to a matrix game, Brown's method can be applied to any linear program. There are many other iterative methods for solving of matrix games and linear programs. For example, J. von Neumann, who was a major contributor to game theory, suggested a continuous time version of Brown's method for symmetric games. One of the more recent methods is Karmarkar's method. It is more complicated than Brown's method but gives faster convergence. See the Appendix for more information about interior methods.

Problem 21.3. Apply one iteration of the fictitious play method to the matrix game with the payoff matrix (in dollars, for the row player)

$$A = \begin{bmatrix} 1 & -1 & 0 & 2 & -3 & 1 \\ 1 & 1 & -1 & 0 & 1 & 1 \\ -1 & 0 & 1 & -2 & 2 & 1 \end{bmatrix}.$$

Solution. Since no initial point was given, we are free to choose it. We use the best pure strategies for both players as the initial point. We name the column player Ann and list her pure strategies as c1, c2, c3, c4,c5, c6. We name the row player Bob and list his pure strategies as r1, r2, r3. Now Ann computes the maximal number in each column and mark them all by *: max $= [1, 1, 1, 2, 2, 1]$. So her best pure strategies are c1, c2, c3, c6, and if Ann chooses any mixture of them she pays Bob at most min max $= \$1$.

Next Bob computes the minimal number in each row and marks them all by $'$: min $= [-3, -1, -2]^T$. So his best pure strategy is r2, and if Bob chooses it his payoff is max min $= -\$1$. Since max min and min max are different, there are no saddle points (no position is marked twice). We know that the value of the game v is between -\$1 and \$1.

Since Ann is not sure which of her pure strategies to take as the initial strategy, she chooses the mixed strategy

$$q^{(0)} = [1, 1, 1, 0, 0, 1]/4.$$

Her corresponding worst-case payoff is

$$-u_0 = -\min(Aq^{(0)T}) = -\min([1/4, 1/4, 1/4]^T) = -1/4.$$

So she pays him $1/4$ in the worst case. Now we know that $-1 \leq v \leq 1/4$. He chooses $p^{(0)} = r2 = [0.1, 0]^T$. Now we do one iteration, using the dot notation $A.q$ for Aq^T and $p.A$ for $p^T A$:

$$p^{(0)} = [0, 1, 0]^T$$
$$A.q^{(0)} = [1, 1, 1]^T/4$$
$$\min(p^{(0)}.A) = -1$$
$$p^{(1)} = [1, 1, 1]^T/3$$
$$\bar{p}^{(1)} = [0.1, 0]^T$$
$$A.\bar{q}^{(1)} = [1, -3, 5]^T/8$$
$$\min(\bar{p}^{(1)}.A) = -1/2$$

$$q^{(0)} = [1, 1, 1, 0, 0, 1]/4$$
$$p^{(0)}.A = [0, 1, -1, 0, 1, 1]$$
$$\max(A.q^{(0)}) = 1/4$$
$$q^{(1)} = [0, 0, 1, 0, 0, 0]$$
$$\bar{q}^{(1)} = [1, 1, 5, 0, 0, 1]/8$$
$$\bar{p}^{(1)}.A = [0, 1, -1, 0, 1, 2]/2$$
$$\max(A.\bar{q}^{(1)}) = 5/8.$$

So now Bob has a strategy $\bar{p}^{(1)}$ where he pays at most $1/2$, and Ann has a strategy $q^{(0)}$ where she pays at most $1/4$. Still there is a gap for the value of game, $-1/2 \leq v \leq 1/4$, and it takes time to continue iterations by hand. If you cannot guess the optimal strategies (which is really easy in this case), try to do Exercise 5 by hand. ∎

Exercises

1-4. Solve the matrix games:

1. $\begin{bmatrix} 3 & 2 & 0 \\ 3 & 1 & 3 \\ 4 & 4 & 2 \end{bmatrix}$

2. $\begin{bmatrix} 3 & 2 & 0 \\ 3 & 1 & 0 \\ 0 & 4 & 2 \end{bmatrix}$

3. $\begin{bmatrix} 3 & 2 & 0 \\ 3 & 1 & 4 \\ 3 & 2 & 1 \\ 4 & 4 & 2 \end{bmatrix}$

4. $\begin{bmatrix} 3 & 2 & 0 \\ 3 & 1 & 3 \\ 0 & 0 & 2 \\ 3 & 2 & 2 \\ 3 & 3 & 1 \\ 5 & 4 & 0 \end{bmatrix}$.

5. Solve the matrix games in Problem 21.3.

6–14. Find the value of the matrix game. If you cannot find the exact value, find the best lower and upper bounds for the value you can compute.

6.
$$\begin{bmatrix} 0 & 9 \\ 5 & 2 \end{bmatrix}.$$

7.
$$\begin{bmatrix} 0 & 1 & 2 & 3 \\ -1 & 0 & 4 & 5 \\ -2 & -4 & 0 & 6 \\ -3 & -5 & -6 & 0 \end{bmatrix}.$$

8.

$$\begin{bmatrix} 2 & 1 & 1 & 1 & 1 \\ 2 & 2 & 1 & 2 & 2 \\ 1 & 0 & 0 & 2 & 2 \\ 1 & 1 & 1 & 1 & 2 \\ 1 & 2 & 1 & 2 & 1 \end{bmatrix}.$$

9.

$$\begin{bmatrix} 0 & 0 & -1 & 0 \\ 0 & 9 & -1 & 8 \\ 5 & 2 & 3 & -2 \end{bmatrix}.$$

10.

$$\begin{bmatrix} 1 & 2 & 0 & 3 \\ 5 & 7 & 5 & 6 \\ 4 & 5 & 1 & 6 \\ 7 & 8 & 3 & 9 \\ 0 & 1 & 5 & 1 \end{bmatrix}.$$

11.

$$\begin{bmatrix} 0 & 1 & 1 & 1 & 1 \\ -1 & 0 & 1 & 1 & 2 \\ -1 & -1 & 0 & 1 & 3 \\ -1 & -1 & 0 & 0 & 4 \end{bmatrix}.$$

12.

$$\begin{bmatrix} 0 & 1 & -1 & 1 & 1 \\ -1 & 0 & 1 & 1 & 2 \\ 2 & -1 & 0 & 1 & 3 \\ -1 & 3 & 1 & 1 & 2 \\ -1 & -1 & 0 & 0 & 4 \end{bmatrix}.$$

13.

$$\begin{bmatrix} 0 & 1 & 0 & 1 & 1 & 1 & 1 \\ 0 & 1 & -1 & 0 & 1 & 1 & -2 \\ -1 & -1 & 0 & 2 & 0 & 1 & 3 \\ -1 & 0 & 0 & 1 & 1 & 1 & 2 \\ 0 & 0 & 1 & -1 & 0 & 0 & 4 \end{bmatrix}.$$

14.

$$\begin{bmatrix} 4 & 4 & 1 & 0 & 1 & 1 & 1 & 1 \\ 0 & 1 & -1 & 4 & 2 & 1 & 1 & -2 \\ 0 & 1 & 4 & -1 & 2 & 1 & 1 & -2 \\ -1 & -1 & -1 & 2 & 0 & 4 & 1 & 3 \\ -1 & 0 & 3 & 1 & 4 & 1 & 1 & 2 \\ 0 & 0 & 2 & -1 & 0 & 4 & 0 & 4 \end{bmatrix}.$$

Chapter 8

Linear Approximation

§22. What Is Linear Approximation?

Before we can start solving a real-life problem using mathematics, we often need to collect numerical data. This is not always an easy task. For example, how we can measure the happiness of a person? The height (stature) of a person is considered a less controversial quantity, but precise measurements reveal that it not a constant during the same day even for an adult person. How about the speed of light? Since it is a physical constant, should not all observations by a skilled observer using the same tools and doing best to eliminate the sources of variation give exactly the same answer? Not at all! Even the most careful experiments produce variable results.

Even counting the passengers in an airplane sometimes gives discrepancies that may delay your flight. But airlines would like to know numbers of passengers not only in the present but also on future flights! (Cf. Exercise 4 in §24.)

A typical approach for a scientist is to observe the quantity several times and then take an average. An average is a single value that summarizes or represents a set of values. It is always between the minimal and maximal values. The averages used most often are the mean, median, and midrange. The differences between the observations and the selected average are called *residuals, discrepancies, vertical deviations,* or *error terms.* The term *vertical deviations* comes from a figure in which the average is represented by a horizontal line.

Now we define these averages and explain in what sense they are optimal. We consider m observations (numbers) $a_1, \ldots, a_m$. The arithmetic *mean* is

$$(a_1 + \cdots + a_m)/m.$$

To define other averages, it is convenient to order the observations in increasing order: $a_1 \leq \cdots \leq a_m$. In particular, $a_1 = \min(a_i)$ and $a_m = \max(a_i)$. Then the *midrange* is

$$(a_1 + a_m)/2.$$

When m is odd, the *median* is defined as $a_{(k+1)/2}$. When m is even, a *median* is any number x such that $a_{m/2} \leq x \leq a_{m/2+1}$. In some textbooks, it is defined to be $(a_{m/2} + a_{m/2+1})/2$ to make it unique. We will call the last number the *sample median* or the *central value*.

Example. For numbers 2, 5, 5, 7, the mean is $19/4 = 4.75$, the midrange is $9/2 = 4.5$, and the median is 5.

Remark. To compute the midrange and the interval of medians, it is not necessary to order the given numbers. It takes $m-1$ comparisons to find $\min(a_i)$. Then it takes $m-2$ comparisons to find $\max(a_i)$. So it takes $2m-3$ comparisons and 2 arithmetic operations to compute the midrange. Also, the medians can be found in time linear in m (see the Appendix). On the other hand, it takes at least $\log_2 m! \geq m(\log m - 1)$ comparisons to order m number. ∎

One reason that these three averages are used so often is the fact that they are optimal (the best fit to the given numbers) in the following three senses.

Theorem 22.1.   The mean x_2 is the optimal solution for the following optimization problem:

$$\sum_{i=1}^{m}(a_i - x)^2 \to \min.$$

Proof. We can write the objective function as

$$\sum_{i=1}^{m}(a_i - x)^2 = m(x - x_2)^2 + C$$

with a constant C. Now it is obvious that $\min = C$ at $x = x_2$ and that this optimal solution is unique. ∎

So the mean is *the least squares fit*, or the best l^2-fit.

Theorem 22.2. The midrange x_∞ is the optimal solution for the following optimization problem:

$$\max_i |a_i - x| \to \min.$$

Proof. We order numbers as before. Then the objective function becomes

$$\max(a_1 - x, x - a_1, a_m - x, x - a_m)$$

$$= \begin{cases} a_m - x & \text{if } x \leq x_\infty \\ x - a_1 & \text{otherwise} \end{cases} = |x - x_\infty| + (a_m - a_1)/2.$$

Now it is obvious that

$$\min = (a_m - a_1)/2 \text{ at } x = x_\infty$$

and that this optimal solution is unique. ∎

So the midrange can be called the best l^∞-fit.

Theorem 22.3. A number x_1 is a median if and only if it is an optimal solution for the following optimization problem:

$$\sum_{i=1}^{m} |a_i - x| \to \min.$$

Proof. We order numbers as before. Then the objective function is affine on each of the following two rays and $m - 1$ intervals:

$$x \leq a_1; a_i \leq x \leq a_{i+1} (i = 1, 2, \ldots, m - 1); x \geq a_m.$$

Its slope is $-m; -m + 2i$ (if $a_i \neq a_{i+1}$); m, respectively. Now it is obvious that the set of optimal solutions is the interval $a_{m/2} \leq x_1 \leq a_{m/2+1}$ when m is even and $a_{m/2} \neq a_{m/2+1}$. Otherwise, there is exactly one optimal solution which is

$$x_1 = \begin{cases} a_{m/2} = a_{m/2+1} & \text{when } m \text{ is even and } a_{m/2} = a_{m/2+1} \\ a_{(k+1)/2} & \text{when } m \text{ is odd.} \end{cases}$$

This agrees with the definition of medians. ∎

Using the integer part function $\lfloor \ \rfloor$ the definition of medians x_1 can be written by one formula:

$$a_{\lfloor (k+1)/2 \rfloor} \leq x_1 \leq a_{\lfloor (k+2)/2 \rfloor}.$$

Theorem 22.3 tells that the medians are the best l^1-fits.

Remark. Similarly, we can define best l^p-fit, but the values $p = 2, 1, \infty$ are most common. One reason for this is that those fits are easiest to compute. ∎

To introduce bivariate models, we consider an example. Are you overweight? Underweight? Just right? A possible answer is, "It is my own business and I do not want to discuss it." Some health experts warn against excessive or insufficient weight (body mass). But what is the normal weight?

There are different points of view on this controversial issue. Some say that a person's ideal (or the setpoint) weight is a matter of genotype, the number of fat cells, health, lifestyle, and personal taste and has nothing to with the weight of others.

There are some situations, however, when one's weight relative to the average in a group is important. For example, someone hoping to play on the offensive line of a football team may want to be heavier than other players in the game. A suma wrestler may want to be heavier than his competition. You may want extra weight if you live in cold climates or compete in endurance tests such as the popular "Survival" TV show. On the other hand, a horse racing jockey may want to be the lightest among his competitors. Some runners may also strive to minimize their weight.

A simple-minded and probably politically incorrect way to judge your own weight is to compare it with an average weight of other persons (your peers). Depending on what average and which peers you use, the answer can be different. However, experts suggest a more sophisticated approach: Compare your weight with your own parameters such as your height.

For example, the Web site

$$\text{http://health.yahoo.com/health}$$

advises the following method:

An easy way to determine your own desirable body weight is to use the following formula:

Women: 100 pounds for the first 5 feet of height, 5 pounds for each additional inch; using this formula, the desirable body weight can be calculated.

Men: 106 pounds of body weight for the first 5 feet of height, 6 pounds for each additional inch.

Writing w for the weight and h for the height, this recipe can be written $w = 5h + 75$ for women and $w = 6h + 76$ for men. Ever wonder where those coefficients 5, 75 and 6, 76 come from? Did a great scientist in an ivory tower compute them using basic laws of nature?

Or did somebody conduct a statistical analysis of real-life data? In the latter case, if you use these formulas, you compare implicitly your weight with the weights of other persons. A more explicit way of comparison is as follows: Assume that the ideal weight is a constant b (so the formula is $w = b$), and evaluate this constant as an average over a group of peers. Recall that we have considered three different concepts of an average.

Although you might like to be the heaviest in your group to make the football team, your doctor might be more concerned about the relation between your weight and your height. So we discuss now your weight relative to your height. This leads to more complicated mathematics. Our goal is to show how the coefficients a, b in the model $w = a + bh$ can be determined. We assume that you are a student in a class of 49 students on linear programming and all agreed to disclose their vital statistics. Consider the heights h_i and weights w_i in the class, $i = 1, 2, \ldots, 49$. We can plot the points (h_i, w_i) in the plane and look for a pattern in this scatterplot. It may happen that points cluster around a straight line in which case we want to find the line $w = a + bh$ that fits our data best.

It is not likely that a line $w = a + bh$ passes through all 49 points. In other words, it is unlikely that the system of 49 linear equations $w_i = a + bh_i$ $(i = 1, 2, \ldots, 49)$ for two unknowns a, b has a solution. So we are looking for an approximate "solution." Obviously, we want the best approximation. But how we can compare two different approximations and decide which is better?

In other words, we have 49 objective functions $|w_i - a - bh_i|$ of two variables a, b to minimize and we want to combine them into one objective function so that our optimization problem would make sense.

There are many ways to do this. Three most common ways are

$$e_1^2 + e_2^2 + \cdots + e_{49}^2 \to \min, \tag{22.4}$$

$$|e_1| + |e_2| + \cdots + |e_{49}| \to \min, \tag{22.5}$$

$$\max(|e_i|) \to \min, \tag{22.6}$$

where $e_i = w_i - a - bh_i$ are called *residuals, vertical deviations, or error terms*. Taking (22.4), (22.5), (22.6) as objective functions, we obtain the best l^p-fits for $p = 2, 1, \infty$ respectively.

Remark. Here is the reason why the objective function $\max|e_i|$ is referred to as the $p = \infty$ case:

$$\|e\|_p = (\sum_{i=1}^{m} |e_i|^p)^{1/p} \to \max(|e_i|)$$

as $p \to \infty$. ∎

A more complicated way to relate the weight and height is $w = a+bh+ch^2+d^3$. Now we set $e_i = w_i-(a+bh+ch^2+d^3)$ and have one of three objective function (22.4), (22.5), (22.6) to minimize. Besides height, other parameters can be brought into model. For example, the U.S. Navy uses a circumference method involving measurements of height, neck, and abdomen for men and height, abdomen, neck, and hip for women.

A simple model to relate h and w is used by CDC (the Centers for Disease Control and Prevention, the lead federal agency for protecting the health and safety of people), NIH (the National Institutes of Health, another federal agency), and AHA (the American Heart Association): $w = ch^2$. When the height h is measured in meters and weight w is in kilograms, the ratio w/h^2, measured in kg/m^2, is known as the body mass index (BMI).

By opinion of the CDC, NIH, and AHA, the BMI value is more useful for predicting health risks than weight alone. A BMI between 19 and 25 was considered to be "healthy" by AHA. These numbers were changed in 2001 to 18.5 to 24 (see AHA's Web site http://www.americanheart.org for updates; other Web sites give similar but different numbers that are changing with time). In a recent study (1996), researchers determined that 49% of women in the United States and 59% of men have a BMI of over 25, which would classify more than half of Americans as overweight. Of people between the ages of 50 and 60, 64% of women and 73% of men were identified as overweight.

Here is how the CDC answers the question "How does BMI relate to health among adults?": A healthy BMI for adults is between 18.5 and 24.9. BMI ranges are based on the effect body weight has on disease and death. In 1998 the NHI adapted the same range for "normal weight."

BMI has its limitations (e.g., for body builders), which are pointed out in the NIH guidelines (1998), where it is also suggested to use waist circumference for BMI between 25 and 34.9 kg/m in addition to BMI.

Thus, we plot points (h_i^2, w_i) and try to approximate them by a straight line passing through the origin. Once we choose (22.4), (22.5), or (22.6) as the objective function, with $e_i = w_i - ah_i^2$, we have an optimization problem in one variable c.

Remark. You should not make decisions about your health based solely on college textbooks. ∎

Problem 22.7. Find the best l^p-fit $w = ch^2$ for $p = 1, 2, \infty$ given the following data:

i	1	2	3
Height h in m	1.6	1.5	1.7
Weight w in kg	65	60	70

Compare the optimal values for c with those for the best fits of the form $w/h^2 = c$ with the same p. Compare the minimums with those for the best fits of the form $w = b$ with the same p.

Solution. *Case $p = 1$.* We could convert this problem to a linear program with four variables and then solve it by the simplex method (see §23 below). But we can just consider the nonlinear problem with the objective function

$$f(c) = |65 - 1.6^2 c| + |60 - 1.5^2 c| + |70 - 1.7^2 c|$$

to be minimized and no constraints. The function $f(c)$ is piecewise affine and convex, with the slope changing at

$$c = 70/1.7^2 \approx 24, c = 65/1.6^2 \approx 25, \text{ and } c = 60/1.5^2 \approx 27.$$

The slopes of $f(c)$ are

$$-1.6^2 - 1.5^2 - 1.7^2 < 0 \text{ for } c \le 70/1.7^2,$$
$$-1.6^2 - 1.5^2 + 1.7^2 \approx -2 \text{ for } 70/1.7^2 \le c \le 65/1.6^2,$$
$$1.6^2 - 1.5^2 + 1.7^2 \approx 3 \text{ for } 65/1.6^2 \le c \le 60/1.5^2,$$

and

$$1.6^2 + 1.5^2 + 1.7^2 > 0 \text{ for } c \ge 60/1.5^2.$$

Now it is clear that $f(c)$ is minimized at

$$c = x_1 = 65/1.6^2 \approx 25.39.$$

This value equals the median of the three observed BMIs

$$65/1.6^2, 60/1.5^2, 70/1.7^2.$$

The optimal value is

$$\min = |65 - c_1 1.6^2| + |60 - c_1 1.5^2| + |70 - c_1 1.7^2| = 6.25.$$

To compare this with the best l^1-fit for the model $w = b$, we compute the median $b = x_1 = 65$ and the corresponding optimal values:

$$\min = |65 - x_1| + |60 - x_1| + |70 - x_1| = 10.$$

So the model $w = ch^2$ is better than $w = b$ for our data with l^1-approach.

Case $p = 2$. Our optimization problem can be reduced to solving a linear equation for c (see §23 below). Here we solve the problem using calculus, taking the advantage of the fact that our objective function

$$f(c) = (65 - 1.6^2 c)^2 + (60 - 1.5^2 c)^2 + (70 - 1.7^2 c)^2$$

is differentiable. We set $f'(c) = 0$, which gives, after division by -2,

$$1.6^2(65 - 1.6^2 c) + 1.5^2(60 - 1.5^2 c) + 1.7^2(70 - 1.7^2 c) = 0,$$

hence the optimal solution is

$$c = x_2 = 2518500/99841 \approx 25.225.$$

This x_2 is not the mean of the observed BMIs, which is about 25.426. The optimal value is $\min \approx 18$.

The mean of w_i is 65, and the corresponding minimal value is $5^2 + 0^2 + 5^2 = 50$. So again the model $w = ch^2$ is better than $w = b$.

Case $p = \infty$. We could reduce this problem to a linear program with two variables and then solve it by graphical method or simplex method (see §23). But we can do a graphical method with one variable. The objective function to minimize now is

$$f(c) = \max(|65 - 1.6^2 c|, |60 - 1.5^2 c|, |70 - 1.7^2 c|).$$

This objective function $f(c)$ is piecewise affine and convex. We plot the function $f(c)$:

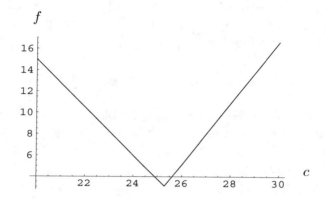

Figure 22.8. The objective function for l^∞-fit

The figure shows the optimal solution $a \approx 25$. Around this point,

$$65 - 1.6^2 c \approx 1, 60 - 1.5^2 c \approx 3.75, 70 - 1.7^2 c \approx -2.25;$$

hence

$$f(c) = \max(60 - 1.5^2 c, -70 + 1.7^2 c).$$

So the exact optimal solution satisfies $60 - 1.5^2 c = -70 + 1.7^2 c$; hence the optimal solution is

$$c_\infty = 6500/257 \approx 25.29.$$

It differs from the midrange of the BMIs, which is about 25.44. The optimal value is ≈ 3.

On the other hand, the midrange of the weights w_i is 65, which gives min $= 5$ for the model $w = b$ with the best l^∞-fit. So again the model $w = ch^2$ is better than $w = b$.

Remark 22.9. The optimal solutions for the values $p = 1, 2, \infty$ are all different in this example. It does not make sense to compare the corresponding optimal values M_p unless we normalize them:

$$M_p \mapsto (M_p/m)^{1/p} \text{ for } p \neq \infty, M_\infty \mapsto M_\infty,$$

where m is the number of observations ($m = 3$ in Problem 22.7). This transformation converts the l^p-norm of the column of residues to an average for the absolute values of the residues. But even after the transformation, the comparison of the averages with different p is difficult to justify. ∎

General Setup for Linear Approximation

In general, we may have data consisting of a column w of m entries and an m by n matrix A. We want to find a column X with n entries such that the column $e = (e_i) = w - AX$ of residuals is smallest in the sense one of the following norms:

$$\|e\|_2 = (\textstyle\sum e_i^2)^{1/2}, \ \|e\|_1 = \textstyle\sum |e_i|, \ \|e\|_\infty = \max(|e_i|). \qquad \blacksquare$$

So we have three optimization problems with m variables and nonlinear objective functions. These kind of problems are typical in statistics. They also arise in other areas of mathematics and in computer science.

Remark 22.10. Geometrically, we want to approximate a vector w by a vector in the column space of A (by definition, the column space consists of all linear combinations of columns of A). In the next section, we will show that the first minimization problem is in fact about solving a system of linear equations, while the other two can be reduced to linear programs.

In the preceding theorems and examples, we considered the case when the matrix A consists of one column (i.e., $n = 1$). In fact, in the theorems A is the column with m ones, and this case was completely solved in the theorems. Finding optimal fits in the case $n = 1$ with general column A are optimization problems with one variable and no constraints. Solving them without a computer may present a challenge, as Problem 22.7 shows.

Examples like $w = a + bh + ch^2 + dh^3$ [so the residuals are $e_i = w_i - (a + bh + ch^2 + dh^3)$] are covered by the general setup for linear approximation with $n = 4$ because the residuals are linear functions of unknowns X. Functions used by scientists for data fitting could be any functions they know (polynomial, rational, trigonometric, exponential, etc.). But to find coefficients we either solve system of linear equations (for least squares approximations) or use linear programming.

Remark 22.11. *Connection with statistics.* Linear approximation, especially with the least squares approach, appears in statistics. In regression analysis, traditionally, the first column of our matrix A consists of m ones. In simple regression analysis, the matrix A consists of $n = 2$ columns. In multiple regression, $n \geq 3$. In time series analysis, the second column of A is an arithmetic progression representing time; typically, this column is $[1, 2, ..., m]^T$.

Deep probability tools are used based on assumptions on normal distributions of residuals, which are considered as random noise masking a systematic pattern. These assumptions justify the least squares approach (l^2-fit), which is the method of choice in statistics. In the case when $n = 1$, this assumption gives preference to the mean over the other averages, but the sample median is also widely used.

Even though the assumptions of multiple regression cannot be tested explicitly, gross violations should be dealt with appropriately. In particular, outliers (i.e., extreme cases) can seriously bias the results by "pulling" or "pushing" the regression line in a particular direction, thereby leading to biased regression coefficients. Often, excluding just a single extreme case can yield a completely different set of results. The general purpose of multiple regression (the term was first used by Pearson, 1908) is to learn more about the relationship among several independent or predictor variables and a dependent or criterion variable. From the statistical point of view, the number m of observations should be much larger than the number n of variables to get reliable estimates for the n unknown coefficients in the column X.

Exercises

1–4 Compute the mean, the median, and the midrange of the following numbers.

1. $2, -7, 0, 2, 1$. **2.** $23, 56, -6, 0, 8, 0, 67$.

3. $2, -7, 0, 2, 1, 0, 0, -1, 8$. **4.** $2, 3, 5, 6, -6, 0, 8, 0, 6, 7$.

5. Construct examples when the mean x_2, the median x_1, and the midrange x_∞ of given numbers satisfy:

(a) $x_2 < x_1 < x_\infty$, (b) $x_1 < x_2 < x_\infty$,
(c) $x_2 < x_\infty < x_1$, (d) $x_\infty < x_2 < x_1$,
(e) $x_\infty < x_1 < x_2$, (f) $x_1 < x_\infty < x_2$.

6. Note that a simple-minded computation of the median of BMIs in Problem 22.7 gives the same result $a_1 = 65/1.6^2 \approx 25.39$. Is it always the case (which would make the computation of the best l^1-fits for all models of the form $w = ax$ much easier) or a coincidence? *Hints*: Try other examples. Since the objective function is piecewise linear, nonconstant, and nonnegative, to find an optimal solution we need to look only at the points where the slope changes—that is, at the given BMIs w_i/h_i^2.

7. Using the data of Problem 22.7, find the best l^p-fit of the form $w = ah$ for $p = 1, 2, \infty$. Compare the results with the best fits in Problem 22.7.

8. Using the data of Problem 22.7, find the best l^p-fit of the form $w = ah^3$ for $p = 1, 2, \infty$. Compare the results with the best fits in Problem 22.7.

Remark. This model was suggested in literature. The quantity w/h^3 was named *normalized body mass* (NBM) and suggested as an alternative to BMI w/h^2. BMI was introduced by the Belgian statistician and anthropologist Lambert-Adolphe-Jacques Quetelet (1796-1874). The same metric units are used for both BMI and NBM. Also, the *ponderal index* $h/w^{1/3} = $ NBM$^{-1/3}$ was suggested in literature. Another suggested alternative is w/h^4.

9. Find the best l^p-fit $w = ah^2$ for $p = 1, 2, \infty$ given the following data:

i	1	2	3	4
Height h in m	1.6	1.5	1.7	1.8
Weight w in kg	65	60	70	80

Compare the results with the best fits (with the same p) for the model $w = b$. Compare the results with the best fits (with the same p) for the model $w = b$.

10. Find the best l^p-fit $w = ah$ for $p = 1, 2, \infty$ given the data in Exercise 9.

11. Find the best l^p-fit $w = ah^3$ for $p = 1, 2, \infty$ given the data in Exercise 9.

12. Here is the list of the first 100 primes p_n: 2, 3, 5, 7, 11, 13, 17, 19, 23, 29, 31, 37, 41, 43, 47, 53, 59, 61, 67, 71, 73, 79, 83, 89, 97, 101, 103, 107, 109, 113, 127, 131, 137, 139, 149, 151, 157, 163, 167, 173, 179, 181, 191, 193, 197, 199, 211, 223, 227, 229, 233, 239, 241, 251, 257, 263, 269, 271, 277, 281, 283, 293, 307, 311, 313, 317, 331, 337, 347, 349, 353, 359, 367, 373, 379, 383, 389, 397, 401, 409, 419, 421, 431, 433, 439, 443, 449, 457, 461, 463, 467, 479, 487, 491, 499, 503, 509, 521, 523, 541. Find the best l^p-fit $p_n = cn$ for $p = 1, 2, \infty$.

12. Using the data in Exercise 12, find the best l^p-fit $p_n = cn \log(n)$ for $p = 1, 2, \infty$.

13. The Fibonacci sequence F_t is defined by the recurrence

$$F_t = F_{t-1} + F_{t-2}, \ F_0 = F_1 = 1.$$

Using the first 50 Fibonacci numbers, compute the best l^p-fit $F_t = 2^{ct}$ for $p = 1, 2, \infty$.

§23. Linear Programming and Linear Approximation

The best l^2-fit is well known as the least squares fit. It is widely used for the following two reasons: It can be justified by some probability assumptions on the residuals and it can be found relatively easily. We remind now how to find it by solving a system of linear equations (which can be considered as a particular case of linear programming).

The Best l^2-Fit

The Euclidean norm $\|e\|_2 = \sum e_i^2$ is the most common way to measure the size of a vector and is used in Euclidean geometry. To find a vector in the column space, we drop a perpendicular from w onto the column space (see Remark 22.10). In other words, we want the vector $w - AX$ to be orthogonal to all columns of A—that is, $A^T(w - AX) = 0$. This gives a system of n linear equations $A^T AX = A^T w$ for n unknowns in the column X. The system always has a solution. Moreover, the best fit AX is the same for all solutions X. In the case when w belongs to the column space, the best fit is w. Otherwise, X is unique. ∎

Example 23.1. In the general setup, let $n = 1$, and let all entries of the column A be ones. Then we want to find a number $X = a$, such that $\sum(w_i - a)^2 \to$ min. In other words, we want to approximate m given numbers w_i by one number a. The equation $A^T AX = A^T w$ becomes $na = \sum w_i$; hence $a = X = \sum w_i/n$ is the arithmetic mean of the given numbers. This agrees with Theorem 22.1. ∎

Problem 23.2. Find the best l^2-fit (up to two decimal points) of the form $w = a + bh$ to the following data:

i	1	2	3	4	5
h	1.5	1.6	1.7	1.7	1.8
w	60	65	70	75	80

Solution. In terms of the general setup, $X = [a, b]^T$,

$$A = \begin{bmatrix} 1 & 1 & 1 & 1 & 1 \\ 1.5 & 1.6 & 1.7 & 1.7 & 1.8 \end{bmatrix}$$

and $w = [60, 65, 70, 75, 80]$. The system of linear equations $A^T AX = A^T w$ takes the form

$$\begin{bmatrix} 5 & 8.3 \\ 8.3 & 13.83 \end{bmatrix} \begin{bmatrix} b \\ a \end{bmatrix} = \begin{bmatrix} 350 \\ 584.5 \end{bmatrix}.$$

Solving this system, we find $a \approx -41.7, b \approx 67.3$.

Problem 23.3. Using the data in Problem 23.2, find the best l^2-fit (up to two decimal points) of the form $w = a + ch^2$.

Solution. In terms of the general setup, $X = [a, c]^T$,

$$A = \begin{bmatrix} 1 & 1 & 1 & 1 & 1 \\ 1.5^2 & 1.6^2 & 1.7^2 & 1.7^2 & 1.8^2 \end{bmatrix}$$

and $w = [60, 65, 70, 75, 80]$. The system of linear equations $A^T A X = A^T w$ takes the form

$$\begin{bmatrix} 5 & 13.83 \\ 13.83 & 38.82 \end{bmatrix} \begin{bmatrix} b \\ a \end{bmatrix} \approx \begin{bmatrix} 350 \\ 979.65 \end{bmatrix}.$$

Solving this system, we find $a \approx 13.37, c \approx 20.47$.

The Best l^1-Fit

We reduce the optimization problem with the objective function $\|e\|_1 \to \min$, where $e = (e_i) = w - Aa$, to a linear program using m additional variables u_i such that $|e_i| \leq u_i$ for all i. We obtain the following linear program with $m + n$ variables a_j, u_i and $2m$ linear constraints:

$$\sum u_i \to \min, \quad -u_i \leq w_i - A_i a \leq u_i \text{ for } i = 1, \ldots, m,$$

where A_i is the ith row of the given matrix A.

Problem 23.4. Using the data from Problem 23.2, find the best l^1-fit of the form $w = bh$.

Solution. Here $A = [1.5, 1.6, 1.7, 1.7, 1.8]^T$. As previously, we can convert this problem to a linear program with six variables and then solve it by the simplex method. But we can just consider the non-linear problem with the objective function $f = \sum |w_i - bh_i| \to \min$ in one variable a and no constraints and solve it graphically or as follows. First we compute $c_i = w_i/h_i$ and obtain

$$c_1 = 40, c_2 = 65/1.6 \approx 40.6, c_3 = 70/1.7 \approx 41.2,$$

$$c_4 = 75/1.7 \approx 44.1, c_5 = 80/1.8 \approx 44.4.$$

If $b \leq c_1 = 40 = \min(c_i)$, then $f = \sum(w_i - ah_i)$ and its slope is

$$-\sum h_i = -h_1 - h_2 - h_3 - h_4 - h_5 = -8.3.$$

On the next interval, $c_1 \leq b \leq c_2$, the slope of f is

$$h_1 - h_2 - h_3 - h_4 - h_5 = -5.3.$$

On the next interval, $c_2 \leq b \leq c_3$, the slope of f is

$$h_1 + h_2 - h_3 - h_4 - h_5 = -2.1.$$

On the next interval, $c_3 \leq b \leq c_4$, the slope of f is

$$h_1 + h_2 + h_3 - h_4 - h_5 = 1.3.$$

For a bigger b, the slope is larger. So it is clear that the slope changes sign at $a = c_3$, hence the minimum is obtained at $b = c_3 = 70/1.7 \approx 41.2$. So our answer is $w = 70h/1.7$.

Remark. The first l^1-approximation problems appeared in connection with data on star movements. Bosovitch (about 1756), Laplace (1789), Gauss (1809), and Fourier (about 1822) proposed methods of solving those problems. In fact, Fourier consider also l^∞-approximation, and he suggested a method of finding feasible solutions for an arbitrary system of linear constraints. Strangely enough, works on l^2-approximation, where finding the best fit reduces to solving a system of linear equations, appeared only in the nineteenth century (Legendre, Gauss).

The best l^∞-fit is also know as the least-absolute-deviation fit and the Chebyshev approximation.

The best l^∞-fit

We reduce the optimization problem with the objective function $\|e\|_\infty \to \min$, where $e = (e_i) = w - Aa$ to a linear program using an additional variable u such that $e_i| \leq u$ for all i. A similar trick was used when we reduced solving matrix games to linear programming. We obtain the following linear program with $n + 1$ variables a_j, u and $2m$ linear constraints:

$$t \to \min, \quad -u \leq w_i - A_i a \leq u \text{ for } i = 1, \ldots, m,$$

where A_i is the i-th row of the given matrix A. ∎

Problem 23.5 Find the best l^∞-fit of the form $w = ah$ using the data from Problem 23.2.

Solution. We can reduce our problem to a linear program with two variables a, u and ten linear constraints and then solve this problem graphically or by the simplex method. Or we can plot the objective function $f = \max(w_i - ah_i)$ (Figure 23.6).

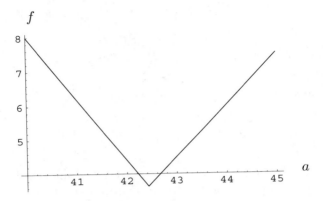

Figure 23.6. $f = \max(w_i - ah_i)$

We see that the optimal solution is $a \approx 42.4$. For a more precise answer, we can plot all five functions $|w_i - ah_i|$ near $a = 42.4$, f being the maximum of these five functions (Figure 23.7):

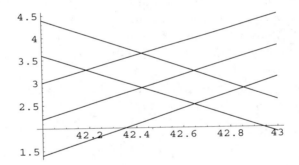

Figure 23.7.

We see that near $a = 42.4$ our objective function f is vskip-5pt

$$\max(1.5a - 60, 80 - 1.8a),$$

vskip-5pt so the optimal solution is vskip-5pt

$$a = 140/3.3 \approx 42.42.$$

Problem 23.8. Find the half-life of a radioactive isotope using the following radiation measurements made every hour:

time in hours t	0	1	2	3	4
radiation level r	100	76	58	45	34.

Solution. We want to find the half-life λ in hours. Without any computations it is clear that $2 < \lambda < 3$. The radiation r in t hours should be $c2^{-t/\lambda}$. The given numbers should form a geometric progression up to measurement errors and round-offs. But the model $r = c2^{-t/\lambda}$ is not linear with respect to λ. Taking log, we can make it linear with respect to $w = \log_2 r, a = \log_2 c$ and $b = 1/\lambda$:

$$w = a - tb. \tag{23.9}$$

There are several options to proceed with. The simplest one is to compute an average of r_t/r_{t-1}. Thus, we obtain the following estimates for 2^b : the mean 1.30964, the midrange 1.30079, the central value 1.31966. They give the following estimates for λ: 2.56958, 2.6358, 2.49896.

Now we will work with the linear model (23.9). The best l^2-fit is obtained by solving the linear system $A^T A X = A^T w$ with

$$w = [\log_2 100, \log_2 76, \log_2 58, \log_2 45, \log_2 34]^T, \tag{23.10}$$

$$X = \begin{bmatrix} a \\ -b \end{bmatrix}, \quad A^T = \begin{bmatrix} 1 & 1 & 1 & 1 & 1 \\ 0 & 1 & 2 & 3 & 4 \end{bmatrix}.$$

So the system is

$$\begin{bmatrix} 5 & 10 \\ 10 & 30 \end{bmatrix} \begin{bmatrix} a \\ -b \end{bmatrix} = \begin{bmatrix} \log_2 674424000 \\ \log_2 31133130272352000 \end{bmatrix} \approx \begin{bmatrix} 29.3291 \\ 54.7893 \end{bmatrix}.$$

Solving this, we obtain $b \approx 0.387$; hence $\lambda \approx 2.58$.

Let us now find the best l^∞-fit for the model (23.9). The corresponding linear program is

$$u \to \min, -u \le w_t - a + bt \le u \text{ for } t = 0, 1, 2, 3, 4,$$

where w_t are the entries of the column w in (23.10). This problem has three unknowns a, b, u and ten linear constraints. The optimal value for b is $b \approx 0.385259$ which gives $\lambda \approx 2.596$.

Let us now find the best l^1-fit for the model (23.9). The corresponding linear program is

$$u_0 + u_1 + u_2 + u_3 + u_4 \to \min,$$

$$-u_t \le w_t - a + bt \le u_t \text{ for } t = 0, 1, 2, 3, 4$$

where w_t are the entries of the column w in (23.10).

This problem has seven unknowns $a, b, u_0, u_1, u_2, u_3, u_4$ and ten linear constraints. The optimal value for b is $b \approx 0.389098$, which gives $\lambda \approx 2.570$.

Unless we know more about how the data were produced, it is hard to decide which method is better. The answer $\lambda = 2.55 \pm 0.05$ looks reasonable.

Problem 23.11. A radiation counter was calibrated using three samples: 1 mg of isotope A, 1 mg of isotope B, 1 mg of isotope B:

time in hours	t	0	1	2	3	4
1 mg of isotope A		4100	510	64	8	1
1 mg of isotope B		1300	320	80	20	5
1 mg of isotope C		160	80	40	20	10.

After this, the counter was used to find contents of the isotopes in different samples. A sample S, consisting of the isotopes A, B, C and nonradioactive components, gave the following readings:

time in hours t	0	1	2	3	4
radiation level r	100	76	58	45	34.

Find the weights (in mg) of the isotopes A, B, C in the sample.

Solution. Let a, b, c be weights of A,B,C in the sample.

Unless we know something about nuclear interactions between isotopes, we assume that the radiation level is

$$100 = 4100a + 1300b + 160c + e_0 \text{ at } t = 0,$$
$$76 = 510a + 320b + 80c + e_1 \text{ at } t = 1,$$
$$58 = 64a + 80b + 40c + e_2 \text{ at } t = 2,$$
$$45 = 8a + 20b + 20c + e_3 \text{ at } t = 3,$$
$$34 = a + 5b + 10c + e_4 \text{ at } t = 4$$

with small errors e_0, e_1, e_2, e_3, e_4.

The best l^p-fits are

$$a \approx 0.12, b \approx -0.65, c \approx 2.74 \text{ for } p = 2,$$
$$a \approx 0.13, b \approx -0.70, c \approx 2.89 \text{ for } p = 1,$$
$$a \approx 0.16, b \approx -0.78, c \approx 2.96 \text{ for } p = \infty.$$

We see that the answer depends on the criterion. But the negative value for b in all three solutions is not acceptable. We could prevent this, imposing the sign restrictions on the unknowns. The problems with l^1- and l^∞-criteria would stay linear programs, but the additional constraints would take the l^2-problem from linear algebra to nonlinear programming.

But since the negative values for b are not so close to 0, this is a strong indication that something is wrong with our solutions, our data, or our assumptions. Speaking about the assumptions, could it be that an unexpected isotope is present in the sample?

Speaking about solutions, if a computer was used, did we introduce the data correctly according to the software specifications? Does the software have a bug that resulted in a wrong solution? Did the computer make a random mistake?

We describe now two different ways to find the best l^2-fit with the software package *Mathematica*. The first way uses the command "FindMinimum." We introduce data (with e0 $= e_0$ etc.):

e0 = -100 + 4100a + 1300b + 160c;
e1 = -76 + 510a + 320b + 80c;
e2 = -58 + 64a + 80b + 40c;
e3 = -45 + 8a + 20b + 20c;
e4 = -34 + a + 5b + 10c;

[the semicolumns are used to suppress printing data back, and they allowe us to enter the data in one block]. Then we type and enter
FindMinimum[e0^ 2 + e1^ 2 + e2^ 2 + e3^ 2 + e4^ 2,
$\{a, 1\}, \{b, 1\}, \{c, 1\}]$
which results in the following response:
$\{158.785, \{a -> 0.124638, b -> -0.653399, c -> 2.74108\}\}$

The first number in the response, 158.785, is the optimal value ($\pm$ 0.005). The numbers 1 in
$\{a, 1\}, \{b, 1\}, \{c, 1\}]$
indicate an initial point in an iterative procedure for searching for optimal solutions.

The second way is to reduce finding the least squares fit to solving a system of linear equations $A^T A X = A^t B$ where $X = [a, b, c]^T$ and

$$[A|B] = \begin{bmatrix} 4100 & 1300 & 160 & |100 \\ 510 & 320 & 80 & |76 \\ 64 & 80 & 160 & |58 \\ 8 & 20 & 20 & |45 \\ 1 & 5 & 10 & |34 \end{bmatrix}.$$

We input A, X, and B into *Mathematica* as follows:
X={a,b,c}; B={100, 76, 58, 45, 34}; A={{4100, 1300, 160}, {510, 320, 80}, {64, 80, 40}, {8, 20, 20}, {1, 5, 10}};

In *Mathematica*, the system $A^T A X = A^t B$ is

Transpose[A].A.X==Transpose[A].B

To solve it we use the command "Solve":

Solve[Transpose[A].A.X==Transpose[A].B,{a,b,c}]

The output is

$$(\{\{a-> \frac{69369996}{556572001}, b-> -\frac{10909908649}{16697160030}, c-> \frac{91536610621}{33394320060}\})$$

To get this in decimals, we input N[%] and get

$\{\{$ a $--> 0.124638$, b $--> -0.653399$, c $--> 2.74108\}\}$.

This agrees with what we found with "FindMinimum." Note that *Mathematica* also has other commands to compute the least squares fits.

For additional checking we solve the problem with *Maple*. Here is how we input matrices into *Maple* :

A := array([[4100, 1300, 160], [510, 320, 80],
[64, 80, 40], [8, 20, 20], [1, 5, 10]]);
B := array([100, 76, 58, 45, 34]);

Here is how to compute the least squares fit:

with(linalg);
leastsqrs(A, B); .

Here is how the output looks:

$$\begin{bmatrix} 69369996 \\ ----- \\ 556572001 \end{bmatrix}, \begin{bmatrix} -10909908649 \\ ------- \\ 16697160030 \end{bmatrix}, \begin{bmatrix} 91536610621 \\ ----- \\ 33394320060 \end{bmatrix}$$

This agrees with the answer by *Mathematica*. So probably something is wrong with our data or our assumptions.

Problem 23.12. Here are data about SAT (the Scholastic Aptitude Test) and GPA (the Grade Point Average) of ten students of Oxbridge University:

SAT1	x	750	720	710	780	700	730	760	770	720	720
SAT2	y	740	730	710	770	720	740	770	760	710	730
GPA	z	3.5	3.4	3.6	3.7	3.2	3.2	3.8	3.7	3.5	3.4.

Find the best l^p-fit of the form $z = ax + by$ for $p = 1, 2, \infty$.

Solution. We set

$$A = \begin{bmatrix} 750 & 720 & 710 & 780 & 700 & 730 & 760 & 770 & 720 & 720 \\ 740 & 730 & 710 & 770 & 720 & 740 & 770 & 760 & 710 & 730 \end{bmatrix}^T ,$$

$$b = [\,3.5, 3.4, 3.6, 3.7, 3.2, 3.2, 3.8, 3.7, 3.5, 3.4\,]^T,$$

$$e = A \begin{bmatrix} x \\ y \end{bmatrix} - b.$$

Our three optimization problems are

$$\|e\|_p \to \min \text{ for } p = 1, 2, \infty.$$

The three optimal solutions are:

$x \approx 0.008, y \approx -0.004$ for $p = 3$;
$x \approx 0.015, y \approx -0.010$ for $p = \infty$;
$x \approx 0.008, y \approx -0.003$ for $p = 1$.

The negative value here is not so impossible as in the previous example, but the conclusion that the high score in Part 2 of the SAT inhibits GPA seems to be questionable. So we can say that our model is not good or our data are not sufficient to come to any conclusion. There are zillions of models and computations that ended up in trash rather than in publications. Some published models and computations also belong to trash.

In general, interpretation of results of computations is a very important part of applied mathematics. Details about collecting and handling data, computing with data, and interpreting results of computations can be found in textbooks on statistics.

Exercises

1. Find the least squares fit $w = a + bh$ for the data in Problem 22.7. Compare the optimal value for this fit with those for the least squares fit of the form $w = ah^2$.

2. Find the least squares solution of the linear system in Example 6.10. Recall that the system has no solutions. In general, the least squares solution $\hat{x}$ of a system $Ax = b$ is not a solution (i.e., $A\hat{x} \neq b$ in general). Rather, $A\hat{x}$ is as close to b as possible.

3. Find the least squares solution of the linear system

$$\begin{array}{cc} a & b \\ \begin{bmatrix} 3 & 4 \\ -1 & 5 \\ 3 & 0 \\ 1 & -7 \end{bmatrix} & \begin{matrix} = 1 \\ = 2 \\ = 4 \\ = 5 \end{matrix} \end{array}$$

4. Find the least squares fit $w = a + bh^2$ for the data in Exercise 9 of §22.

5. Find the best l^1-fit of the form $w = ah^2$ to the data in Problem 23.2.

6. Find the best l^1-fit of the form $w = ah^3$ to the data in Problem 23.2.

7. Find the best l^1-fit of the form $w = ah + b$ to the data in Problem 23.2.

8. Compare the minimal value in Problem 23.2 with those in Exercises 5 and 6 and find which model gives us the best fit for our data.

9. Find the best l^∞-fit of the form $w = ah^2$ to the data in Problem 23.2.

10. Find the best l^∞-fit of the form $w = ah^3$ to the data in Problem 23.2.

11. Find the best l^∞-fit of the form $w = ah + b$ to the data in Problem 23.2.

12. Find the best l^2-fit of the form $p_n = an + b$ to the data in Exercise 12 in §22. Compare the result with the fit $p_n = n \log(n)$, where log means the natural logarithm (which has no parameters).

Remark. It is known that $p_n/(n \log(n)) \to 1$ as $n \to \infty$.

13. Find the best l^2-fit of the form $F_n = a2^{bn}$ to the data in Exercise 13 in §22. Compare the result with the fit $F_n = \alpha^{n+1}/\sqrt{5}$ where $\alpha = (\sqrt{5}+1)/2$, the golden section ratio.

Remark. It is known that $F_t = (\alpha^{t+1} + 1/\alpha^{t+1})/\sqrt{5}$ for all t.

14. Rewrite as a linear program:

$$|e_1| + 2|e_2| + \max[|e_3|, |e_4|] \to \min$$

subject to
$$e_1 = 2x_1 + 3x_2 - 1.$$
$$e_2 = x_1 - 2x_2 - 2.$$
$$e_3 = -x_1 + x_2 + 3.$$
$$e_4 = x_1 - x_2 - 4.$$
$$e_1 \le |e_3| \le 4.$$

Solve the program.

§24. More Examples

In the following examples we discuss what model to use and how to interpret the results of computations, and we pay little attention to how to solve the corresponding optimization problem , which can be reduced to linear programs as explained in §23. The data are not made up. They either come from Web sites or from mathematical research problems.

Example 24.1.   *Time series.* Suppose you are interested in per capita chocolate consumption w (in grams) in Japan in 1995, but you know only what it was in 10 preceding years:

1985	1986	1987	1988	1989	1990	1991	1992	1993	1994
1253	1313	1394	1535	1590	1535	1648	1626	1585	1499

How can you use these numbers (instead of more traditional things like tarot cards, which you would not expect in this book) to predict the number for 1995? Anybody who could find a good way to answer this kind of questions could make a lot of money playing the stock market.

We will try a simple model, $w = ah+b$, where h is year and w is the per capita chocolate consumption in grams. Maybe w depends in fact on the weather, per capita income, chocolate price, and/or health concerns, but suppose that we do not have any data about those factors. We decide to use the best l^p-fit for $p = 1, 2, \infty$ rather than other fits. Here are answers obtained by a computer together with the prediction $w_{11} = 1995a + b$ for 1995:

$a \approx 46.4, b \approx -90802.8, w_{11} \approx 1765.2$ for $p = 1$,

$a \approx 33.7, b \approx -65548.4, w_{11} \approx 1683.2$ for $p = 2$,

$a \approx 27.333, b \approx -52888.2, w_{11} \approx 1641.83$ for $p = \infty$.

It is up to the reader to pass judgment on whether those fits predicted sufficiently well the value 1566 g for 1995. All data were taken from the Web site

http://202.167.121.158/ebooks/jetro/November.html#01.

Figure 24.2 plots $w - 1253$ versus $h - 1984$ for 11 years together with the best l^∞-fit.

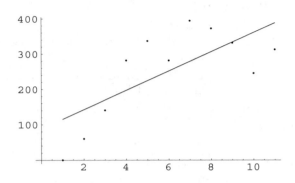

Figure 24.2. Per capita chocolate consumption w
(in grams, reduced by 1253) in Japan in 1985–1995.

Looking at the figure (which is usually a good thing to do before
any computations) reveals that the model $w = ah + b$ does not work
well. It seems that the trend in consumption is not a gradual growth
with a constant rate a, but the rate $w_i - w_{i-1}$ of growth goes down
and becomes negative in 1992. Since the trend does appear to be
monotonous, a better model could be $w_i - w_{i-1} = a'h + b'$—that is,
$w = ah^2 + bh + c$, where $a = a'/2, b = b' + a'/2$.

Since the model $w = ah^2 + bh + c$ is more general than $w = bh + c$,
the fits for 1985–1994 must be better or the same, but the predictions
for 1995 need not be better. Here they are: $w_{11} \approx 1487.2$ for $p = 1$,
$w_{11} \approx 1453.1$ for $p = 2$, $w_{11} \approx 1641.8$ for $p = \infty$.

It is always possible to find a model that predicts exactly the
answer you want. But we would like a simple model which predicts
an answer we do not know. See

http://202.167.121.158/ebooks/jetro/November.html#01

for the data interpretation.

Example 24.3. A statistician is interested in how often the digit 1
occurs in the number $\pi = 3.14159\ldots$. She computed 2 billions digits
and recorded the positions after the decimal point where 1 occurs:

$2, 4, 38, 41, 50, 69, 95, 96, 104, 111, 139, 149, 154, 155, 156, \ldots$.

She asks us to approximate the position w_i as ai with unknown
constant a. Her conjecture is that a must be close to 10.

What can we do with the given 15 numbers using the tools we
learned? First we can compute the three averages of the numbers
w_i/i ($1 \le i \le 15$).

The mean is

$$2793047/270270 \approx 10.3343;$$

the midrange is
$$109/14 \approx 7.78571;$$

and the central value is

$$415/36 \approx 11.5278.$$

Thus, we found the best l^p-fits of the form $w_i/i = a$ for $p = 1, 2, \infty$. The mean is closest to 10.

Now we compute the best l^∞-fits for the model $w_i = ai$. They are $a \approx 11.56$ for $p = 1$, $a = 11.5$ for $p = 2$, $a = 11$ for $p = \infty$.

Should we try now to repeat these computations for 2 billion digits? We will have to wait a few years for more powerful computers.
Remark. How can we try to confirm or refute this conjecture? Notice that if you change or drop finitely many members of an infinite sequence, you do not change the limit (if it exists).

Example 24.3. *One-sided fits.*
Your monthly paycheck \$5K is deposited electronically (available on first day of the month) to your money market account (MMA) where you get 3% interest. The interest is computed monthly on the minimal balance and credited to your account on December 31. You also have a checking account (CA) at the same bank with no interest paid. You use the CA to pay all your bills by mail and never use cash, except cash for post stamps, which you withdraw from your MMA. So you have to go to the bank often to transfer money from from the MMA to the CA to pay for the mortgage, telephone, cable TV, cellular phone, Internet access, car insurance, credit cards, and other bills every month, totaling 10 checks, \$3K. On the top of this, you pay five bills every quarter (in March, June, September, December) totaling \$2K per quarter and tax bills, \$2K and \$5K in April and \$4K in August. Finally, each year, on December 31 you write your last seven checks this year for the rest of your annual income, \$5K minus postage total, and send them to your favorite mutual funds and charities. We assume that the intitial balance at your MMA is sufficiantly large so you need not worry about overdrawing.

Now your bank offers to change your routine: They will pay all your bills from your CA, without any fee. So you do not need to write 150 checks and addresses every year and you save on postage and checks. On the top of this, the bank offers to set up an automatic monthly transfer (on the first day of each month) from your MMA to your CA and one automatic yearly transfer (on January 1) at no cost to you. If you accept, you do not need to go to the bank or post office ever again. You have to decide about the amounts a and b of your monthly and annual transfers.

If you set the transfer amount to be all your salary, \$5K (that is, $a = 5, b = 0$) everything works well accept that you do not get any interest from your MMA account. For small b you get fines and other troubles for an insufficient balance in your CA. So what is the optimal solution? To see this we put all data in a table:

Month t	To pay in \$K	Transfer in \$K	CA balance e_t in \$K
1	3	$a + b$	$a + b - 3$
2	3	a	$2a + b - 6$
3	$3 + 2$	a	$3a + b - 11$
4	$3 + 7$	a	$4a + b - 21$
5	3	a	$5a + b - 24$
6	$3 + 2$	a	$6a + b - 29$
7	3	a	$7a + b - 32$
8	$3 + 4$	a	$8a + b - 39$
9	$3 + 2$	a	$9a + b - 44$
10	3	a	$10a + b - 47$
11	3	a	$11a + b - 50$
12	$3 + 2 + 5$	a	$12a + b - 60$

Table 24.5. Data and variables for Example 24.4

You are ready to state your optimization problem. You have two variables, a and b, subject to the conditions

$$a \geq 0, b \geq 0, 12a + b = 60.$$

In addition, you have 12 constraints $e_i \geq 0$ (see Table 24.5). You can state your objective without the MMA balances since maximizing interest in your MMA is equivalent to minimizing the total balance $e_1 + \cdots + e_{12}$ in your CA, where you get no interest.

We leave solving this particular problem to the reader (see Exercise 3 on the next page). We observe that the problem is similar to finding the best l^1-fit $\sum |e_i| \to$ min with $e_i = ah_i + b - w_i$ but with additional constraints $e_i \geq 0$ and $a, b \geq 0$. These sign restrictions $a, b \geq 0$, $e_i \geq 0$ help to rewrite the optimization problem as a liner program in canonical form—namely, $\sum_i ah_i + b - w_i \to$ min, $ah_i + b - w_i \geq 0$ for all i.

Exercises

1. Here is the U.S. fresh strawberry production w (in millions of pounds) for 9 years. Predict the production in 1993 using the model $w = ah + b$ and the best l^p-fits for $p = 1, 2, \infty$. The date are from

http://www.nalusda.gov/pgdic/Strawberry/ers/ers.htm.

Compare your predictions with actual production 987.6. *Hint*: Some computer software does not like big numbers. Replace the year h by $h - 1988$ and the production w by $w - x_2$, where x_2 is its mean over 9 years.

1984	1985	1986	1987	1988	1989	1990	1991	1992
748.2	754.1	734.8	780.4	855.5	861.6	864.2	971.5	980.3

2. A student is interested in the number w of integer points [x,y] in the disc $x^2 + y^2 \leq r^2$ of radius r. He computed w for some r:

r	1	2	3	4	5	6	7	8	9
w	5	13	29	45	81	113	149	197	253

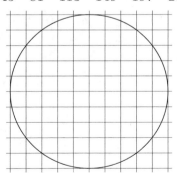

Figure 24.6. 81 integer points
in the disc of radius 5

The student wants to approximate w by a simple formula $w = ar + b$ with constants a, b. But you feel that the area of the disc, πr^2 would be a better approximation, and hence the best l^2-fit of the form $w = ar^2$ should work even better for the numbers above.

Compute the best l^2-fit (the least squares fit) for both models, $w = ar + b$ and $w = ar^2$ and find which is better. Also compare both optimal values with

$$\sum_{i=1}^{9}(w_i - \pi i^2)^2.$$

3. Solve the optimization problem in Example 24.3. *Hint*: Use the graphical method.

4. Some passengers with confirmed reservations were denied boarding ("bumped") from their flights because the flights were oversold. The airlines oversell because they cannot be sure how many passengers will show up. Here are the year t and the numbers x and y of boarded (the second row, in millions) and bumped (the third row, in thousands) passengers for the domestic nonstop scheduled flights by the 10 largest U.S. air carriers. Data are taken from

http://www.bts.gov/btsprod/nts/Ch1_web/1-55.htm:

1990	1991	1992	1993	1994	1995	1996	1997	1998	1999
421	429	445	449	457	460	481	503	514	523
628	646	764	683	824	843	957	1072	1126	1070

Find the best l_p-fits with $p = 1, 2, \infty$ of the form

$$y = at + bx + c$$

for the data for 1990–1998 and use these fits to predict the number 1070 in 1999.

Appendix

Guide to MP

A1. Mathematical Programming

We mentioned in §1, that a mathematical program is to minimize (or maximize) a function f of n variables $x_1, \ldots, x_n$ over a set S:

$$f(x) \to \min, \ x \in S, \tag{A1.1}.$$

The decision variables $x_1, \ldots, x_n$ are also called control, planning, or strategic variables. We write them in a column x, so the feasible region S is a subset of R^n, all columns of n real entries. The set S is often given by a system of constraints of the following three types: $g(x) = 0, g(x) \geq 0$, or $g(x) \leq 0$, where $g(x)$ is a function on R^n. Note that the same set S can be given by different systems of constraints. Theoretically, it can always be given by one constraint of the type $g(x) = 0$ as well as one constraint of the type $g(x) \leq 0$. Using a penalty method, all constraints can be eliminated.

The real-valued function f is called the objective function or *minimand*. It always can be arranged to be linear. Namely, given program (A1.1), we can introduce a new variable z and consider the equivalent problem

$$z \to \min, \ x \in S, \ f(x) - z \leq 0 \tag{A1.2}$$

with $n + 1$ variables and the objective function being a linear form. The optimal values for (A1.1) and (A1.2) are the same.

In this generality, we cannot develop much of theory, but we can give a couple of definitions. We endow R^n with the Euclidean norm

$$|x| = \|x\|_2 = (x^T x)^{1/2}.$$

Definition A1.3. A point a in S is called a *local* (or *relative*) *optimum point*, or *local optimizer*, or a *locally optimal solution* if there is $\varepsilon > 0$ such that a is an optimal solution after we add the constraint $|x - a| \leq \varepsilon$.

Remark. If we used the $\| \; \|_1$- or $\| \; \|_\infty$-norm instead of the $\| \; \|_2$-norm, then the constraint $|x - a| \leq \varepsilon$ would be equivalent to a system of linear constraints. An advantage of the $\| \; \|_2$-norm is that it is smooth (all partial derivatives of every order exist). The choice of norm in Definition A1.3 does not matter because

$$\|e\|_2 / n^{1/2} \leq \|e\|_\infty \leq \|e\|_1 \leq \|e\|_2 n^{1/2}$$

for all vectors e in R^n. ∎

For example, any optimal solution (which can be called a global optimum point) is also a local optimum point. In linear programming, the converse is true, see A3 below.

Definition A1.4. A *feasible direction* at a point a in S is a nonzero vector b with the following property: there is $\varepsilon_0 > 0$ such that $a + b\varepsilon$ belongs to S for all ε in the interval $0 \leq \varepsilon \leq \varepsilon_0$. ∎

Note that this concept is independent of the objective function. When $S = R^n$, every direction at every point is feasible.

Many methods in mathematical programming involve choosing a feasible direction at a point such that small movement away from the point in this direction improves the objective function. Then there is a line search to decide how far we should move until we get the next point in our iterative procedure. The choice of a feasible direction at a point x may depend on the values of the objective function and the functions in the constraints at x as well as on the values of derivatives. Some iterative methods also use the values at previous points, in which case the method starts with more than one initial point.

In parallel and genetic programming, "time" could be a more complicated then the sequence $0, 1, \ldots$. In some cases, an arbitrary or random choice is involved in finding the next point; otherwise we have a deterministic algorithm.

Notice that iterative procedures in mathematical programming usually do not give exact optimal solutions. Moreover, the final answer is not always even feasible. One reason is that the exact answer could involve irrational numbers even if all data are rational.

Linear programs are exceptional in this respect, because theoretically the simplex method is guaranteed to give an exact optimal solution in finitely many pivot steps. But suppose the answer involves integers with 10,000,000,000 digits. Can we really compute all these digits? If so, can we use such an answer and is it worth our time to compute the exact answer? Does it make sense to compute many digits of the answer if the data are not precise?

The answer with 10^{10} digits may look somewhat far-fetched, because an optimal solution for a LP is a unique solution for a system of linear equations (once you know the basis), and such systems have been solved for many years. However, in practice they are usually solved approximately.

Some iterative methods produce an infinite sequence

$$x^{(t)} = [x_1^{(t)}, \dots, x_n^{(t)}], \quad t = 0, 1, \dots$$

in R^n, and others are augmented by various stopping rules. The simplest stopping rules are as follows: Stop after a certain number of iterations or a certain amount of time. A natural termination occurs when $x^{(t)} = x_s$ for all $s > t$. If it is easy to recognize an optimal solution (or a local optimizer), a good stopping rule is to stop when $x^{(t)}$ is optimal (respectively, locally optimal). Other rules involve a a small positive number ε, called *tolerance* [e.g., stop when $f(x^{(t)})$ is within ε from the optimal value or stop when $x^{(t)}$ is within of ε from an optimal solution (assuming that the closeness can be easily judged)]. Some methods stop when $|x^{(t+1)} - x^{(t)}| \le \varepsilon$ or when $|f(x^{(t+1)}) - f(x^{(t)})| \le \varepsilon$.

An iterative method usually works on a class of mathematical programs. For example, methods using derivatives assume their existence. If we know nothing about the function $f(x)$ besides its values $f(x^{(t)})$ and the sequence $x^{(t)}$ does not exhaust S, we do not know how close we come to the optimal value. To guarantee the convergence of a method or to get estimates of how close it comes to the optimum, further restrictions on the class of mathematical programs could be needed. A typical condition on $f(x)$ is the *Lipschitz condition* with a *Lipschitz constant K*:

$$|f(x) - f(y)| \le K|x - y| \text{ for all } x, y. \tag{A1.5}$$

This condition implies that $f(x)$ is continuous. When the gradient

$$f'(x) = \nabla f(x) = [\partial f(x)/\partial x_1, \dots, \partial f(x)/\partial x_n] \tag{A1.6}$$

exists and is continuous for all x, (A1.5) is equivalent to the following: $|f'(x)| \le K$ for all x.

Another useful condition is the convexity of f which guarantees that the function is continuous and that a local optimizer is optimal. Any affine function $f(x) = cx + d$ is convex, smooth, and admits Lipschitz constant $K = |f'(x)| = |c|$.

A desirable property of an iteration method for solving (A1.1) is the *descent property*

$$f(x^{(t+1)}) < f(x^{(t)}) \text{ unless } x^{(t+1)} = x^{(t)}$$

or the *strong descent property* for $f(x)$: $f((1-\alpha)x^{(t)} + \alpha x^{(t+1)})$ is a decreasing function of α when $0 < \alpha \le 1$ unless $x^{(t+1)} = x^{(t)}$.

The simplex method has the strong descent property.

Now we discuss conditions for a point x to be an optimal solution in the case when the objective function $f(x)$ and the constraint functions are differentiable. First we consider the unconstrained optimization (i.e., $S = R^n$.) Then a point $x = x^* = [x_1^*, \ldots, x_n^*]^T$ is locally optimal only if it is *critical* or *stationary* [i.e., the gradient vanishes: $f'(x^*) = 0$.]

Consider now the case when S is given by a system of constraints $g_i(x) = 0$, $i = 1, \ldots, k$ with differentiable functions g_i. In this case it is well-known that if x^* is a local optimizer, then the gradient vectors

$$\nabla f(x^*), \nabla g_1(x^*), \ldots, \nabla g_k(x^*) \text{ are linearly dependent.} \quad (A1.7)$$

In the *regular* case, when $\nabla g_1(x^*), \ldots, \nabla g_k(x^*)$ are linearly independent, the condition (A1.7) can be written as

$$\nabla f(x^*) = \sum_{i=1}^{k} \lambda_i \nabla g_i(x^*) \quad (A1.8)$$

where the coefficients λ_i are known as *Lagrange multipliers*.

Suppose now that S is given by a system of constraints

$$g_i(x) = 0 \text{ for } i = 1, \ldots, k \text{ and } g_i(x) \le 0 \text{ for } i = k+1, \ldots, l \quad (A1.9)$$

with differentiable functions g_i. A feasible solution x is called *regular* if the gradients $\nabla g_i(x)$ with $g_i(x) = 0$ (i.e., *active* gradients) are linearly independent. Here are the Karush-Kuhn-Tucker (KKT) conditions for a regular feasible solution x^* to be locally minimal:

$$-\nabla f(x^*) = \sum_{i=1}^{l} \lambda_i \nabla g_i(x), \ \lambda_i g_i(z^*) = 0 \ \forall i, \ \lambda_i \ge 0 \text{ for } i > k.$$

The condition $\lambda_i g_i(z^*) = 0 \ \forall i$ means that only the active gradients are involved. The KKT conditions are a system of linear constraints for λ_i, which becomes a system of linear equations (A1.8) in the case $l = k$. The fact that the KKT conditions are necessary follows easily from the duality in linear programming combined with an implicit function theorem. In the the case when the constraints are linear, the condition of regularity can be dropped. (This is also true in the case $l \le 1$.) Moreover, in this case the KKT conditions are sufficient for x^* to be optimal provided that $f(x)$ is convex, cf. A3 below.

A2. Univariate Programming

This is mathematical programming with a single decision variable ($n = 1$ in notations of A1). The feasible region most often is an interval, a ray, or the whole line R (unconstrained optimization). By various methods, mathematical programs with several variables (multivariate programs) are reduced to or use univariate programs. For instance, we can try to approximate the feasible region S in R^n by a "space-filling" curve. Or we can choose a feasible direction at a point and then do a line search along this direction to find the next point.

Direct Search Methods

These methods are used when derivatives are expensive to calculate or do not exist. However, there is no sharp distinction from methods using derivatives, because derivatives can be approximated using values of $f(x)$.

A simple-minded way to minimize a function $f(x)$ on an interval $a \leq x \leq b$ is to choose a large natural number N, compute

$$f(a + i(b - a)/N) \text{ for } i = 0, 1, \ldots, N,$$

and find the minimal value. When $f(x)$ admits a Lipschitz constant K, this value differs from the optimal value not more than by $K/(2N)$. More refined methods, *grid-based methods*, involve subdivision some of the intervals $a + i(b-a)/N \leq x \leq a + (i+1)(b-a)/N$ into smaller intervals in search for better solutions.

As an alternative, we can compute $f(x)$ at N points chosen at random in the interval. When $f(x)$ satisfies a Lipschitz condition, the minimal value of $f(x)$ at N points converges to the optimal value with probability 1 as $N \to \infty$.

We call a function $f(x)$ on an interval $a \leq x \leq b$ *unimodal* if it is continuous and has exactly one locally optimal solution x^*.

Fibonacci search for the minimum of a unimodal function uses the Fibonacci sequence F_t, see Exercise 13 in §22. We fix a number N and we want to restrict the unknown optimal solution x^* to the smallest possible interval by evaluating the objective function at N judicially chosen points. When $N = 0$ or 1, we cannot restrict x^* to a smaller interval. When $N = 2$ we can restrict x^* to the interval $a \leq x \leq (a + b)/2 + \varepsilon$ or $(a + b)/2 - \varepsilon \leq x \leq b$ by comparing $f((a+b)/2+\varepsilon)$ and $f((a+b)/2-\varepsilon)$ where $0 < \varepsilon < (a-b)/2$. [When $f'((a + b)/2)$ exists, we can do bisection instead, saving ε.]

For any $N > 2$, we compare $f(a + (1 - F_{N-1}/F_N)(b - a))$ and $f(a + (b - a)F_{N-1}/F_N)$ and reduce the interval $a \leq x \leq b$ to either the subinterval $a \leq x \leq a + (b - a)F_{N-1}/F_N$ or the subinterval $a + (1 - F_{N-1}/F_N)(b - a) \leq x \leq b$. The length of the subinterval is $(b - a)F_{N-1}/F_N$ in both cases. After $N - 2$ steps, we restrict x^* to a subinterval of length $2/F_n$. The last two evaluations reduce the length to $(1 + \varepsilon)/F_n$ with arbitrary small positive ε.

When $N \to \infty$, the Fibonacci search becomes the *search by golden section*, when we compare $f(a\alpha + (1 - \alpha)b)$ and $f((1 - \alpha)a + \alpha b)$ and reduce the interval $a \leq x \leq b$ to either $a \leq x \leq (1 - \alpha)a + \alpha b)$ or $a\alpha + (1 - \alpha)b \leq x \leq b$, where $\alpha = 2/(1 + \sqrt{5}) \approx 0.618$. The number $1/\alpha$ is known as *golden section ratio*. After k evaluations, the uncertainty interval for x^* is reduced to a subinterval of length $(b - a)\alpha^{k-1}$.

For a unimodal function, the search by golden section produces a sequence $x_t = x^{(t)}$ convergent to the optimal solution x^*. Applied to an arbitrary continuous function, the method produces a sequence x_t convergent to a point x^* such that

Either $x^* = a$ or there are infinitely many t such that $x_t < x^*$ and $f(x_t) \geq f(x^*)$;

Either $x^* = b$ or there are infinitely many t such that $x_t > x^*$ and $f(x_t) \geq f(x^*)$.

Such a point x^* is critical if $f'(x^*)$ exists and $x^* \neq a, b$.

Splitting the feasible interval into smaller intervals allows us to find more such points.

Better methods can be devised if we use additional information about the objective function. For example, it helps if the function is differentiable, in which case a is locally optimal if $f'(a) > 0$, b is locally optimal if $f'(b) < 0$, and a point c inside $(a < c < b)$ is locally optimal only if it is critical [i.e., $f'(c) = 0$.] A critical point c is locally minimal if $f(x)$ is convex or if $F''(c)$ exists and is positive.

Bisection Method

This is a method for finding a local minimizer using the derivative $f'(x)$. Assume that $f'(x)$ exists in the interval $a \leq x \leq b$ and that $f'(a) \leq 0 \leq f'(b)$ (otherwize, we have a local optimizer). We set $a_0 = a$, $b_0 = b$, $x_0 = (a_0 + b_0)/2$.

Given a_t, b_t, and $x_t = (a_t + b_t)/2$, we define a_{t+1}, b_{t+1}, and $x_{t+1} = (a_{t+1} + b_{t+1})/2$ as follows. If $f'(x_t) > 0$, then $a_{t+1} = a_t, b_{t+1} = x_t$. Otherwise, $a_{t+1} = x_t, b_{t+1} = b_t$.

Note that $f'(a_t) \leq 0 \leq f'(b_t)$, $a_{t+1} \leq a_t < b_{t+1} \leq b_t$, and $b_t - a_t = (b - a)/2^t$ for $t = 0, 1, \ldots$. It follows that the sequence $x_t = (a_t + b_t)/2$ converges, say $x_t \to x^*$. If $f'(x)$ is continuous at $x = x^*$ then $f'(x^*) = 0$, so x^* is a good candidate for a local optimizer. It is a local optimizer if, for example, $f'(x)(x - x^*) \geq 0$ for x close to x^*.

To look for more than one local optimizer, we can start with splitting the initial interval $a \leq x \leq b$ into many smaller intervals and apply the bisection method to each small interval.

If the derivatives do not exist or are hard to compute, *dichotomous search* can be used, where we replace computation of $f'(c)$ by evaluating $f(c - \varepsilon)$ and $f(c + \varepsilon)$ for a small positive ε. However, the search by golden section is more efficient because it reduces the interval of uncertainty $(1 + \sqrt{5})^2/4 \approx 2.61803$ times using two evaluations of $f(x)$.

In the case of unconstrained program $(S = R)$ we can decide that we do not care about solutions x with $|x| > M$ for some large M. Then we have a program where S is a bounded interval, so we can use the bisection method or other methods devised for optimization over intervals. We also can cover R by a sequence of finite intervals. However, there are methods that are simpler in the unconstrained case.

Gradient Methods

Let $S = R$, and let $f'(x)$ exist and be continuous for all x. We start with an initial point x_0. Given any point x_t, the next point in the gradient method is $x_{t+1} = x_t - f'(x_t)$. A modified gradient method is given by

$$x_{t+1} = x_t - \alpha f'(x_t) \tag{A2.1}$$

with some $\alpha > 0$.

The method terminates naturally at a critical point. If $f'(x)$ admits the Lipshitz constant $K < 2/\alpha$, then

$$f(x_t) - f(x_{t+1}) \geq \alpha(1 - K\alpha/2)f'(x_t)^2 > 0$$

so we have the descent property for $f(x)$.

Assume now that $f'(x)$ admits the Lipshitz constant $K \leq 1/\alpha$. Then we have the strong descent property for $f(x)$ (see A1). Moreover, we have the strong descent property for $|f'(x)|$.

The last property implies that the sequence x_t is strictly monotone for all t or until we hit s such that $f'(x_s) = 0$. When $f'(x_0) < 0$ [respectively, $f'(x_0) > 0$], then either the sequence converges to the first critical point x^* on the right (respectively, on the left) of x_0 or there are no critical points on the right (respectively, left) and $x_t \to \infty$ (respectively, $x_t \to -\infty$).

Under the additional condition that

$$f'(x) - f'(y) \geq (x - y)K_2 \text{ for all } x, y$$

for some $K_2 > 0$ [e.g., $f''(x) \geq K_2$ for all x] the function $f(x)$ is convex (so every critical point is optimal) and we can estimate the rate of convergence:

$$f'(x_{t+1}) \leq f'(x_t)(K - K_2)/K \leq f'(x_0)(1 - K_2/K)^{t+1}$$

hence

$$|x_{t+1} - x_t| \leq \alpha|f'(x_0)|(1 - K_2/K)^t$$

and

$$|x_t - x^*| \leq \alpha|f'(x_0)|(1 - K_2/K)^t K/K_2.$$

If no Lipshitz constant for $f'(x)$ exists or available, we can use the *damped gradient method*

$$x_{t+1} = x_t - \alpha_t f'(x_t) \tag{A2.2}$$

with a *damping sequence* $\alpha_1, \alpha_2, \ldots$ such that $\alpha_t > 0$ for all t, $\alpha_t \to 0$ as $t \to \infty$, and $\sum \alpha_t = \infty$. (See [V2] for a discussion of damping sequences.) We get convergence of x_t to a critical point, ∞, or $-\infty$, provided that there are only finitely many critical points.

To find more than one critical point, we can try different initial points. For example, if the critical point x^* we found is not a local optimizer, we can start again with a point y_0 close to x^* such that $f(y_0) < f(x^*)$. More sophisticated approaches modify $f(x)$ in a way to exclude the critical points that are already found.

In the case when S is a finite interval $a \leq x \leq b$, any iterative method can be adjusted by replacing infeasible x_{t+1} by a or b whichever is closer.

To avoid calculation of $f'(x_t)$ in the gradient method, we can replace it by, say, $(f(x_t) - f(x_{t-1})/(x_t - x_{t-1})$ and start with two distinct initial points x_0, x_1.

Newton Methods

More sophisticated methods than the gradient methods are Newton methods. They are designed to work in regions where $f(x)$ is convex. The standard version assumes that

$$f''(x) \text{ exists and is positive for all } x. \qquad (A2.3)$$

However, there are versions that do not use the first and/or second derivatives. We start with a version which uses $g(x) = f'(x)$ but not $f''(x)$.

Assume that the derivative $g(x) = f'(x)$ exists and is continuous. We want to find a zero of this function $g(x)$. We start with two initial points $x_0 \neq x_1$. Then we draw the straight line through the points $(x_0, g(x_0)), (x_1, g(x_1))$ and intersect this line, which approximates the function $g(x)$ and whose slope is

$$s = (g(x_1) - g(x_0))/(x_1 - x_0),$$

with the horizontal axis. So we want $s = (0 - g(x_1)/(x_2 - x_1)$, hence $x_2 = x_1 - g(x_1)/s$ for our next point x_2 when $s \neq 0$. When $s = 0$, we take $x_2 = (x_0 + x_1)/2$. Then we repeat the procedure with x_1, x_2 instead of x_0, x_1, and so on. We obtain a sequence x_t that (we hope) converges to a zero of $g(x)$.

For the rest of this section, we assume the condition (A2.3). So the objective function $f(x)$ is convex. In its pure form, the Newton method for finding a minimizer of $f(x)$ [i.e., a zero of $g(x) = f'(x)$] starts with one initial point x_0 and defines the sequence x_t by

$$x_{t+1} = x_t - g(x_t)/g'(x_t). \qquad (A2.4)$$

To guarantee the convergence of the Newton method (A2.4) even for a real analytic function, we need additional conditions.

In the case when $g'(x) = f''(x)$ admits a Lipshitz constant K_3 and $|g(x)/g'(x)^2| \leq K_1$ for all x, we have

$$|g(x_{t+1})| \leq (g(x)/g'(x))^2 K_3/2 \leq |g(x)| K_1 K_3/2;$$

hence (A2.4) converges if $K_1 K_3/2 = q < 1$ for all x. Namely, under this condition, $g(x_{t+1}) \leq g(x_0) q^{n+1}$. Also the method converges very fast [with the ascent property for $|g(x)|$] if $g'(x) \geq K_2 > 0$ for all x and we start with x_0 such that $|g(x_0)| K_3/(2K_2^2) = q_0 < 1$.

Namely, we have the descent property for $|g(x)|$ and $q_0|g(x_{t+1})| \leq (q_0|g(x_t)|)^2$; hence

$$q_0|g(x_t)| \leq q_0^{2^t}$$

for all t.

A modified Newton method is

$$x_{t+1} = x_t - \alpha g(x_t)/g'(x_t) \qquad \text{(A2.5)}$$

with some $\alpha > 0$. It converges and has the descent property for $f(x)$ when $\alpha g'(y)/g'(x) \leq 1$ for all x, y (see the preceding discussion of the gradient method). There is a whole spectrum of quasi-Newton methods which generalize and fill up the gap between the gradient methods and the Newton methods.

For example, here is a damped Newton method:

$$x_{t+1} = x_t - \alpha_t g(x_t)/g'(x_t)$$

where α_t is a sequence such that $\alpha_t \to 0$ (e.g., $\alpha_t = 1/t$).

The Newton method is based on approximation of $f(x)$ by a quadric $a_0 + a_1x + a_2x^2$ using information at a single point or at two distinct points. It converges in one step for a strictly convex quadric ($a_2 > 0$). More complicated methods use approximation by a cubic

$$a_0 + a_1x + a_2x^2 + a_3x^3.$$

Self-Concordant Functions

An important analysis of the modified Newton method was done in [NN]. Instead of traditional bounds on $f''(x)$ to select α in (A2.5) and analyze convergence, we assume that the function

$$f''(x)^{-1/2} \text{ admits a Lipschitz constant } \kappa_2. \qquad \text{(A2.6)}$$

In the case when $f'''(x)$ exists, this Lipschitz condition is equivalent to the following *self-concordant* condition which was introduced in [NN] and used in many subsequent publications:

$$|f'''(x)|/f''(x)^{3/2} \leq 2\kappa_2 \text{ for all } x. \qquad \text{(A2.6')}$$

In the case when $\kappa_2 = 0$, the Newton method terminates in one step with the optimal solution, so we can assume that $\kappa_2 > 0$.

On one hand, we want α in (A2.5) to be small so we have the strong descent property for $f(x)$ that prevents overshooting the local minimum. Since we assume that $f''(x) > 0$ for all x, the strong descent property for $f(x)$ is equivalent to that for $|g(x)|$. On the other hand, taking α smaller than necessary slows down the convergence. So what is the smallest possible α under the condition (A2.6) such that $g(x_{t+1}) = 0$?

The Lipschitz condition (A2.6) means that

$$|g'(x)^{-1/2} - g'(y)^{-1/2}| \le \kappa_2 |x - y| \text{ for all } x, y.$$

Taking here $x = x_t$ and $a = g'(x_t)^{-1/2}$, we obtain that

$$g'(y) \ge 1/(a + \kappa_2 |y - x_t|)^2 \text{ for all } y \qquad (A2.7)$$

and

$$g'(y) \le 1/(a - \kappa_2 |y - x_t|)^2 \text{ when } |y - x_t| < a/\kappa_2. \qquad (A2.8)$$

We set $b = |g(x_t)|$. Assume now that $g(x_t) < 0$ [the case $g(x_t) > 0$ is similar]. Then we integrate (A2.8) from x_t to $x_t + z$ with $0 \le z < a/\kappa_2$ and obtain

$$g(x_t + z) \le -b + 1/(\kappa_2(a - \kappa_2 z)) - 1/(\kappa_2 a) \text{ for } 0 \le x < a/\kappa_2 \quad (A2.9)$$

On the interval $0 \le z < a/\kappa_2$, the right hand side of (A2.8) increases from $-b$ to ∞ taking the zero value at $z^* = a^2 b/(1 + ab\kappa_2)$. So $g(x_t + z) \le 0$ for $0 \le z \le z^*$. From the equation

$$z^* = -\alpha g(x_t)/g'(x_t) = \alpha a^2 b,$$

we find

$$\alpha = \alpha_t = 1/(1 + ab\kappa_2) = 1/(1 + g'(x_t)^{-1/2}|g(x_t)|\kappa_2).$$

Thus, we have the strong descent property for (A2.5) with this α. [The same α works in the case $g(x_t) > 0$.]

Now we estimate the decrease rate for $|g(x_t)|$. Integrating (A2.7) we obtain

$$g(x_t + x) \ge -b - 1/(\kappa_2(a + \kappa_2 x)) + 1/(\kappa_2 a) \text{ for all } x \ge 0. \quad (A2.10)$$

Taking here

$$x = x^* = a^2b/(1 + ab\kappa_2),$$

we obtain that

$$|g(x_{t+1})| \leq 2ab^2\kappa_2/(1 + 2ab\kappa_2) = |g(x_t)|(1-\alpha)/(1-\alpha/2). \quad \text{(A2.11)}$$

[The same is true in the case $g(x_t) \geq 0$.]

Assume now that

$$ab = |g(x_t)|/g'(x_t)^{1/2}$$

is bounded, that is,

$$|f'(x)| \leq \kappa_1 f''(x)^{1/2} \text{ for all } x. \quad \text{(A2.12)}$$

We set

$$\kappa = \kappa_1\kappa_2, \ \alpha^* = 1/(1+\kappa), \ \beta^* = (1-\alpha^*)(1-\alpha^*/2) = 2\kappa/(1+2\kappa). \quad \text{(A2.13)}$$

Then $\alpha^* \leq \alpha_t \leq 1$ for all t and

$$|g(x_{t+1})| \leq |g(x_t)|\beta \leq |g(x_t)|\beta^*$$

for all t hence

$$|g(x_t)| \leq |g(x_0)|\beta^{*t} \to 0$$

as $t \to \infty$.

We have the strong descent property for $|g(x)|$, which implies the monotone convergence $x_t \to x^*$ as $t \to \infty$, where x^* is a number or $\pm\infty$. When $|x^*| < \infty$, we have $g'(x) \geq K_2 > 0$ for all x between x_0 and the minimizer x^*. This low bound on $g'(x)$) implies that

$$|x_t - x^*| \leq |g(x_t)|/K_2 \leq |g(x_0)|\beta^t/K_2.$$

Moreover,

$$|g(x_{t+1}| \leq |g(x_t)ab| \leq cg(x_t)^2$$

with $c = \kappa_2/K_2^{1/2}$ for all t. Once $|cg(x_s)| = q < 1$ for some s, we have

$$|cg(x_{s+t})| \leq q^{2^t}$$

for all t.

Example A2.14. Let $f(x) = -c_1 \log(x) - c_2 \log(1-x)$ with $c_1, c_2 > 0$. On the interval $0 < x < 1$ we have the bound

$$f''(x) \geq K_2 = (c_1^{1/3} + c_2^{1/3})^3$$

as well as (A2.6) with $\kappa_2 = (\min[c_1, c_2])^{-1/2}$ and (A2.12) with $\kappa_1 = (\max[c_1, c_2])^{1/2}$; hence

$$\kappa = \kappa_1 \kappa_2 = (\max[c_1, c_2] / \min[c_1, c_2])^{1/2}.$$

If we start with x_0 in the interval and use the modified Newton method with $\alpha = \alpha^* = 1/(1 + \kappa)$, then we stay in the interval by the strong descent property and we have $|f'(x_t)| \leq |f'(x_t)|\beta^{*t}$ and $|x_t - x^*| \leq |f'(x_t)|\beta^t / K_2$ with

$$\beta^* = 2\kappa/(1 + 2\kappa) = (1 - \alpha)/(1 - \alpha/2) < 1.$$

Using the variable step size

$$\alpha = \alpha_t = 1/(1 + g'(x_t)^{-1/2}|g(x_t)|\kappa_2) \geq \alpha^* \qquad \text{(A2.15)}$$

we obtain even faster convergence (of the type $|x_t - x^*| \leq cq^{2^t}$ with $0 \leq q < 1$) preserving the strong descent property for $f(x)$.

Example A2.16. This is a generalization of the previous example. Let

$$f(x) = -\sum_{i=1}^{m} c_i \log(b_i - a_i x)$$

with all $c_i > 0$. This function is defined on the set S given by the constraints $a_i x < b_i$. The set S can be given as follows: $a < x < b$ with a being a number or $-\infty$ and b being a number or ∞. Assume now that $a < b$ (i.e., S is not empty).

We have (A2.12) with $\kappa_1 = (\sum c_i)^{1/2}$ and (A2.6) with $\kappa_2 = \min(c_i)^{-1/2}$ for all $x \in S$.

In the particular case when all $c_i = 1$, we have $\kappa_1 = m^{1/2}$, $\kappa_2 = 1$ and $\kappa = m^{1/2}$. So we can take $\alpha^* = 1/(1 + m^{1/2})$.

When S is bounded, we also have a bound $f''(x) \geq K_2 > 0$ for all $x \in S$. If S is unbounded, we do not have such a bound. When $a = -\infty$ and $f'(x_0) > 0$, we have min $= -\infty$ and $x_t \to -\infty$. When $b = \infty$ and $f'(x_0) < 0$, we have min $= -\infty$ and $x_t \to \infty$. In all other cases we have

$$|x_t - x^*| \leq c_1(2m^{1/2}/(1 + 2m^{1/2}))^t$$

if we use the modified Newton method with fixed

$$\alpha = \alpha^* = 1/(1 + m^{1/2}).$$

We have even faster convergence if we use the variable step size (A2.15).

A3. Convex and Quadratic Programming

This section is about minimizing a convex function $f(x)$ over a convex set S. Convex programming generalizes linear programming.

Theorem A3.1. For any convex program, any locally optimal solution is optimal.

Proof (cf. [B1], [FV]). Let x^* be a locally optimal solution and y is a feasible solution. We have to prove that $f(x^*) \leq f(y)$. Suppose that $f(x^*) > f(y)$.

Since S is convex, $z = (1 - a)x^* + ay$ is a feasible solution for $0 \leq a \leq 1$. Since f is convex $f(z) \leq (1 - a)f(x^*) + af(y) < f(x^*)$ for $0 < a \leq 1$. For a close to 0, z is close to x^*, which contradicts x^* being locally optimal. ∎

If desirable, by an easy trick (A1.2), $f(x)$ can be arranged to be a linear form while keeping the feasible set convex. The set S is often given by a constraint or constraints of the form $g(x) \leq 0$, with a convex function g. Besides being convex, such a set S is *closed* (i.e. contains its limit points). Conversely, any closed convex set can be given by a constraint $g(x) \leq 0$ with convex $g(x)$. In the case when S is given by a system of convex constraints $g_i(x) \leq 0$ for $i = 1, \dots, m$, such a convex function $g(x)$ can be written as $g(x) = \max[g_1(x), \dots, g_m(x)]$.

Some convex sets S are given by constraints involving *quasiconvex functions*.

Definition A3.2. A function $g(x)$ is called *quasiconvex* if the set $g(x) \leq c$ is convex for every number c. ∎

Equivalently, $g(ax + (1 - a)y) \leq \max(f(x), f(y))$ for all x, y and all a in the interval $0 \leq a \leq 1$. Every convex function is quasiconvex. A function $g(x)$ of one variable x is quasiconvex if it is unimodal.

Note that several quasiconvex constraints $g_i(x) \leq 0$ for $i = 1, \dots, m$ can be replaced by a single quasiconvex constraint $g(x) = \max[g_1(x), \dots, g_m(x)]$.

In the case when $S = R^n$ and the function f is differentiable, x is optimal if and only if $\nabla f(x) = 0$. In general, we have the following result.

Theorem A3.3. Let x^* be a feasible solution for the convex program

$$f(x) \to \min, g_i(x) \leq 0 \text{ for } i = 1, \dots, m, \tag{A3.4}$$

where $f(x)$ is convex and differentiable at x^* and all $g_i(x)$ are quasiconvex and differentiable at x^*. If x^* satisfies the KKT conditions (see A1 above), then x^* is optimal.

Proof. We write

$$-\nabla f(x^*) = \sum_{i=1}^{m} \lambda_i \nabla g_i(x)$$

with $\lambda_i g_i(z^*) \geq 0$ for all i and $\lambda_i = 0$ when $g_i(x^*) = 0$.

We have to prove that $f(x^*) \leq f(x')$ for any feasible solution x'. Since $g_i(x) \leq 0 = g_i(x^*)$ for any active constraint and the restriction of $g_i(x)$ on the line segment connecting x^* and x' is quasiconvex, we conclude that $\nabla g_i(x^*)(x' - x^*) \leq 0$. So

$$\nabla f(x^*)(x' - x^*) = -\sum_{i=1}^{m} \lambda_i \nabla g_i(x)(x' - x^*) \geq 0$$

Since the restriction of $f(x)$ on the segment is convex, we conclude that $f(x^*) \leq f(x')$. ∎

Corollary A3.5. Let x^* be a regular feasible solution for the convex program $f(x) \rightarrow \min$, $g_i(x) \leq 0$ for $i = 1, \ldots, k$, where $f(x)$ is convex and differentiable at x^* and all $g_i(x)$ are quasiconvex and differentiable at x^*. Then x^* is optimal if and only if it satisfies the KKT conditions. ∎

In the case when a function $f(x)$ has continuous second derivatives, $f(x)$ is convex if and only if its Hessian, that is, the matrix

$$\nabla^2 f(x) = [\nabla_{i,j}^2 f(x)] = f''(x) = [\frac{\partial^2 f(x)}{\partial x_i \partial x_j}] \qquad (A3.6)$$

$= Q$ is *positive semidefinite* (i.e., $y^T Q y \geq 0$ for all $y \in R^n$). Equivalently, the quadratic form $y^T Q y$ is a sum of squares of linear forms.

Traditionally, *quadratic programming* is a part of convex programming. It is about minimizing a function $f(x)$ that is a sum of an affine function and squares of linear forms, subject to a finite set of linear constraints. The objective function can be written in the form $f(x) = d + cx + x^T ax$, where x is a column of n decision variables, d a given number, c a row of n given numbers, and a is an $n \times n$ symmetric positive semidefinite matrix. As a sum of convex functions, our $f(x)$ is convex, so quadratic programming is a particular case of convex programming. When a quadratic program is bounded, it has an optimal solution. This is not true for convex programs, as the univariate unconstrained example $e^x \rightarrow \min$ shows. When $a = 0$, our quadratic program becomes a linear program.

When all constraints are equations, an optimal solution can be found by solving two systems of linear equations. Namely, we can solve our system of constraints and excluding variables to reduce our problem to a quadratic program without any constraints. Then we can find all optimal solutions by setting all partial derivatives to be zero. The most often used methods for unconstrained convex programs are Newton and quasi-Newton methods, see the next section. In the quadratic case, the Newton method gives the exact optimal solution in one step.

In the constrained case, the interior methods are most often used, see A5. Many methods of constrained optimization work particularly well in the case of a convex program, see A4. In fact, for nonconvex programs those methods are often *heuristic*, i.e., they seek a solution but do not guarantee they will find one. By contrast, while applied to convex programs some methods have good convergence properties. Moreover, under additional restrictions on the Hessian, we can estimate the convergence rate, cf., A2 and A4.

One approach to solving a convex program is to approximate the convex objective function by a piecewise linear function (i.e., by the maximum of a finite set of affine functions). Also, we approximate the feasible region S by a finite system of linear constraints. When S is given by constraints of the form $g_i(x) \leq 0$ with convex $g_i(x)$, this can be achieved by approximating each $g_i(x)$ by a piecewise linear function. The point here is that the convex program (A3.4) with piecewise linear functions $f(x), g_i(x)$ is equivalent to a linear program. There are special modifications of the simplex method to handle linear program arising this way. Some of them work with tableaux of variable size, where columns and rows are added when necessary and redundant rows and columns are dropped.

Duality in linear programming has been extended to convex programming. For example, for a quadratic program
$$x^T Q x + cx \rightarrow \min, Ax \leq b, x \leq 0$$
with $Q = Q^T$, its dual according to [D] is
$$-y^t Q y + bu \rightarrow \max, \ u \geq 0, \ A^T u - 2Qy \leq c.$$
Finally, there are tricks to reduce nonconvex programs to convex programs. For example, consider the program $f(x) \rightarrow \min$ with a twice differentiable $f(x)$. If $K(f'(x)u)^2 \geq -u^T f''(x)u$ (respectively, $K(f'(x)u)^2 \geq -u^T f''(x)uf(x))$ for some $K \geq 0$ and all $x, u \in R^n$ then our program has the same optimal solutions as the convex program $\text{sign}(f(x))|f(x)|^{K+1} \rightarrow \min$ (respectively, $e^{Kf(x)} \rightarrow \min$).

A4. Multivariate Programming

Methods for univariate optimization (see A2) have been generalized or used in many ways for multivariate optimization. Some methods are simpler when the feasible region is the whole space R^n so we start with unconstrained optimization and discuss the constrained case later.

Coordinate Descent Method

We start with an initial point $x^{(0)} = [x_1^{(0)}, \ldots, x_n^{(0)}]$. To find the next point, we choose a *descent coordinate* and optimize $f(x)$ with respect to this variable keeping other variables fixed. Then we do this with other coordinates to get a sequence $x^{(0)}, x^{(1)}, \ldots$ with $f(x^{(t+1)}) \le f(x^{(t)})$. A possible choice of descent coordinates is cyclic:

$$x_1, x_2, \ldots, x_n; x_1, x_2, \ldots.$$

If $f(x^{(t+n)}) = f(x^{(t)})$ for some t and $f'(x)$ exists and is continuous at $x = x^{(t)}$, then $f'(x^{(t)}) = 0$.

When the gradient $f'(x)$ of $f(x)$ is available, it can be used for a better choice of descent coordinate. This method is simpler than the gradient method (see below), but the convergence is usually poorer.

Grid-Based Methods

Grid-based methods involve a derivative-free search for a local optimizer over successively refined meshes. See [CP1].

Simplex-Based Methods

Simplex-based methods construct an evolving pattern of $n + 1$ distinct points in R^n that are viewed as the vertices of a simplex (in the case $n = 1$ we have an evolving pair of points and the simplex is an interval; cf. A2). In some methods, the next simplex is obtained by using simple rules such as reflecting away from a vertex with the largest value of $f(x)$ and contracting to toward a vertex with the smallest value of $f(x)$.

In other methods, evolution is more complicated, with the main goal being strict improvement of the best objective function value at each iteration (cf. [K2], [SMY]).

Gradient Methods (a.k.a. Methods of Steepest Descent)

We want to find a critical point of a continuously differentiable function $f(x)$ (to be minimized) starting with an initial point x_{0}. The next point $x^{(t+1)} = F(y, t)$ depends on the previous point $x^{(t)} = y$ as follows:

$$F(y, t) = y - \alpha_t f'(y)^T \tag{A4.1}.$$

The numbers $\alpha_t > 0$ here are called *step sizes*, and $f'(x) = \nabla f(x)$ is the gradient [see (A1.6)].

The method terminates naturally when it hits a critical point y in which case $F(y, t) = y$. Until this happens, for sufficiently small α_t, we improve our objective function at each step (i.e., we have the strong descent property).

The *step length* is $b_t = |x^{(t+1)} - x^{(t)}| = |\alpha_t f'(y)|$. For a method to converge to a point we want $b_t \to 0$. (This corresponds to cooling in simulated annealing.) To reach an optimal solution that could be far away from the initial point, we would like to either have $\sum 1/b_t = \infty$ or try many different initial points. If $x^{(t)}$ converges to a point z, then $f'(z) = 0$ provided that $f'(x)$ is continuous.

The perfect choice for α_t would be the value α that is optimal for the univariate minimization $f(y - \alpha f'(y)^T) \to \min$, where $y = x^{(t)}$. The perfect line search can be done exactly in some cases [e.g., when $f(x)$ is a polynomial of total degree 2]. In general, line search is done approximately (see A2). For example, we can set $a_t = 1/t$ or $b_t = 1/t$ (cf., Brown's method).

When $f(x)$ is not differentiable or computation of its derivatives is not so easy, we replace the direction $-f'(y)$ in the gradient method by any direction taking $f(x)$ down (method of feasible directions). On the other hand, when the second derivatives of $f(x)$ are available, the Newton method (discussed subsequently) is preferable.

For instance, let $f(x) = x^T Q x + cx$ with a positive definite symmetric matrix Q (i.e., $z^T Q z > 0$ for all $0 \neq z \in R^n$). Then there is a unique optimal solution $x^* = -Q^{-1} c^T / 2$ and the gradient method with perfect line search gives the global convergence $|x_t, - x^*| \leq \alpha \beta^t$ with some α and $0 < \beta < 1$. (A similar bound holds in the more general case when $f(x)$ is a convex function with continuous second derivatives and $y^T f''(x) y / z^T f''(x) z$ is bounded when $x, y, z \in R^n$ and $|y| = |z| = 1$, see [L].) The Newton method in this case gives the global convergence in one step: $x_{,1} = x^*$.

Trust Region Methods

Given a point w we find the next point $F(w)$ by minimizing a *merit function* $\tilde{f}(x$, which approximates $f(x)$, over a *trust region* S'. We *trust* our approximation in S'. See [CGT].

For example, the *linear approximation* of a differentiable function $f(x)$ at a point w is the affine function

$$f(w) + f'(w)(x - w) \tag{A4.2}$$

of x. Taking S' to be the ball $|x - w| \leq \alpha_t |f'(w)|$, we obtain the gradient method (A4.1) above.

Using the same linear approximation but $S' = S$ given by a finite set of linear constraints, we obtain a method for finding local optimizers in concave problems, see below.

In the unconstrained case, taking the linear approximation as $\tilde{f}(x$, and S' given by $(x - w)^T Q(x - w) \leq 1$ with symmetric positive definite matrix Q, we obtain the classical *variable metric method*.

Using again the linear approximation but a different trust region, depending on constraints, we obtain the affine scaling method (see A5).

Assume now that the Hessian $f''(x)$ (see (A3.6)) exists. The *quadratic approximation*

$$\tilde{f}(x) = f(w) + f'(w)(x - w) + (x - w)^T f''(w)(x - w)/2 \quad \text{(A4.3)}$$

of $f(x)$ at w is a polynomial in x of total degree ≤ 2. Suppose that the matrix $Q = f''(w)$ is positive definite. Using the quadratic approximation and $S' = R^n$, we obtain the Newton method (A4.4) below.

Newton Methods

We assume that the matrix $f''(x)$ exists and is continuous and invertible for all x. The next point $F(w)$ in the Newton method is the critical point of the quadratic approximation $\tilde{f}(x)$ (see (A4.3)) which can be computed explicitly:

$$F(w) = w - f''(w)^{-1} f'(w)^T. \quad \text{(A4.5)}$$

Note that $F(w)$ is the minimizer for $\tilde{f}(x)$ if and only if the matrix $f''(w)$ is positive definite, i.e., $\tilde{f}(x)$ is convex.

To get convergence we usually imposes Lipschitz conditions on second derivatives or bounds on third derivatives of $f(x)$. For example, suppose that $f''_d(x) \geq K_0 > 0$ for all $x, d \in R^n$ with $|d| = 1$ and $f''_d(x)$ admits a Lipschitz constant K_3 for every $d \in R^n$ with $|d| = 1$. Here f''_d is the directional derivatives of f, i.e., the second derivative d^2/dz^2 of the restriction of f onto a line $x = w + dz$ with $d \in R^n$. Then

$$|f'(F(w))| \leq |f'(F(w))|^2 K_3/(2K_0^2). \quad \text{(A4.6)}$$

Thus, the Newton method converges well to a local minimizer x^* if $f''(x^*)$ is positive definite and we start sufficiently close to x^*.

We do not need to assume that $f(x)$ is convex everywhere in R^n. The damped version

$$F(w) = w - \alpha f''(w)^{-1} f'(w)^T \qquad \text{(A4.7)}$$

with $0 < \alpha \leq 1$ can be used to increase the region of convergence.

For large n, the matrix $f''(w)^{-1}$ is usually computed approximately, using its previous value as a starting point. This leads to *the conjugate direction methods* and *quasi-Newton methods*, which fill the gap between the gradient methods and the Newton methods, (see [L, Ch.8 and 9]).

Self-Concordant Functions
Assume that $f''(x)$ exists, is continuous, and is positive definite. Assume also that the restriction of $f(x)$ onto every line satisfies the condition (A2.6), that is,

$$(e^T f''(x)e)^{-1/2} u - (e^T f''(x+e)e)^{-1/2} \leq \kappa_2 \text{ for all } x, e \in R^n, e \neq 0. \qquad \text{(A4.8)}$$

As we saw in A2, the modified Newton method (A4.7) has the strong descent property for $f(x)$ if

$$\alpha \leq \frac{1}{1 + \kappa_2 |f'(w)u||u|/(u^T f''(w)u)^{1/2}}$$

where $u = -f''(x)^{-1} f'(w)^T$ is the Newton direction at w hence $f'(w)u = -f'(w)f''(x)^{-1} f'(w)^T < 0$.

But to get better decrease in $|f'(x)|$ [rather than in $|f'(x)u/|u||$] we want a smaller

$$\alpha \leq \alpha_t = \frac{1}{1 + \kappa_2 |f'(w)|^{1/2} |u|^{-1/2} (-f'(w)^T f''(w)u)^{-1/2}}, \qquad \text{(A4.9)}$$

where

$$0 < -f'(w)^T f''(w)u = f'(w)^T f''(w)f'(w)$$

$$\leq (u^T f''(w)u)^{1/2} (f'(w)^T f''(w)f'(w))^{1/2}.$$

With such α we have (assuming that the third derivatives exist and continuous as in [NN])

$$|f'(F(w))| \leq |f'(w)|(1 - \alpha)/(1 - \alpha/2).$$

Now we introduce an n-dimensional version of (A2.12):

$$f'(x)f''(x)^{-1}f'(x)^T \le \kappa_1 |f'(x)|^{1/2} |f''(x)^{-1}f'(x)^T|^{1/2}| \ \forall \ x \in R^n.$$
(A4.10)

Under this condition, we set

$$\kappa = \kappa_1 \kappa_2, \alpha = 1/(1+\kappa), \beta = 2\kappa/(1+2\kappa) = (1-\alpha)/(1-\alpha/2).$$

Then $\alpha^* \le \alpha_t$ for all t and the method (A4.7) has the strong descent property and $f'(x^{(t)}) \le f'(x^{(0)})\beta^t$ for all t where $x^{(t)} = F^t x^{(0)}$.

It follows that either $|x^{(t)}| \to \infty$ or $x^{(t)} \to x^* \in R^n$. Under the additional condition

$$u^T f''(x)u \ge K_2 |u|^2 \ \forall \ u \in R^n$$
(A4.11)

for some $K_2 > 0$, we obtain that $|x^{(t)} - x^*| \le \beta^t/K_2$. Taking $\alpha = \alpha_t \ge \alpha^*$ as in (A4.9), we obtain faster convergence near x^*. ∎

Now we discuss the constrained optimization. Even finding a feasible solution in this case could be difficult.

In some cases, constraints defining S can be explicitly solved for some variables which allows us to eliminate those constraints and variables. For example, if all our constraints are linear equations, we can eliminate all constraints and obtain an unconstrained program.

Some iterative methods can be adapted to the constrained case by restricting, if necessary, the step size to stay in F. If we hit the boundary, more sophisticated methods, like the gradient projection [finding a feasible direction closest to $-f'(x)$] are used to stay in S [see subsequent discussion and [L]. So it could be a good idea to avoid the boundary using barrier methods (see the next section).

Penalty Methods

A widely used way to remove constraints in (A1.1) involves a *penalty function*. This function $p(x)$ should be zero on S and positive elsewhere. So the feasible solutions for (A1.1) are the optimal solutions for the unconstrained program $p(x) \to \min$.

The program (A1.1) is replaced by a sequence $f(x) + c_k p(x) \to \min$ of unconstrained programs P_k with a sequence $0 < c_k \to \infty$ of large penalties for violating the constraints.

Assuming that both $f(x)$ and $p(x)$ are continuous and that the program P_k has an optimal solution $x^{(k)}$ [which is automatic when $f(x)/p(x) \to 0$ as $|x| \to \infty$], every limit point of the sequence $x^{(k)}$ is an optimal solution for the program $f(x) \to \min$, $x \in S$.

When S is given by constraints (A1.8) a penalty function can be given as

$$p(x) = \sum_{i=1}^{k} g_i(x)^2 + \sum_{i=k+1}^{l} (\max[0, g_i(x)])^2. \qquad (A4.12)$$

When all $g_i(x)$ are differentiable, so is $p(x)$.

Note that the same S can be given by different systems of constraints, which results in different penalty functions. A good judgement should be exercised to get $p(x)$ such that minimization of $f(x) + c_k p(x)$ can be done efficiently by a chosen method (e.g., a Newton method). Another point to be decided on is how to coordinate the degree of precision for solving P_k with a choice of the sequence c_k.

When we have a program (A3.4) with convex $g_i(x)$ the function $p(x)$ becomes $p(x) = \sum_{i=1}^{m}(\max[0, g_i(x)])^2$ which is a convex function, so each P_k is a convex program. Another convex penalty function in this case is $p(x) = \max[0, g_1(x), \dots, g_m(x)]$.

A penalty function $p(x)$ is called *exact* if for some number c a local minimizer of the unconstrained problem $f(x) + cp(x) \to \min$ is also a local minimizer for the original program (A1.1). Search for such functions leads to duality and the KKT conditions. See [L]. Some iterative methods bridge or combine the ideas of duality and penalty.

Linear Constraints

Suppose that S is given a finite system of linear constraints, like in linear programming. If all constraints are linear equations then we can solve this linear system. If there are no feasible solutions or there is only one feasible solution, we are done. Otherwise, we can exclude some variables and get an equivalent unconstrained optimization problem (with a smaller number of variables).

In the case of linear program, finding a feasible solution in general is as difficult as finding an optimal solution. One method is the simplex method. The ellipsoid method can be used for a linear or a convex program; see [H1]. Also, there are *infeasible interior methods*, which are similar to interior methods but are exterior methods working with infeasible solutions and intended to produce a feasible solution.

When S is given by linear constraints and $f'(x) = \nabla f(x)$ exists, the following method using linear programming has been suggested.

Given a feasible solution w, we minimize the linear approximation (A4.1) of $f(x)$ at a point w, which is an affine function of x,
to find the next point $F(w)$. Note that we have the strong descent
property in the case when $f(x)$ is convex. However, when no vertex
is optimal, line search on the segment connecting w and $F(w)$ is
needed for convergence. Here is an example of damping:

$$x^{(t+1)} = x^{(t)} + (F(x^{(t)}) - x^{(t)})/t.$$

To find an initial point $x^{(0)}$ (or to find that the program is infeasible)
we can minimize an arbitrary linear form over S. We terminate the
procedure when $f'(x^{(t)}) = 0$. In the convex case, it is clear that either
$x^{(t)}$ converges to an optimal solution or $|x^{(t)}| \to \infty$. In general, every
limit point of the sequence is critical.

The method with perfect line search was suggested in [FW] for
quadratic programming. In this case the line search is accomplished
by the Newton method in one iteration. Also in this case an optimal
solution exists if S is not empty and $f(x)$ is bounded from below.

Here is what can be done if the direction d at a feasible solution
y improves $f(x)$ but is not feasible. We consider the linear system
$Ax = b$ corresponding to the active constraints. Then we project d
onto the subspace $Ax = 0$, i.e., replace d by the closest vector pd in
the subspace, see §23. The explicit formula for the matrix p in the
case when y is regular (i.e., AA^T is invertible) is

$$p = 1_n - A^T(AA^T)^{-1}A.$$

The direction pd is feasible and improves $f(x)$ unless $pd = 0$. In the
case when $d = -f'(x)$ we obtain a *gradient projection* method.

In the case when the objective function is *concave* [i.e., $-f(x)$
is convex], we have the corner principle, so no damping is necessary.
Concave programs appear in some applications [VCS]. Once a local
optimizer is found, an additional linear constraint can be used to
exclude it (*cutting plane* methods, see [L, Chapter 13]). After this
the procedure is repeated to get a new local minimizer. It is possible
to obtain an optimal solution (if it exists) in finitely many steps. See
[HT], [P].

Note that more general feasible regions, especially convex ones,
can be approximated by regions given by linear constraints. This
extends possible applications of methods mentioned in this subsec-
tion.

A5. Interior Methods

Given a set S in R^n, a point a in S is called *interior* if there is an $\varepsilon > 0$ such that the ball $|x - a| \leq \varepsilon$ is contained in S.

An interior (point) method for solving a mathematical program $f(x) \to \min, x \in S$ starts with an interior point $x^{(0)}$ and generates a sequence of interior points $x^{(0)}, x^{(1)}, \ldots$ convergent to an optimal solution (or finds that there are no optimal solutions). In some cases such global convergences is too much to hope, and we are satisfied with one of the following: $f(x^{(t)}) \to \min$; a subsequence of $x^{(t)}$ converges to a local optimizer.

Brown's fictitious play method is an interior point method when it starts with a mixed strategy $x^{(0)}$ with nonzero entries. Karmarkar [K1] made a breakthrough in mathematical programming when he suggested an interior point algorithm with good convergence for linear programming. Since that time many improvements and generalizations were suggested in thousands of publications. Most of these publications deal with convex programming.

The set $\mathrm{Int}(S)$ of interior points in S can be empty even for a nonempty S. The *boundary* of S is defined to be the set $a \in R^n$ such that for every $\varepsilon > 0$, the ball $|x - a| \leq \varepsilon$ contains a point in S and a point outside S. Note that when the objective function is affine but not constant, every optimal solution (if any exists) belongs to the boundary of the feasible region.

Some mathematical programs with empty $\mathrm{Int}(S)$ can be transformed into those with nonempty interiors and then solved by interior methods. For example, when S is given by constraints $g_i(x) \leq 0$ for $i = 1, \ldots, m$ with continuous functions, we can use an exterior method and obtain a point y such that $g_i(x) < \varepsilon$ for all i with a small $\varepsilon > 0$. Then we relax the constraints $g_i(x) \leq 0$ to $g_i(x) \leq \varepsilon$ to enlarge S to the set S_ε including this point into the $\mathrm{int}(S_\varepsilon)$.

Another way works for linear programs. Solving some linear equations and excluding some variables in a LP with the feasible region S, we can obtain an equivalent (by affine transformations) linear program with the feasible set S' such that either both programs are infeasible, or both problems have exactly one feasible solution, or $\mathrm{int}(S')$ is nonempty. A more sophisticated way, which does not require solving any equations, involves connection with matrix games. Solving any linear program can be reduced to solving a symmetric matrix game, which in its turn can be reduced to solving a linear program with a known optimal value (namely, 0) and a known feasible solution in the interior.

Many methods for unconstrained optimization or univariate optimization can be adjusted to become interior methods. For example, this is clear for the coordinate descent method. In the gradient method or the Newton method we decrease the stepsize, if necessary, to stay in $\text{int}(S)$.

Even when $\text{int}(S)$ is not empty it could be a difficult problem, the *feasibility problem* or Phase 1, to find an interior point. Different modifications of the interior method, called infeasible interior methods, are suggested to handle this problem.

Now we consider some methods that became particularly important after [K1].

Affine Scaling Methods

Let the feasible region S be convex and given as in (A3.4). Let $\text{int}(S)$ be given by $g_i(x) < 0$ for $i = 1, \ldots, m$. We start with a point $x^{(0)} \in \text{int}(S)$. Given any point $w = x^{(t)} \in \text{int}(S)$, the constraints giving S can be rewritten as follows:

$$(g_i(w) - g_i(x))/g_i(w) \leq 1 \text{ for } i = 1, \ldots, m.$$

We define the region $S_t \subset S$ containing x_t by

$$\sum_{i=1}^{m} ((g_i(w) - g_i(x))/g_i(w))^2 \leq 1. \tag{A5.1}$$

Next we define $F(w) = x^{(t+1)}$ as a minimizer of $f(x)$ over S_t. In the unlikely case when $F(w)$ hits the boundary, we can do some damping [i.e., replace $F(w)$ by $w + \alpha(F(w) - w)$ with positive $\alpha < 1$.]

Actually the method is useless unless minimization of $f(x)$ over S_t is easier than that over S. In general, it is not easier. For example, when $S_t = S$ for all t when $m = 1$. When $m \geq 2$ the set S_t need not be convex. However, there is an important case when the minimization over S_t is easier. Namely, assume that the functions $f(x)$ and $g_i(x)$ are affine and that $f(x) = cx + d$ is not constant (i.e., $c \neq 0$). Then the constraint (A5.1) defining S_t has a polynomial $g(x)$ of total degree ≤ 2 on the left-hand side. Therefore, we can find $x^{(t+1)}$ easily. One way to do this is to make an affine change of variables and bring $g(x)$ to one of the following two standard forms: $g(x) = z_1^2 + \cdots + z_m^2 + d_0$ or $g(x) = z_1^2 + \cdots + z_k^2 + d_1 z_{k+1} + d_0$ with $k \leq m - 1$.

In the first case, S_t is the ball $|z|^2 \leq 1 - d_0$ in the new coordinate (an ellipsoid in the original coordinates), and it is easy to minimize $f(x) = cx + d = \tilde{c}z + \tilde{d}$ over S_t: The unique optimal solution is $z^{(t+1)} = -(1 - d_0 1)^{1/2} \tilde{c}/|c|$.

In the second case, since the program is bounded, $\tilde{c}_i = 0$ for $i > k$, so the first k components of an optimal solution $z^{(t+1)}$ are unique and given a similar formula, the kth component is arbitrary when $d_1 = 0$ or is subject to a linear constraint, while the other components (if they exist) are arbitrary.

Instead of changing variables, we can just solve a system of linear equations (the KKT conditions; see A1).

Note that the condition that $f(x)$ is affine can be satisfied easily (see A1), and that any convex set S can be approximated by a system of linear constraints. So the method of affine scaling can be used for more general convex programs, at least in principle. For example, we can replace the constraints $g_i(x) \leq 0$ by linear constraints $g_i(x^{(t)}) + g_i'(x^{(t)})(x - x^{(t)}) \leq 0$. Damping could be used to stay in $\text{int}(S)$.

Practical computations showed that the method is sensitive to the choice of an initial interior point x_0. A good tip is to stay away from the boundary. There is some evidence [H2] to indicate good results for convex programs provided that we start close to an optimal solution or, more generally, to the *central path* (see subsequent discussion).

Barrier Methods

Let S be a subset of R^n with nonempty interior $\text{int}(S)$. A *barrier* function $B(x)$ for S is a continuous function on $\text{int}(S)$ such that $B(x) \geq 0$ for all $x \in \text{int}(S)$ and $B(z^{(k)}) \to \infty$ for every sequence $z^{(k)} \in \text{int}(S)$ that converges to a point outside $\text{int}(S)$. Note that unless $S = R^n$ a barrier function cannot be extended to a continuous function on R^n.

If S is given as in (A3.4), here are some barrier functions $B(x)$:

$$-\sum_i \log(-g_i(x)); \; -\sum_i -1/g_i(x); \; -1/\max_i[g_i(x)].$$

Keeping S intact, changes in $g_i(x)$ generate more examples and erase the difference between these three examples.

Given a bounded mathematical program (A1.1) with a continuous $f(x)$, a barrier function $B(x) \geq 0$ on nonempty $\text{Int}(S)$, and a sequence $\delta_t > 0$ such that $\delta_t \to 0$, we approximate the program by a sequence of programs

$$f(x) + \delta_t B(x) \to \min, x \in \text{int}(S), \tag{A5.2}.$$

If we start in $\text{int}(S)$ and use a method with the strong descent property to solve (A5.2) ignoring the constraint $x \in \text{int}(S)$, then we stay in $\text{int}(S)$.

Theorem A5.3. Suppose that $v = \inf_{x \in \text{int}(S)}(f(x)) > -\infty$. Set $v_t = \inf_{x \in \text{int}(S)}(f(x) + \delta_t B(x))$. Then $v_t \to v$ as $t \to \infty$.

Proof. Clearly $v_k \geq v_{k+1} \geq v$. For any $\varepsilon > 0$, we find $y \in \text{int}(S)$ such that $f(y) - v \leq \varepsilon/2$. Next we find k such that $\delta_t B(y) \leq \varepsilon/2$ for $t \geq k$. Then

$$|v_k - v| = v_k - v \leq (f(y) + \delta_t B(y)) - v \leq \varepsilon/2 + \varepsilon/2 = \varepsilon$$

for $t \geq k$. ∎

Assume now that $f(x) = cx$ is linear, $B(x)$ is convex and that the program

$$f(x) + \delta B(x) \to \min, x \in \text{int}(S)$$

has a unique optimal solution $x^*(\delta)$, The points $x^*(\delta)$ form the so called *central path*, so barrier methods are also known as central path methods. The sequence $x^{(t)} = x^*(\delta_t)$ in Theorem A5.3 follows this path, hence the term a *path-following method*.

The point $x^{(t+1)}$ is usually found by one or more steps of the modified Newton method starting from $w = x^{(t)}$. So first we find the Newton direction $u = -B''(w)^{-1}(c/\delta_t + B'(w))^T$ for $f(x) + \delta_t B(x)$. Then we set $x^{(t+1)} = w + \alpha u$ where the numbers α are chosen to improve $f_k(x)$ The strong descent property would keep $x^{(t+1)}$ interior automatically. According to A2, $\alpha = 1/(1 + \kappa)$ is a good choice where κ is an upper bound for $-(cu/\delta_t + b'(0)|b'''(0)|/b''(0)^2$ and $b(s) = B(w + us)$.

Larger step size α may take us outside S. An alternative choice for finding α that does not require $f'''(x)$ is $\alpha = \beta_t \alpha_t$, where α_t is the maximal value for the univariate program $\alpha \to \max, x^{(t)} + \alpha z \in S$ and β_t is a damping sequence (say, $\beta_t = 1/t$).

For good convergence we want $B(x)$ to be self-concordant with $f''_d(x) \geq K_0 > 0$. Then $f_k(x)$ is also self-concordant. The existence of self-concordant barriers in convex programming was proved in [NN]. Moreover, for several classes of programs such barriers were constructed explicitly [after making the objective function linear as in (A1.2)].

For S given by a finite system of linear constraints $g_i(x) \leq 0$, a good self-concordant barrier is the logarithmic barrier $B(x) = -\sum_i \log(-g_i(x))$.

One purpose of the barrier method is to introduce a penalty for approaching the boundary and hence stay in the feasible region while using methods for unconstrained optimization.

For this, we usually require that $B(z^{(k)}) \to \infty$ for any sequence $z^{(k)}$ in int(B) which converges to a point at the boundary. However, this may inhibit our approach to optimal solutions at the boundary, so the parameter $\delta_t \to 0$ is used. In some cases we can find and use a barrier function such that $B(z^{(k)}) \to \infty$ for any sequence $z^{(k)}$ in $\text{int}(S)$ that converges to any nonoptimal point at the boundary, with a fixed value δ_t.

Here is a version of the path following method that can be called the sliding objective method or the cutting plane method. Given a convex program (A1.1) with linear $f(x) = cx$, we use a convex barrier function $B(x)$ for S and a positive sequence $\alpha_t \to 0$.

As initial point, we take $x^{(0)}$ to be the minimizer of $B(x)$ over S. Given $x^{(t)}$, we find the next point $x^{(t+1)}$ by applying a step (or several steps) of the modified Newton method to the objective function $B(x) - \log(f(x^{(t)}) - f(x) + \alpha_t)$ which is a convex barrier function for $\{x \in S, f(x) \le f(x^{(t)}) + \alpha_t\}$.

It was shown in [NN] that for any convex S with nonempty bounded $\text{int}(S)$ there is a convex barrier function $B(x)$ which is self-concordant in the sense of (A4.8) with κ_2 depending on n and which also admits k_1 in the sense of (A4.10). When $f(x)$ is affine, $f(x) + cp(x)$ is self-concordant with the same κ_2 and admits a different κ_1.

For some S, self-concordant barriers are constructed explicitly in [NN], [H2]. In [NN] self-concordant functions are allowed to have noninvertible Hessian, which requires some changes in definitions and in the Newton method. However, after excluding the variables which do not effect $f(x)$, we are back in the case of an invertible Hessian.

Example A5.4. Let S be given by linear constraints

$$g_i(x) \le 0, \ i = 1, \ldots, m.$$

We assume that $\text{int}(S)$ is not empty. Then

$$B(x) = \sum_{i=1}^{m} \log(-g_i(x))$$

is a convex barrier function.

When $\text{int}(S)$ contains a whole straight line, a variable can be excluded from our program. Otherwise, $f''(x)$ is invertible for all x. When S is bounded, the condition (A4.11) holds for some $K_2 > 0$.

By our computation in Example A2.16, $p(x)$ is self-concordant with $\kappa_2 = 1$ and the condition (A4.10) holds with $\kappa_1 = m^{1/2}$.

Potential Reduction Methods

Given a mathematical program (A1.1), a *potential function* $h(x)$ is a function on int(S) satisfying the following condition: for any sequence $x^{(k)}$ in int(S), $f(x^{(k)}) \to -\infty$ if and only if $f(x^{(k)}) \to f_0$, where f_0 is the optimal value. So minimization of $f(x)$ is equivalent to that of $h(x)$ or $e^{h(x)}$.

For example, let $B(x)$ be a barrier function. Assume that $B(x) + C_0 \log(f(x) - f_0)) \to -\infty$ when x converges to an optimal solution. Then $-C_0 \log(f(x) - f_0)) + B(x)$ is a potential function.

Now we use setting in (A3.4) with affine functions $f(x), g_i(x)$. In [K1], $e^{h(x)} = (f(x) - f_0)^m / \prod_{i=1}^m g_i(x)$ with such a function $h(x)$ was used to measure the progress of descent but it did not enter explicitly in the determination of the direction of movement. This is the traditional concept of *merit function*. However nothing is wrong with minimizing the potential or merit function instead of the objective function (except possible confusion with terminology).

However, replacement of $f(x)$ by $h(x)$ or $e^{h(x)}$ makes sense only if some optimization method works better for $h(x)$ or $e^{h(x)}$ than for $f(x)$. In [K1], $h(x)$ is not convex, but $e^{h(x)}$ is convex and self-concordant.

Linear Complementary Problem (LCP)

The problem is to find two vectors $x, y \in R^n$ such that

$$y = Ax + b, \; x \geq 0, \; y \geq 0, \; x_i y_i = 0 \text{ for all } i$$

with given $n \times n$ matrix A and a vector $b \in R^n$.

An LCP is *monotone* if the matrix A is positive semidefinite. By the KKT conditions, any quadratic program (in particular, any linear program) can be cast as a monotone LCP.

Interior methods were developed for solving LCP (cf. [FMP]).

Semidefinite Programming (SDP)

We consider the set $M_n(R)$ of all $n \times n$ real matrices. Given two matrices $A = [A_{i,j}], B = [B_{i,j}] \in M_n(R)$ we denote by $A \bullet B$ the number $\sum_{i=1}^n \sum_{j=1}^n A_{i,j} B_{i,j}$. A semidefinite program is a mathematical program of the form $C \bullet X \to \min$, subject to

$$A_{(i)} \bullet X = b_i \text{ for } i = 1, \ldots, m, X \text{ is positive semidefinite,}$$

where X is a symmetric $n \times n$ matrix of variables and the data consist of symmetric matrices $C, A_{(i)} \in M_n(R)$ and numbers b_i.

SDP has wide applications in convex programming, combinatorial optimization, and control theory. Interior methods are used in SDP, [W1], [K4], [R].

A6. Pertubation

Pertubation was the first and still is the easiest way to show that cycling can be avoided; that is, there is a simplex method that always
works. It is not practical because it requires additional computations. However, the concept of perturbation is useful in other areas
of mathematical programming (see A7). In Phase 2 of the simplex
method (see §10), we replace the column $b = [b_1, \ldots, b_n]^T$ in the
tableau

$$\begin{array}{cc} x & 1 \end{array}$$
$$\begin{bmatrix} A & b \\ c & d \end{bmatrix} \begin{array}{l} = u \\ = z \to \min, \quad x \geq 0, \quad u \geq 0 \end{array}$$

by the column $b(\epsilon) = [b_1 + \epsilon, \ldots, b_n + \epsilon^n]^T$. Formally, we work with
polynomials in ε. Informally, $b(\varepsilon)$ is a small perturbation of b; ε is
considered a small positive number. We compare polynomials in ε
as follows:

$a_0 + a_1\varepsilon + a_2\varepsilon^2 + \cdots \geq f_0 + f_1\varepsilon + f_2\varepsilon^2 + \cdots$ if either
$a_0 > f_0$, or
$a_0 = f_0$, and $a_1 > f_1$, or
$a_0 = f_0$, $a_1 = f_1$, and , $a_2 > f_2$, or
$\ldots$
$\ldots$
$\ldots$

Now, when we apply the simplex method, stage 2, to the tableau

$$\begin{array}{cc} x & 1 \end{array}$$
$$\begin{bmatrix} A & b(\varepsilon) \\ c & d(\varepsilon) \end{bmatrix} \begin{array}{l} = u \\ = z \to \min, \quad x \geq 0, \quad u \geq 0 \end{array}$$

where $d(\varepsilon) = d$ in the initial tableau, we obtain tableaux of the same
form, and no entry of $b(\varepsilon)$ is ever 0. This is because the entries of
$b(\varepsilon)$ are always linearly independent over the real numbers (they were
linearly independent in the initial tableau, and then pivot steps result
in addition and multiplication operations on these entries with real
coefficients, so they remain linearly independent). So the entry $d(\varepsilon)$
decreases its value after each pivoting step; that is, $d(\varepsilon)$ cannot stay
the same. This makes cycling impossible. If we obtain an optimal
ε-tableau, then, by setting $\varepsilon = 0$, we obtain an optimal tableau for
the original problem. If we obtain an ε-tableau with a bad column,
then, by setting $\varepsilon = 0$, we obtain a tableau for the original problem
with a bad column.

A7. Goal Programming

In real life, we can have several goals or objectives to optimize. It does not happen often that an optimal solution for one objective function is also optimal for another objective function. To apply mathematical programming, we need to combine those goals into one objective function and possibly take into account some goals in additional constraints. There are many ways to do this, so there are several books on goal, vector, or multiobjective programming (cf. [CVL], [K5], [S]). A similar problem occurs in game theory, where goals of different players could be different.

Suppose that we have several functions $f_1(x), \ldots, f_m(x)$ to minimize. We can set a numerical goal b_i for each objective and then minimize a convex combination of our functions:

$$f(x) = \sum_{i=1}^{m} c_i f_i(x) \to \min, f_i(x) \le b_i \text{ for all } i, \ x \in S$$

where $c_i > 0$ for all i.

Every optimal solution x^* of this program is *Pareto optimal* or *efficient* [i.e., x^* is feasible, and there is no other feasible point, y, such that $f_i(y) \le f_i(x^*)$ for all i and $f_i(y) < f_i(x*)$ for some i]. The set of such points is called the efficient frontier or the Pareto boundary. In some situations, especially when S is convex, every Pareto optimal solution can be obtained as an optimal solution for a convex linear combination of $f_i(x)$.

In the case when $k = n = 2$, S is convex, $f_1(x) = x_1, f_2(x) = x_2$, Nash suggested using

$$f(x) = -(b_1 - f_1(x))(b_2 - f_2(x))$$

instead of

$$f(x) = c_1 f_1(x) + c_2 f_2(x).$$

He did this in the contest of two-person cooperative games, when x_i was the payoff for player i.

The convexity of S gives uniqueness for optimal solution. The optimal solution x^* exists if there is a feasible solution and S is bounded and closed, which is automatic in the game theory context. Moreover, x^* is Pareto optimal if $f(x^*) \ne 0$.

A natural generalization of Nash's approach is the following program:

$$f(x) = -\sum_{i=1}^{m} \log(b_i - f_i(x)) \to \min, \ f_i(x) \le b_i \text{ for all } i, \ x \in S.$$

To get a convex program in the terms of new variables $y_i = f_i(x)$, we would like the image S' of S under the mapping

$$x \mapsto [f_1(x), \dots, f_m(x)] \in R^m$$

to be convex. In the game theory context this is achieved by allowing the joint mixed strategies. This amounts to replacing S' by its convex hull. So our problem takes the form

$$g(y) = -\sum_{i=1}^{m} \log(b_i - y_i) \to \min, \ y_i \le b_i \ \text{ for all } i, \ y \in S'$$

with a convex self-concordant objective function, which makes the modified Newton method, with an appropriate barrier function, work well.

These approaches are used when the goals are of roughly comparable importance. In the case of *preemptive goal programming* the functions are partially ordered by priority. We can combine goals at the same priority level as before. However, goals at different levels are treated differently.

For example, let $f_i(x)$ be sorted by priority with $f_1(x)$ being most important. We can use the following sequential procedure to solve the overall problem by a sequence of k mathematical programs. First we minimize $f_1(x)$ and obtain the optimal value c_1^*. Then we minimize $f_2(x)$ with the additional constraint $f_1(x) = c_1^*$ (this makes sense only if the first program has more than one optimal solution). And so on. A nice way to state the overall problem is

$$f(x) = \sum_{i=1}^{k} \varepsilon^{i-1} f_i(x) \to \min, f_i(x) \le b_i \text{ for all } i, \ x \in S,$$

where the objective function is a polynomial in ε and the polynomials are compared as in A6. Note that the objective function now is not real-valued. Instead of polynomials, we can consider the rows with k entries that are ordered lexicographically.

Note also that any optimal solution in preemptive goal programming is also Pareto optimal.

A very general way to combine our m objectives $f_i(x)$ is by a function $h(y)$ of m variables that is nondecreasing with respect to every variable in the region $y \ge 0$ [in the differentiable case, this means that $g'(y) \ge 0$) in the region]. Then our program is

$$-h(b_1 - f_1(x), \dots, b_m - f_m(x)) \to \min, f_i(x) \le b_i \ \text{ for all } i, \ x \in S.$$

Every optimal solution for this program is also Pareto optimal. In the above examples, $h(y) = cy$ with $c \ge 0$ and $g(y_1, y_2) = y_1 y_2$.

A8. Linear Programming in Small Dimension

Recent interior methods beat simplex methods for very large problems. On the other hand, it is known that for LPs with small number of variables (or, by duality, with a small number of constraints), there are faster methods than the simplex method, cf. [C2]. We will demonstrate this for linear programs with two variables, say x and y.

We write the program in canonical form

$$cx + c'y \to \min, ax + a'y \le b, x \ge 0, y \ge 0$$

where c, c' are given numbers and a, a', b are given columns of m entries each. The program can be written in the standard row tableau

$$\begin{array}{ccc} x & y & 1 \\ \left[\begin{array}{ccc} -a & -a' & b \\ c & c' & 0 \end{array} \right] & \begin{array}{c} = * \ge 0 \\ \to \min \end{array} \end{array} \qquad (A8.1)$$

The feasible region S has at most $m + 2$ vertices. Phase 2, starting with any vertex, terminates in at most $m + 1$ pivot steps.

At each pivot step, it takes at most two comparisons to check whether the tableau is optimal or to find a pivot column. Then in m sign checking, at most m divisions, and at most $m - 1$ comparisons we find a pivot entry or a bad column.

Next we pivot to compute the new $3m+3$ entries of the tableau. In one division we find the new entry in the pivot row that is not the last entry (the last entry was computed before) and in $2m$ multiplications and $2m$ additions we find the new entries outside the pivot row and column. Finally, we find the new entries in the pivot column in $m + 1$ divisions. So a pivot step, including finding a pivot entry and pivoting, can be done in $8m + 3$ *operations*—arithmetic operations and comparisons.

Thus, Phase 2 can be done in

$$(m + 1)(8m + 3)$$

operations, While small savings in this number are possible [e.g., at the $(m + 1)$-th pivot step we need to compute only the last column of the tableau] it is unlikely that any substantial reduction of this number for any modification of the simplex method can be achieved (in the worst case).

Concerning the number of pivot steps, for any $m \geq 1$ it is clearly possible for S to be a bounded convex $(m+2)$-gon, in which case for any vertex there is a linear objective function such that the simplex method requires exactly $\lfloor 1+n/2 \rfloor$ pivot steps with only one choice of pivot entry at each step. It is also possible to construct an $(m+2)$-gon, an objective point ,and an initial vertex in a way that m pivot steps with unique choice are required (or with two choices at the first step such that the first choice leads to the optimal solution while the second choice leads to m additional pivot steps with unique choice).

Now we outline a method to find an optimal solution, assuming that (A8.1) is feasible, into $\leq 100m + 100$ operations. We proceed by induction on m. For $m \leq 12$ we can use the simplex method and finish in

$$(m+1)(8m+3) < 100m + 100$$

operations. So we assume that $m \geq 13$.

Assume that the objective function is nonconstant (otherwise, we are done in two comparisons). We set $u = -cx - c'y$. Set $v = x$ when $c' \neq 0$ and $v = y$ otherwise.

We rewrite all $m + 1$ conditions (not including $v \geq 0$) in the form

$$u \leq a_i v + b_i, \ i = 1, \ldots, l,$$

$$u \geq a_{-i} v + b_{-i}, \ i = 1, \ldots, m + 1 - l.$$

This requires at most $3(m + 1)$ operations. If $l = 0$, then the program is unbounded, and we are done in $\leq 3(m + 1) \leq 100m + 100$ operations. So we assume that $l \geq 1$.

If the numbers $a_{2i-1} - a_{2i}$ and $b_{2i-1} - b_{2i}$ have different signs or both are zero for some integer i such that either $1 \leq i \leq l/2$ or $1 \leq -i \leq (m + 1 - l)/2$, then we can drop one of these two constraints. This involves at most $m+1$ subtractions and $0.5m+0.5$ sign comparisons. We denote by $l' \leq l$ and $m' - l' - 1$ the remaining numbers of constraints of type $\leq$ and $\geq$. Note that $l' \geq 1$. If $m' \leq 12$, then we are done in

$$3(m + 1) + (m + 1) + (0.5m + 0.5) + 100 \cdot 12 + 100$$

$$= 4.5m + 4.5 + 100 \cdot 12 + 12 < 100m + 100$$

operations. So we assume that $m' \geq 13$.

We write the remaining constraints as

$$u \leq a_i' v + b_i', \ i = 1, \ldots, l',$$

$$u \geq a'_{-i}v + b'_{-i}, \ i = 1, \ldots, m' + 1 - l'$$

and we remember the computed numbers $q_i = a_{2i-1} - a_{2i}, p_i = b'_{2i-1} - b'_{2i}$ with $\text{sign}(p_i) = \text{sign}(q_i) \neq 0$.

Now we compute $v_i = p_i/q_i$ for $1 \leq i \neq l'/2$ and $1 \leq -i \neq (m' + 1 - l')/2$. This gives

$$k = \lfloor l'/2 \rfloor + \lfloor (m + 1 - l')/2 \rfloor \leq (m' + 1)/2$$

numbers. The number of arithmetic operations is

$$k \leq (m' + 1)/2 \leq 0.5m + 0.5.$$

Then we compute a median v_0 of these $k \geq 0.5m' - 1.5$ numbers. It requires at most $18k + 18 \leq 9m + 27$ comparisons (ee A10).

Next we find the maximum u'' of $a'_i v_0 + b'$ with $1 \leq -i \leq m' + 1 - l'$. This requires $m' - l'$ comparisons.

We also find the minimum u' of $a'_i v_0 + b'$ with $1 \leq l'$ computing at the same time the set Y of $\text{sign}(a'_i)$ with $a'_i x_i + b'_i = u'$. This requires at most $2l' - 1$ comparisons. The total number of operations is at most

$$(4.5m + 4.5) + (0.5m + 0.5) + (9m + 27) + (m' - l') + (2l' - 1)$$

$$\leq 16m + 31.$$

If $u'' \leq u'$ and Y contains either 0 or both 1 and -1, then v_0 is an optimal solution and u' is the optimal value. So we are done in $(16m + 31) + 1 < 100m + 100$ operations.

If either $u'' > u'$ or $Y = \{-1\}$, then $v^* < v_0$ for each optimal solution v^*. In the remaining case, when $u'' \leq u'$ and $Y = \{1\}$, then there is an optimal solution $v^* \leq v_0$. In both cases, we know what side of x_0 to search for v^*.

Now we can drop one constraint for every v_i (including v_0) that is on the other side of v_0 than v^*. This means that at least $k/2$ constraints drop, and at most

$$m' - k/2 \leq m' - (m' - 3)/4 = 0.75m' + 0.75 \leq 0.75m + 0.75$$

stay. By the induction hypothesis, we can finish in at most

$$100(0.75m + 1.75)$$

operations. Thus, in at most

$$(16m + 31) + 1 + 100(0.75m + 1.75) = 91m + 207 < 100m + 100$$

operations, we can solve our linear program.

The number $100m + 100$ can be improved by a more careful accounting.

A9. Integer Programming

This section is about linear programs with additional conditions that some variables are integers. A particular case, is boolean, binary, or combinatorial programming when all variables are required to be 0 or 1. The general case with all variables being bounded integers can be reduced to binary case by writing variables in base 2 (binary representation). In the case of combinatorial programming, a solution can be thought as a set of variables (taking value 1). If all variables are required to be integers, we have a *pure* integer program.

Example A9.1. Maximize $f(n)$ subject to $1 \leq n \leq N$, n an integer, where N is a given positive integer.

This is an univariate integer program. We want to find a maximal term in the finite sequence $f(1), f(2), \ldots, f(N)$. The problem can be solved by $N - 1$ comparisons (see A10 below).

Example A9.2. *Job Assignment Problem* (see p. 90).

This is a boolean program. It can be reduced to a linear program, see p. 192.

Example A9.3. *Knapsack Problem:*

$$cx \rightarrow \max, ax \leq b, x = [x_1, \ldots, x_n]^T$$

with integers $x_i \geq 0$, where $a, c \geq 0$ are given rows and b is a given number.

This integer program models the maximum value of a knapsack that is limited in weight by b, where x_j is the number of items of type i with value c_i and weight a_i. In a bounded knapsack problem, we have additional constraints x_i e_i. In a 0-1 knapsack problem, all $e_i = 1$.

Example A9.4. *Traveling Salesman Problem* (TSP, cf. [GP]).
Given an $n \times n$ cost matrix $[c(i,j)]$, a *tour* is a permutation of $[\sigma(1), \ldots, \sigma(n)]$ of *cities* $1, 2, \ldots, n$. A tour means visiting each city exactly once, and then returning to the first city (called home). The cost of a tour is the total cost

$$c(\sigma(1), \sigma(2)) + c(\sigma(2), \sigma(3)) + \ldots + c(\sigma(n-1), \sigma(n)) + c(\sigma(n), \sigma(1)).$$

The TSP is to find a tour of minimum total cost. It can be stated as an integer program using (like the job assignment problem) the binary variables x_{ij}. ∎

Using binary variables we can reduce logically complicated systems of constraints to usual systems where all the constraints are required to be satisfied. For example, the program

minimize $f(x)$ satisfying any two of the following
three constraints $g_1(x) \leq 0, g_2(x) \leq 0, g_3(x) \leq 0$

can be written as

$f(x) \to \min, y_1 g_1(x) \leq 0, y_2 g_2(x) \leq 0, y_3 g_3(x) \leq 0,$
$y_1 + y + 2 + y_3 = 2, y_i$ binary.

Now we discuss some methods of solving integer programs.

Rounding Linear Solutions

Dropping the conditions that the variables are integers, we obtain a linear program, the LP-*relaxation* of the integer program, IP. If an optimal solution of the LP satisfies the integer restriction (as for the job assignment problem), it is optimal for IP. In general, rounding the optimal solution for LP, we obtain a "solution" for IP. This solution need not be feasible, and if it is feasible, it need not be optimal. However, it is used sometimes in real life when better solutions are hard to find.

Exhaustive and Random Enumeration

When the variables restricted to be integers are bounded, the IP is reduced to a finite set of LPs by running over all possible values for those variables. This approach is practical only when the number of LPs is small. In general, we can choose the values at random. The more choices tried, the closer to 1 is the probability of finding an optimal value for IP.

Reduction of IP to an LP Using Convex Hull

Consider the feasible set S for our IP and its convex hull S' (i.e., the set of all convex combinations of all points in S). Then S' can be given by a finite set of linear constraints, and optimization of our objective function $f(x)$ over S is equivalent to optimization of $f(x)$ over S', which is a linear program. Moreover, every extreme point of S' that is optimal belongs to S and hence is optimal for the IP. Although theoretically this approach reduces any IP to a LP, this method works only when S' can by described by a small number of linear constraints. See §3 for examples.

Branch-and-Bound Method

We outline this method for integral programs (with binary variables $x_1, \ldots, x_n$) using the best-first branching rule although it can be used for any mathematical program where some variables are integers, and there are different branching rules. We start with the LP-relaxation LP_0 of our IP minimization program IP_0 (i.e., the linear program obtained by replacing the conditions $x_i^2 = x_i$ in IP_0 by the conditions $0 \leq x_i \leq 1$). If an optimal solution of LP_0 is integral, we are done. In any case the optimal value for LP_0 is a lower bound for the optimal value of IP_0. Next we split IP_0 into two IPs, IP_1 and IP_2 fixing $x_1 = 0$ or 1. We solve the corresponding linear programs obtaining lower bounds v_1 and v_2 for the optimal values of IP_1 and IP_2. Next we branch the program with lower bound into two programs, setting $x_2 = 0$ and $x_2 = 1$. Continuing the process, we have at step t a tree with t nodes. At each node we have a lower bound for the corresponding IP obtained by solving the corresponding LP-relaxation. If no optimal solution for the nodes with minimal bound is integral, we branch one of these nodes into two. The process terminates in at most $2^n - 1$ branchings.

Cutting Plane Method

We start by solving the LP-relaxation of our IP problem. If the optimal solution satisfies the integrability constraints, we are done. Otherwise, an additional linear constraint is constricted, which cuts out the optimal solution but is satisfied by all feasible solutions (or at least by all optimal solutions) of IP. The process is repeated, so we obtain a sequence of linear programs with a decreasing sequence of the feasible regions. Many ways to construct the cutting constraint were suggested for which it was proved that the process terminates in finitely many steps. However, it is common that the number of steps is too large. *Polyhedral annexation* is a cutting plane approach to finding an optimal solution known to lie at an extreme point of a polyhedron, P. The general algorithm is to start at some extreme point and solve the polyhedral annexation problem. This will result in ascertaining that the extreme point is (globally) optimal, or it will generate a recession direction from which a convexity cut is used to exclude the extreme point. The approach generates a sequence of shrinking polyhedra by eliminating the cut region. Its name comes from the idea of annexing a new polyhedron to exclude from the original, homing in on the extreme point solution.

Many textbooks on linear programming contain chapters on integer programming. Also, there are books on integer programming (cf. [ES], [W2], [KV], [S2]).

A10. Sorting and Order Statistics

A lot of sorting is done by humans and computers. Sorting is a significant part of data processing. So finding efficient ways to sort is important. Sorting here means ordering items in a linear list, like making a list of students in class. More precisely, given $n \geq 1$ numbers $a_1, \ldots, a_n$, let $b_1 \leq \cdots \leq b_n$ be the same numbers sorted.

We want to find these numbers b_i as fast as possible. Usually, any method of sorting also finds a permutation σ such that $b_{\sigma(i)} = a_i$.

Sorting involves comparison of numbers and possibly moving them around. We will count only the number of comparisons, which we call steps for short. So the problem is how to sort a given set of numbers in a smallest number of steps.

Thus, we are interested only in the cost of collecting information (sufficient to order the numbers) about the relative size of numbers and ignoring costs of storing and using this information as well as the complexity of the algorithm.

Since there are $n!$ permutations of n numbers, it is well known that any sorting method requires at least $\lceil \log_2 n! \rceil$ steps, where $\lceil x \rceil$ denotes the least integer $t \geq x$. Here we prove the following well-known result (cf. [K3]).

Theorem A10.1. We can sort s numbers in

$$F_0(n) = \sum_{i=1}^{n} \lceil \log_2 n \rceil = m \lceil \log_2 n \rceil - 2^{\lceil \log_2 n \rceil} + 1$$

steps. ∎.

Proceeding by induction on n, it suffices to prove the following.

Lemma A10.2. Given a sorted list of numbers $b_1 \leq \cdots \leq b_{n-1}$ and another number a_n, it takes at most $\lceil \log_2 n \rceil$ steps to sort all n numbers.

Proof. In other words, we have to prove that k steps are sufficient to insert a new number b_n into an ordered list of $n = 2^k - 1$ numbers. The case $n = k = 1$ requiring one step is trivial. Let $k \geq 2$. We proceed by induction on k, using a bisection method that is similar to that in A2. Namely, we compare a_n with the median $b_{(n+1)/2}$. After this we have the task of inserting a_n into a sorted list of $(n-1)/2 = 2^{k-1} - 1$ numbers, which can be done in $k - 1$ steps by the induction hypothesis. ∎

It is clear that $k - 1$ steps are necessary and sufficient to find min $= \min[a_1, \ldots, a_n]$. The same is true for the maximal number.

If we have some additional information on the sequence a_i this can be used to find min faster. For example, suppose that the sequence is *unimodal* (i.e., there is i^* such that a_i strictly decreases from $i \leq i^**$ and strictly increases from $i \geq i^*$). A method for finding min $= a_{i^*}$ using the minimal number of evaluations, similar to Fibonacci search, is called *lattice search.* It is more efficient than the bisection method which finds the minimum in $\lceil \log_2 n \rceil$ steps.

Suppose again that we know nothing about a_i. While it takes about $\log_2 n!$ steps to find the ordered list b_i it turns out that any particular b_i (an order statistic), including the median $b_{\lceil n/2 \rceil}$, may be computed faster—namely, in Cn steps with C bounded over all n, k; see [K3].

The known proofs with small C are quite long, so we give a couple of examples and then sketch a simple proof with $C = 18$.

For instance, $b_1 = \min(a_i)$ can be computed by $n - 1$ comparisons.

As another example, consider the particular case $n = 5$. Sorting takes at least $\lceil \log_2(5!) \rceil = 8$ steps. On the other hand, as mentioned previosly it takes $n - 1 = 4$ steps (comparisons) to find b_1. After this, it takes three more steps to find b_2, the minimum of the remaining numbers. So b_2 can be found in seven steps. Similarly, b_5 and b_4 can be found in four or seven steps. Finally, the median b_3 can be found in seven steps as follows.

First we sort a_1, a_2, a_3, a_4 by the insertion method in $1+2+2 = 5$ steps and obtain $c_1 \leq c_2 \leq c_3 \leq c_4$. Then we compare a_5 with the medians c_2 and c_3 (two more steps). If $c_2 \leq a_5 \leq c_3$, then $b_3 = a_5$. If $a_5 \leq c_2$, then $b_3 = a_2$. If $a_5 \geq c_3$. then $b_3 = c_3$. Thus, each b_i can be found in seven steps.

It is known [DZ1] that any statistics b_i can be found in at most $2.95n + o(n)$ steps and that at least $2.01n + o(1)$ steps are required in the worst case [DZ2]. Here $o(n)$ [respectively, $o(1)$] stands for a sequence such that $o(n)/n \to 0$ [respectively, $o(11) \to 0$] as $t \to \infty$.

Now we sketch a proof that $18(n-1)$ steps are sufficient to find the kth smallest number b_k. We proceed by induction on n. The cases $n \leq 34$ are trivial because then we can sort n numbers in $18(n-1)$ steps. So let $n \geq 35$.

We write $n = 10l + 5 + r$ with $0 \le r \le 9$. We partition the first $10l + 5$ numbers a_i into $2l + 1$ 5-tuples and find the medians $c_1, \ldots, c_{2l+1}$ in each 5-tupple. This can be done in $7(2l + 1)$ steps. By the induction hypothesis, we can find the median d of c_j in $36l$ steps.

Now we have $3l + 2$ numbers a_i on the right of d and $3l + 2$ numbers a_i on the left of d. In

$$n - 6l - 5 = 4l + 5$$

steps we place the remaining $n - 6l - 5$ numbers a_i on the right or left of d.

Now we have to find a certain statistic among $n' \le n - 3l - 3$ numbers on the right of d or among $n'' \le n - 3l - 3$ numbers on the left of d. By the induction hypothesis, we can do this in

$$18(n - 3l - 4)$$

steps.

Thus, the total number of steps is at most

$$7(2l + 1) + 36l + 4l + 5 + 18(n - 3l - 4) = 18n - 60 \le 18n - 18. \blacksquare$$

Finally we discuss the problem of finding saddle points in a given $m \times n$ matrix $[a_{i,j}]$ (or proving that they do not exist). This problem appears when we try to solve a matrix game, and it can be stated as an integer program.

Given i, j it takes $m + n - 2$ steps to check whether this position is a saddle point. On the other hand it takes $n(m - 1)$ steps to find all maximal entries in every column. Similarly, it takes $m(n - 1)$ steps to find all minimal entries in every row. So after $2mn - m - n$ steps we are done (the positions selected twice are the saddle points). No faster method is known.

However, there is a faster method for finding a *strict* saddle point [i.e., (i, j) such that $a_{i,k} > a_{i,j} > a_{l,j}$ for all $k \ne j$ and $l \ne i$ (or proving nonexistence)]. The method [BV] starts with sorting the numbers $a_{i,i}$ on the main diagonal in $F_0(m)$ steps (assuming that $m \le n$), and terminates in

$$F_0(m) + F_)(m - 1) + n + m - 3 + (n - m)\lceil \log_2(m + 1) \rceil$$

steps.

A11. Other Topics and Recent Developments

We have not mentioned several important topics in mathematical programming including special classes of mathematical programs and special methods devised for solving those programs.

Besides explicitly given numbers, the data may include parameters or external variables (e.g., coefficients of linear functions) and/or random data. In §14, we mentioned parametric programming in the context of linear programming. *Stochastic programming* (cf. [BL], [KW]) deals in particular with uncertainty in data which is also the main concern in *fuzzy programming* (cf. [C1], [RI], [RV]). Fuzzy sets are also used to represent preferences which is connected with goal programming.

To solve large problems, special sophisticated tricks are needed to handle data, and methods are modified to obtain a good solution in a reasonable time. Here are some recent books on large-scale optimization: [B2], [T]. For large linear programs there are revised versions of the simplex method, where special attention is paid to handling data. For example, in the case of a very large number of columns in tableaux, they are generated during pivoting. Column generation is dual to the cutting plane method. Also, the simplex method has been modified to handle upper bounds on variables in a special way.

Some large linear programs can be split into subprograms that are weakly related between themselves. This leads to decomposition methods (nested programming) like Dantzig-Wolfe methods. Similarly, aggregation methods reduce solving a large LP to solving a sequence of smaller LPs.

In *fractional programming*, the objective function $f(x)$ and the constraint functions are sums of ratios of the form $a(x)/b(x)$ with affine functions $a(x), b(x)$ such that $b(x) > 0$ over S. In the one-term case, the problem looks like

$$a_0(x)/b_0(x) \to \min, a_i(x)/b_i(x) \le d_i \text{ for } i = 1, \ldots, m$$

with affine functions $a_i(x), b_i(x)$ and constants f_i. This case is special because then the program is equivalent to a linear program [assuming that all $b_i(x) \le 0$ in the feasible region]. See [C-M].

In *separable programming*, the objective function $f(x)$ and every constraint function is a sum of univariate functions $f_i(x_i)$. Piecewise approximations and linear programming are used to solve such programs (cf. [S3]).

In polynomial programming, the objective function $f(x)$ and the constraint functions $g_i(x)$ in (A1.9) are polynomials. The linear programming correspond to the case of total degree 1. It is easy to operate polynomial functions (e.g., to compute their derivatives), and some questions, like the existence of feasible or optimal solutions, can be answered theoretically in finitely many arithmetic operations with rational data (but we do not know how to do it efficiently in higher-degree multivariate cases). However, solving polynomial programs cannot be much easier than solving programs with continuous functions because continuous functions over bounded regions can be approximated by polynomials.

Transportation problem (see Chapter 6) is a particular case of various *network problems.* Some of them can be reduced to transportation problem, and some share the property that integral data result in integral optimal solutions. Many textbooks on linear programming treat network problems, and there are special books on network problems, including nonlinear ones (cf. [ES]). Interior point methods are used nowadays for solving large network problems.

Dynamical programming (cf. [DL]) is concerned with optimal decisions over time. For continuous time, it is used in optimal control and variational calculus. For discrete time, we have multistage (multiperiod) models. The main idea in a multiperiod decision process is solving the problem from the end, going back in time. Position games (a.k.a. games in extensive form) use this approach, too (cf. [FSS], [M]).

One popular application of linear programming is *data envelopment analysis* ; see

http://www.banxia.com/, http://www.deazone.com/.

Neural networks, which imitate biological neural systems, are used to solve some optimization problems ("learning" algorithms), and mathematical programming is used for designing efficient neural networks.

An iterative method may depend on several parameters. Choosing parameters for a mathematical program with an objective function which is difficult to compute or/and we know little about or/and a program with a complicated feasible region could be a daunting task. One approach is *genetic* or *evolutionary programming.* We start with several algorithms, called strings, with parameters and initial points chosen at random. After a few iterations for each string, strings are sorted according to improvement in the objective function. Bad strings are eliminated. Good strings are paired up, and their "offsprings" appear with some values of parameters exchanged (crossover), changed a little bit at random (mutations), or combined in some ways. This approach is particularly attractive combined with parallel computing (many CPUs). (See [CVL], [LP].)

In general, progress in hardware (such as advances in parallel computers, quantum computers, and DNA computers) stimulates new approaches in mathematical programming (cf. [CP2], [DPW], [G], [NC], [LP], [MCC]).

Bibliography

[B1] Berkovitz, L. D., *Convexity and Optimization in R^n*. J. Wiley, New York , 2002.

[B2] Biegler, L. T., et al., *Large-Scale Optimization with Applications*. Springer-Verlag, New York, 1997.

[BL] Birge, J. R. and F. Louveaux, *Introduction to Stochastic Programming*. Springer Series in Operations Research. Springer-Verlag, New York, 1997.

[BV] Byrne C. and L. N. Vaserstein, An improved algorithm for finding saddlepoints of two-person zero-sum games, *Game Theory*, 20:2 (1991), 149–159.

[C1] Carlsson, C. *Fuzzy reasoning in decision making and optimization*. Physica-Verlag, Heidelberg/New York, 2002.

[C2] Chazelle, B., *The Discrepancy Method. Randomness and Complexity*. Cambridge University Press, Cambridge, U.K., 2000.

[CGT] Conn, A. R., N. I. M. Gould, and P. L. Toint, *Trust-Region methods*. SIAM, Philadelphia, 2000.

[CP1] Coope, I. D. and C. J. Price, On the convergence of grid-based methods for unconstrained optimization. *SIAM J. Optim.* 11 (2000), no. 4, 859–869

[CP2] Calude, C. S. and G. Paun, *Computing with Cells and Atoms: An Introduction to Quantum, DNA, and Membrane Computing*. Taylor & Francis, London/New York, 2001.

[CVL] Coello, C. C. A., D. A. Van Veldhuizen, and G. B. Lamont, *Evolutionary algorithms for solving multi-objective problems*. Kluwer Academic, New York, 2002.

[D] Dorn, W. S., Duality in quadratic programming, *Quarterly of Applied Mathematics* 18 (1960), no. 2, 155–162.

[DL] Dreyfus, S. E.and A. M. Law, *The Art and Theory of Dynamic Programming*. Mathematics in Science and Engineering, Vol. 130. Academic Press [Harcourt Brace Jovanovich, Publishers], New York/London, 1977.

[DPW] Du, D.-Z., P. M. Pardalos, and W. Wu (ed.), Mathematical Theory of Optimization. Kluwer Academic, Dordrecht/Boston, 2001.

[DZ1] Dor, D. and U. Zwick, Selecting the median, *SIAM J. Comput.* 28 (1999), no. 5, 1722–1758.

[DZ2] Dor, D. and U. Zwick, Median selection requires $(2 + \epsilon)N$ comparisons. *SIAM J. Discrete Math.* 14 (2001), no. 3, 312–325.

[ES] Eiselt, H. A. and C.-L. Sandblom, *Integer Programming and Network Models.* With contributions by K. Spielberg, E. Richards, B. T. Smith, G. Laporte and B. T. Boffey. Springer-Verlag, Berlin, 2000.

[FMP] Ferris, M.C., O. L. Mangasarian, and J.-S. Pang (editors), *Complementarity: Applications, Algorithms, and Extensions.* Kluwer Academic, Boston, 2001.

[FV] Florenzano, M. and C. L. Van in cooperation with P. Gourdel, *Finite Dimensional Convexity and Optimization* . Springer-Verlag, Berlin/New York, 2001.

[FSS] Forg, F., J. Szp, and F. Szidarovszky, *Introduction to the Theory of Games. Concepts, Methods, Applications.* Revised and expanded version of the 1985 original. Nonconvex Optimization and its Applications, 32. Kluwer Academic, Dordrecht/Boston, 1999.

[FW] Frank, M. and P. Wolfe, An algorithm for quadratic programming. *Naval Res. Logist. Quart.* 3 (1956), 95–110.

[G] Gramss, T. et al., Non-Standard Computation: Molecular Computation, Cellular Automata, Evolutionary Algorithms, Quantum Computers. Wiley-VCH, Weinheim/New York, 1998.

[GP] Gutin G. and A. P. Punnen (editors), *The Traveling Salesman Problem and its Variations* Kluwer Academic, Dordrecht/Boston, 2002.

[H1] Hačijan, L. G., A polynomial algorithm in linear programming. (Russian) *Dokl. Akad. Nauk SSSR* 244 (1979), no. 5, 1093–1096.

[H2] Hertog, D. den., *Interior Point Approach to Linear, Quadratic, and Convex Programming: Algorithms and Complexity.* Kluwer Academic, Dordrecht/Boston, 1994.

[HT] Horst, R. and H. Tuy, On the convergence of global methods in multiextremal optimization. *J. Optim. Theory Appl.* 54 (1987), no. 2, 253–271.

[K1] Karmarkar, N., A new polynomial-time algorithm for linear programming. *Combinatorica* 4 (1984), no. 4, 373–395.

[K2] Kelley, C. T., *Iterative Methods for Optimization* (Frontiers in Applied Mathematics, 18) SIAM, Philadelphia, 1999.

[K3] Knuth, D. E., *The art of computer programming. Volume 3. Sorting and searching.* Addison-Wesley Series in Computer Science and Information Processing. Addison-Wesley Publishing Co., Reading, Mass., 1973.

[K4] Klerk, E. de, *Aspects of Semidefinite Programming: Interior Point Algorithms and Selected Applications.* Kluwer Academic, Dordrecht/Boston, 2002.

[K5] Kalyanmoy D. , *Multi-Objective Optimization Using Evolutionary Algorithms.* John Wiley & Sons, Chichester/ New York, 2001.

[KV] Korte, B. H. and J. Vygen, *Combinatorial Optimization: Theory and Algorithms,* 2nd ed. Springer-Verlag, New York, 2002.

[KW] Kall, P. and S. W. Wallace, *Stochastic programming.* Wiley-Interscience Series in Systems and Optimization. John Wiley & Sons, Ltd., Chichester, 1994.

[L] Luenberger, D.G., *Introduction to Linear and Nonlinear Programming.* Addison-Wesley, Reading, Mass., 1984.

[LP] Langdon, W. B. and R. Poli, *Foundations of Genetic Programming.* Springer-Verlag, New York, 2002.

[M] Morris, P., *Introduction to Game Theory.* Springer-Verlag, New York, 1994.

[MCC] Marathe, A., A. E. Condon, and R. M. Corn, On combinatorial DNA word design. *DNA Based Computers, V* (Cambridge, Mass, 1999), 75–89, DIMACS Ser. Discrete Math. Theoret. Comput. Sci., 54, Amer. Math. Soc., Providence, RI, 2000.

[NC] Nielsen, M. A. and I. L. Chuang, *Quantum Computation and Quantum Information.* Cambridge University Press, Cambridge, U.K., 2000.

[NN] Nesterov, Yu. and A. Nemirovskii, *Interior-Point Polynomial Algorithms in Convex Programming.* SIAM, Philadelphia, 1994.

[P] Porembski, M., Finitely convergent cutting planes for concave minimization, *Journal of Global Optimization,* 20, no. 2 (June 2000), 113–136.

[R] Renegar, J., A Mathematical View of Interior-Point Methods in Convex Optimization. SIAM/Mathematical Programming Society, Philadelphia, 2001.

[RI] Ramík, J. and M. Inuiguchi, *Fuzzy Mathematical Programming.* Papers from the session of the 7th Congress of the International Fuzzy Systems Association held in Prague, June 25–29, 1997. Edited by Jaroslav Ramk and Masahiro Inuiguchi. Fuzzy Sets and Systems 111 (2000), no. 1. North-Holland Publishing Co., Amsterdam, 2000.

[RV] Ramík, J. and M. Vlach, *Generalized Concavity in Fuzzy Optimization and Decision Analysis.* Kluwer Academic, Boston, 2002.

[S1] Schniederjans, M. J., *Goal Programming: Methodology and Applications.* Kluwer Academic, Dordrecht/Boston, 1995.

[S2] Sierksma, G., *Linear and Integer Programming: Theory and Practice.* 2nd ed. Marcel Dekker, New York, 2002.

[S3] Stefanov S.M., *Separable Programming: Theory and Methods.* Kluwer Academic, Dordrecht/Boston, 2001.

[S-M] Stancu-Minasian, I. M., *Fractional Programming: Theory, Methods and Applications.* Kluwer Academic, Dordrecht/Boston, 1997.

[SMY] Sarker R., M. Mohammadian, and X. Yao (editors), *Evolutionary Optimization* . Kluwer Academic Publishers, Boston, 2002.

[T] Tsurkov, V., *Large-Scale Optimization: Problems and Methods.* Applied Optimization, 51. Kluwer Academic, Dordrecht/Boston, 2001.

[VCS] Vaserstein, L. N., V.I. Chmil', and E. B. Sherman, A multiextremal problem for the growth and location of production with a concave objective function. (Russian), in *Mathematical Methods in Economic Research* (Russian), pp. 138–143. Izdat. Nauka, Moscow, 1974.

[V] Vaserstein, L. V., On the best choice of a damping sequence in iterative optimization methods, *Publ. Matem. Univ. Aut. Barcelona* 32 (1988), 275–287.

[W1] Wolkowicz, H. et al (editors), *Handbook of Semidefinite Programming: Theory, Algorithms, and Applications.* International Series in Operations Research & Management Science, 27. Kluwer Academic, Dordrecht/Boston, 2000.

[W2] Wolsey, L. A. *Integer Programming.* Wiley-Interscience Series in Discrete Mathematics and Optimization. John Wiley & Sons, New York, 1998.

Answers to Selected Exercises

§1. What Is Linear Programming?

1. True

3. True

5. True. This is because for real numbers any square and any absolute value are nonnegative.

7. False. For $x = -1$, $3(-1)^3 < 2(-1)^2$.

8. False (see Definition 1.5).

9. False (see Example 1.9 or 1.10).

11. False. For example, the linear program

$$\text{Minimize } x + y \text{ subject to } x + y = 1$$

has infinitely many opotimal solutions.

13. True. It is a linear equation. A standard form is $4x = 8$ or $x = 2$.

15. No. This is not a linear form, but an affine function.

17. Yes if a and z do not depend on x, y.

18. No (see Definition 1.1).

19. No. But it is equivalent to a system of two linear constraints.

21. Yes. We can write $0 = 0 \cdot x$, which is a linear form.

22. True if y is independent of x and hence can be considered as a given number; see Definition 1.3.

23. Yes if a, b are given numbers. In fact, this is a linear equation.

25. No. We will see later that any system of linear constrains gives a convex set. But we can rewrite the constraint as follows $x \geq 1$ OR $x \leq 1$. Notice the difference between OR and AND.

27. See Problem 6.12.

29. $x = 3 - 2y$ with an arbitrary y.

31. min $= 0$ at $x = y = 0, z = -1$. All optimal solutions are given as follows: $x = -y, y$ is arbitrary, $z = -1$.

33. min $= 1$ at $x = 0$.

35. min $= 0$ at $x = -y = 1/2, z = -1$.

37. No. This is a linear equation.

39. No

41. Yes

43. No
45. Yes
47. Yes
49. No
51. Yes
53. Yes. In fact, this is a linear equation.
55. No. This is not even equivalent to any linear constraint with rational coefficients.
57. No.
59. $\min = 2^{-100}$ at $x = 0, y = 0, z = \pi/2, u = -100, v = -100$. In every optimal solution, x, y, u, v are as before and $z = \pi/2 + n\pi$ with any integer n such that $-32 \leq n31$. So there are exactly 64 optimal solutions.

§2. Examples of Linear Programs
2. $\min = 1.525$ at $a = 0, b = 0.75, c = 0, d = 0.25$
4. Let x be the number of quarters and y the number of dimes we pay. The program is
$$25x + 10y \rightarrow \min,$$
$$\text{subject to}$$
$$0 \leq x \leq 100, \ 0 \leq y \leq 90, 25x + 10y \geq C \text{ (in cents)}, x, y \text{ integers.}$$

This program is not linear because the conditions that x, y are integers. For $C = 15$, an optimal solution is $x = 0, y = 2$. For $C = 102$, an optimal solution is $x = 3y = 3$ or $x = 1, y = 8$. For $C = 10000$, the optimization problem is infeasible.

5. Let x, y be the sides of the rectangle. Then the program is
$$xy \rightarrow \min,$$
$$\text{subject to}$$
$$x \geq 0, \ y \geq 0, 2x + 2y = 100.$$

Since $xy = x(50 - x) = 625 - (x - 25)^2 \leq 625$, $\max = 625$ at $x = y = 25$.

7. We can compute the objective function at all 24 feasible solutions and find the following two optimal matchings: Ac, Ba, Cb, Da and Ac, Bb, Ca, Dd with optimal value 7.

8. Choosing a maximal number in each row and adding these numbers, we obtain an upper bound $9 + 9 + 7 + 9 + 9 = 43$ for the objective function. This bound cannot be achieved because of a conflict over c (the third column). So $\max \leq 42$. On the other hand, the matching aa, Bb, Cc, De, Ed achieved 42, so this is an optimal matching.

9. Choosing a maximal number in each row and adding these numbers, we obtain an upper bound $9+9+9+9+8+9+6 = 59$ for the objective function. However looking at B and C, we see that they cannot get $9 + 9 = 18$ because of the conflict over g. They cannot get more than $7 + 9 = 16$. Hence, we have the upper bound max ≤ 57. On the other hand, we achieve this bound 57 in the matching Ac, Bf, Cg, De, Eb, Fd, Ga.

11. Let c_i be given numbers. Let c_j be an unknown maximal number (with unknown j). The linear program is
$$c_1 x_1 + \cdots + c_n x_n \to \text{max, all } x_i \geq 0, x_1 + \cdots + x_n = 1.$$
Answer: $\max = c_j$ at $x_j = 1, x_i = 0$ for $i \neq j$.

§3. Graphical Method

1. Let SSN be 123456789. Then the program is
$$-x \to \text{max}, 7x \leq 5, 13x \geq -8, 11x \leq 10.$$
Answer: $\max = 8/13$ at $x = -8/13$.

3. Let SSN be 123456789. Then the program is
$$x - 3y \to \text{min}, |6x + 4y| \leq 14, |5x + 7y| \leq 8, |x + y| \leq 17.$$
Answer: $\min = 242/11$ at $x = 65/11, y = -59/11$.

5. $\max = 135$ at $x = -9, y = 18$

7. $\min = -1/4$ at $x = 1/2, y = -1/2$ or $x = -1/2, y = 1/2$.

6. The problem is unbounded ($\min = -\infty$).

9. $\max = 1$ at $x = y = 0$

11. $\max = 22$ at $x = 4, y = 2$

13. The program is unbounded.

15. $\max = 3$ at $x = y = 9, z = 1$. See the answer to Exercise 11 of §2.

§4. Logic

1. False. For $x = -1$, $|-1| = 1$.

3. False. For $x = -10$, $|-10| > 1$.

5. True. $1 \geq 0$.

7. True. $2 \geq 0$.

9. True. The same as Exercise 7.

11. False. $1 \geq 1$.

13. True. $5 \geq 0$.

15. False. For example, $x = 2$.

17. True. Obvious.

19. False. For example, $x = 1$.

21. True. $1 \geq 0$.

22. Yes, we can.
23. Yes. $10 \geq 0$.
25. No, it does not. $(-5)^2 > 10$.
27. True.
29. False. The first condition is stronger than the second one.
31. True.
33. (i) $\Rightarrow$ (ii), (iii), (iv)
35. (i) $\Rightarrow$ (iii)
37. (i) $\Leftrightarrow$ (ii) $\Rightarrow$ (iv) $\Rightarrow$ (iii)
41. "only if"
42. This depends on the definition of it linear function.
43. No. $x \geq 1, x \leq 0$ are two feasible constraints, but the system is infeasible.
44. False.
45. False. Under our conditions, $|x| > |y|$.
49. No, it does not follow.
51. Yes, it does. It is the sum of the first two equations.

§5. Matrices

1. $[2, 1, -6, 6]$
3. -14
5. -14
7. 2744
9. No. $1 \neq 4$.

10. $A = \begin{bmatrix} 0 & 1 \\ 0 & 0 \end{bmatrix}$

11. $A = \begin{bmatrix} 1 & 0 \\ 0 & 0 \end{bmatrix}, B = \begin{bmatrix} 0 & 0 \\ 1 & 0 \end{bmatrix}$

12. $\begin{bmatrix} 5 & 2 & 3 & -1 \\ 1 & -1 & -3 & 0 \\ 0 & 0 & 3 & 0 \end{bmatrix} \begin{bmatrix} a \\ b \\ c \\ d \end{bmatrix} = \begin{bmatrix} 1 \\ 2 \\ -1 \end{bmatrix}$

15. $b = a - 1, c = -1/3, d = 7a - 4, a$ arbitrary
19. $AB^T = 5$
21. $A^T B = 5$
23. $(A^T B)^2 = 25$
25. 5^{1000}
27. 3
29. 3

31. 9

33. s^{1000}

35. $E_1 C = \begin{bmatrix} 3 & 6 & 9 \\ -8 & -10 & -12 \end{bmatrix}$ $E_2 C = \begin{bmatrix} 21 & 27 & 33 \\ 4 & 5 & 6 \end{bmatrix}$

$E_1^n = \begin{bmatrix} 3^n & 0 \\ 0 & (-2)^n \end{bmatrix}$ $E_2^n = \begin{bmatrix} 1 & 5n \\ 1 & 0 \end{bmatrix}$

37. $\begin{bmatrix} \alpha & 0 \\ 0 & \delta - \gamma\alpha^{-1}\beta \end{bmatrix}$

41. $\begin{bmatrix} 1 & 0 & 0 & 0 \\ 0 & 1 & 0 & 0 \end{bmatrix}$

43. $\begin{bmatrix} 1 & 0 & 0 \\ 0 & 1 & 0 \\ 0 & 0 & 1 \\ 0 & 0 & 0 \\ 0 & 0 & 0 \end{bmatrix}$

§6. Systems of Linear Equations

1. $\begin{bmatrix} 1 & 0 \\ 0 & -4 \end{bmatrix}$ is invertible; $\det(A) = -4$

3. The matrix is invertible if and only if $abc \neq 0$; $\det(A) = abc$.

5. $\begin{bmatrix} -1 & 0 & 0 & 0 \\ 0 & -1 & 0 & 0 \\ 0 & 0 & -1 & 0 \\ 0 & 0 & 0 & 2 \end{bmatrix}$ is invertible; $\det(A) = -2$

7. $\begin{bmatrix} 1 & 0 & 0 \\ 0 & 13/7 & 0 \\ 0 & 0 & 7 \end{bmatrix}$ is invertible; $\det(A) = 13$

9. $0 = 1$ (no solutions)

11. $x = -z - 3b + 9$, $y = -z + 2b - 6$.

13. If $t \neq 6 + 2u$, then there are no solutions. Otherwise, $x = -2y + +v + 3$, y arbitrary.

15. If $t = 1$, then $x = 1 - y$, y arbitrary.
If $t = -1$, there are no solutions.
If $t \neq \pm 1$, then $x = (t^2 + t + 1)/(t+1)$, $y = -1/(t+1)$.

17. $y = 5b + x - 16$, $z = -3b - x + 10$

19. No. The half-sum of solutions is a solution.

21. $A^{-1} = \begin{bmatrix} 7/25 & 4/25 & -1/25 \\ 19/25 & -7/25 & 8/25 \\ -18/25 & 4/25 & -1/25 \end{bmatrix}$

23. $A^{-1} = \begin{bmatrix} -3/22 & -1/22 & -41/22 & 3/11 \\ -15/22 & -5/22 & -51/22 & 4/11 \\ 5/22 & 9/22 & 61/22 & -5/11 \\ 15/22 & 5/22 & 73/22 & -4/11 \end{bmatrix}$

25. $A = \begin{bmatrix} 1 & 2 \\ 6 & 8 \end{bmatrix} = \begin{bmatrix} 1 & 0 \\ 6 & 1 \end{bmatrix} \begin{bmatrix} 1 & 2 \\ 0 & -4 \end{bmatrix}$

27. This cannot be done. We have $0 = A_{1,1} = L_{1,1}U_{1,1} \neq 0$ since A is invertible, hence U, V are invertible.

29. $A = \begin{bmatrix} 1 & 0 & -1 \\ 5 & 1 & 3 \\ 2 & 4 & 5 \end{bmatrix} = \begin{bmatrix} 1 & 0 & 0 \\ 5 & 1 & 0 \\ 2 & 4 & 1 \end{bmatrix} \begin{bmatrix} 1 & 0 & -1 \\ 0 & 1 & 8 \\ 0 & 0 & -25 \end{bmatrix}$

30. $A = \begin{bmatrix} 1 & -2 & -1 \\ 5 & 1 & 3 \\ 2 & 4 & 5 \end{bmatrix} = \begin{bmatrix} 1 & 0 & 0 \\ 5 & 1 & 0 \\ 2 & 8/11 & 1 \end{bmatrix} \begin{bmatrix} 1 & -2 & -1 \\ 0 & 11 & 8 \\ 0 & 0 & 13/11 \end{bmatrix}$

33. $x = -3(19 + 2d)/8, y = (15 + 2d)/8, z = -(3 + 2d)/8$

35. $x = (15u + 4v)/16, y = (11u + 4v)/16, z = -3u/4$

37. $x = y = 1, z = 1$

39. $x = y = 0, z = 100$

§7. Standard and Canonical Forms for Linear Programs

1. Set $u = y + 1 \geq 0$. Then $f = 2x + 3y = 2x + 3v - 3$ and $x + y = x + u - 1$. A canonical form is
$$-f = -2x - 3v + 3 \to \min, x + u \leq 6, u, x \geq 0.$$
A standard form is
$$-f = -2x - 3v + 3 \min, x + u + v = 6, u, v, x \geq 0$$
with a slack variable $v = 6 - x - u \geq 0$.

2. Excluding $y = x + 1$ and using $y \geq 1$, we obtain the canonical form
$$-x \to \min, \ 2x \leq 8, x \geq 0.$$
Introducing a slack variable $z = 8 - 2x$, we obtain the standard form
$$-x \to \min, \ 2x + z = 8, x \geq 0, z \geq 0.$$

3. We solve the equation for x_3:
$$x_3 = 3 - 2x_2 - 3x_4$$
and exclude x_3 from the LP:

$$x_1 - 7x_2 + 3 \to \min, \ x_1 - x_2 + 3x_4 \geq 3, \text{ all } x_i \geq 0.$$

A canonical form is

$$x_1 - 7x_2 + 3 \to \min, \ -x_1 + x_2 - 3x_4 \leq -3, \text{ all } x_i \geq 0.$$

A standard form is

$$x_1 - 7x_2 + 3 \to \min, \ -x_1 + x_2 - 3x_4 + x_5 = -3, \text{ all } x_i \geq 0$$

with a slack variable $x_4 = x_1 - x_2 + 3x_4 - 3$.

5. Set $t = x + 1 \geq 0, u = y - 2 \geq 0, f = x = y + z = t + u + z + 1$ (the objective function). Then a standard and canonical form for our problem is

$$x + u + z + 1 \to \min; t, u, z \geq 0.$$

7. Using standard tricks, a canonical form is

$$-x \to \min, x \leq 3, -x \leq -2, x \geq 0.$$

A standard form is

$$-x \to \min, x + u + 3, -x + v = -2; x, u, v \geq 0$$

with two slack variables.

9. One of the given equations reads

$$-5 - x - z = 0,$$

which is inconsistant with given constraints $x, z \geq 0$. So we can write very short canonical and standard forms:

$$0 \to \min, 0 \leq -1; x, y, z \geq 0 \text{ and } 0 \to \min, 0 = 1; x, y, z \geq 0.$$

11. Set $x = [x_1, x_2, x_3, x_4, x_5, x_6, x_7, x_8, x_9, x_{10}, x_{11}^T]$ and $c = [3, -1, 1, 3, 1, -5, 1, 3, 1]$. Using standard tricks, we obtain the canonical form

$$cx \to \min, Ax \leq b, x \geq 0$$

with

$$A = \begin{bmatrix} 1 & -1 & -1 & -1 & -1 & 1 & 2 & -3 & -1 \\ -1 & 1 & 1 & 1 & 1 & -1 & -2 & 3 & 1 \\ 2 & -2 & -2 & 2 & 3 & -1 & -2 & 1 & 1 \\ -2 & 2 & 2 & -2 & -3 & 1 & 2 & -1 & -1 \\ 1 & 0 & 0 & 0 & 3 & -1 & -2 & 0 & -1 \\ -1 & 0 & 0 & 0 & -3 & 1 & 2 & 0 & 1 \end{bmatrix}$$

and $b = [-3, -1, 2, -2, 0, 0]^T$.

Excluding a couple of variables using the two given equations, we would get a canonical form with two variables and two constraints less. A standard form can be obtained from the canonical form by introducing a column u of slack variables:

$$cx \to \min, Ax + u = b, x \geq 0, u \geq 0.$$

§8. Pivoting Tableaux

1.

$$
\begin{array}{cccccc}
a & b & c & d & e & 1 \\
\end{array}
$$

$$
\left[
\begin{array}{cccccc}
.3 & 1.2 & .7 & 3.5 & 5.5 & -50 \\
73 & 96 & 20253 & 890 & 279 & 4000 \\
9.6 & 7 & 19 & 57 & 22 & -100 \\
10 & 15 & 5 & 60 & 8 & 0
\end{array}
\right]
\begin{array}{l}
= u_1 \\
= u_2 \\
= u_3 \\
= C \to \min
\end{array}
$$

3. $A = \begin{bmatrix} 3 & -1 & 2 & 2 \\ -1 & 0 & 0 & 2 \\ -1 & 0 & 2 & -2 \\ 0 & 0 & 1 & -1 \end{bmatrix}, b = \begin{bmatrix} 0 \\ -1 \\ 0 \\ -2 \end{bmatrix}$

$$
\begin{array}{cccc}
z & a & 3 & x \\
\end{array}
$$

11. $\begin{bmatrix} 1 & 2 & b+3 & a+1 \\ -1 & 2 & 3 & 1 \end{bmatrix} \begin{array}{l} = y \\ = 1 \end{array}$

$$
\begin{array}{cccc}
1 & a & 3 & z \\
\end{array}
$$

12. $\begin{bmatrix} 1+a & -2a & b-3a & a \\ 1 & -2 & -3 & 1 \end{bmatrix} \begin{array}{l} = y \\ = x \end{array}$

$$
\begin{array}{c}
2 \\
\end{array}
$$

13. $\begin{bmatrix} 1/5 \end{bmatrix} = x$

§9. Standard Row Tableaux

$$
\begin{array}{ccc}
x & y & 1 \\
\end{array}
$$

2. $\begin{bmatrix} -4 & -5 & 7 \\ -2 & -3 & 0 \end{bmatrix} \begin{array}{l} = u \\ = -P \to \min \end{array}$

with a slack variable $u = 7 - 4x - 5y$

$$
\begin{array}{ccc}
x & y & 1 \\
\end{array}
$$

3. $\begin{bmatrix} -1 & -1 & 1 \\ 1 & 1 & 1 \\ 1 & -1 & 1 \\ -1 & 1 & 1 \\ -1 & 0 & 0 \end{bmatrix} \begin{array}{l} = u_1 \\ = u_2 \\ = u_3 \\ = u_4 \\ \to \min \end{array}$

with slack variable u_i

§10. Simplex Method, Phase 2

   4. False. The converse is true.

   5. True

   6. True

13. If the row without the last entry is nonnegative, then the tableau is optimal; else the LP is unbounded.

§11. Simplex Method, Phase 1
   1. The second row (v-row) is bad, so the LP is infeasible.
   2. The tableau is optimal, so
   $\min(w) = 1$ at $x = y = z = z = 0, v = 0$.
   3. This is a feasible tableau with a bad column (the z-column). So the LP is unbounded (z and hence w can be arbitrary large).
   9. True
   10. False

§12. Geometric Interpretation
   1. The diamond can be given by four linear constraints $\pm x \pm y \leq 1$.
   2. Any convex combination of convex combinations is a convex combination.
   7. Both $x = 1$ and $x = -1$ belong to the feasible region, but $0 = x/2 + y/2$ does not.
   8. $2tx + (1 - t^2)y \leq 1 + t^2$, where t ranges over all rational numbers.
   9. A set S is called closed if it contains the limit points of all sequences in S. Any system of linear constraints gives a closed set, but the interval $0 < x < 1$ is not closed. Its complement is closed.
   10. The rows of the identity matrix 1_6.
   11. One

§13. Dual Problems
   1.

$$
\begin{array}{c}
\begin{array}{cccc}
x \\ y \\ z \\ -1
\end{array}
\left[
\begin{array}{ccc}
0 & -1 & 5 \\
1 & -1 & 6 \\
0 & 0 & 2 \\
7 & -3 & 0
\end{array}
\right] \\
\begin{array}{ccc}
\| & \| & \downarrow \\
-v_1 & -v_2 w & \min
\end{array}
\end{array}
$$

   5. Let $cx + d, cy + d$ be two feasible values, where x, y are two feasible solutions. We have to prove that
   $$\alpha(cx + d) + (1 - \alpha)(cy + d)$$
   is a feasible value for any α such that $0\alpha \leq 1$. But
   $$\alpha(cx + d) + (1 - \alpha)(cy + d) = c(\alpha x + (1 - \alpha)y) + d,$$

where $\alpha x + (1 - \alpha)y$ is a feasible solution because the feasible region is convex.

§14. Sensitivity Analysis and Parametric Programming
   3. min $= 0$ at $d = e = 0, a \geq 0$ arbitrary

§15. More on Duality
   1. No, it is not redundant.
   2. Yes, it is 2· (first equation) + (second equation).
   3. No, it is not redundant.
   4. No, it is not redundant.
   5. No, it is not redundant.
   7. Yes, it is.

§16. Phase 1
   1.

20	10	5		35
		5	15	20
20	10	10	15	

   3.

30				35
90				90
11	91	9		111
		1	19	20
140	91	10	19	

§17. Phase 2

1.

	1	2	2	
	1	2	3	
0	175	25	(1)	200
	1	2	2	
0	(0)	100	200	300
	175	125	200	

§18. Job Assignment Problem

1. min = 7 at $x_{14} = x_{25} = x_{32} = x_{43} = x_{51} = 1$, all other $x_{ij} = 0$.

3. min = 7 at $x_{12} = x_{25} = x_{34} = x_{43} = x_{51} = x_{67} = x_{76} = 1$, all other $x_{ij} = 0$.

5. max = 14 at $x_{15} = x_{21} = x_{34} = x_{43} = x_{52} = 1$, all other $x_{ij} = 0$.

7. max = 29 at $x_{15} = x_{26} = x_{33} = x_{41} = x_{57} = x_{62} = x_{74} = 1$, all other $x_{ij} = 0$.

§19. What are Matrix Games?

1. max min = −1. min max = 0. There are no saddle points.
$[1/3, 2/3, 0]^T$ gives at least −2/3 for the row player.
$[1/2, 0, 0, 0, 1/2]$ gives at least 1/2 for the column player.
So −2/3 ≤ the value of the game ≤ −1/2.

3. max min = −1. min max = 2. There are no saddle points.
(second row + 2·third row)/3 ≥ −1/3.
(third column + sixth column)/2 ≤ 1.
So −1/3 ≤ the value of the game ≤ 1.

5. We compute the max in each column (marked by *) and min in each row (marked by •).

$$
\begin{array}{c}
\\
-4 \\ -2 \\ -4 \\ 0^* \\ -9
\end{array}
\begin{array}{c}
\begin{array}{ccccccccc}
4 & 2 & 3 & 5 & 4 & 3^• & 7 & 6 & 3^• \\
\end{array} \\
\left[
\begin{array}{ccccccccc}
4^* & -4^• & 3^* & 0 & 0 & 0 & -1 & 1 & -2 \\
-1 & 0 & 2 & 1 & -2^• & -2^• & 1 & 0 & -2 \\
-4^• & 0 & -2 & -2 & 1 & -1 & 1 & 6^* & 2 \\
1 & 2^* & 2 & 5^* & 3 & 3^* & 7^* & 2 & 0^• \\
-4 & -9^• & -8 & 0 & 4^* & 2 & 2 & 0 & 3^* \\
\end{array}
\right]
\end{array}
$$

Thus, max min = 0. min max = 3. There are no saddle points.

§20. Matrix Games and Linear Programming

1. The optimal strategy for the row player is $[0, 2/3, 1/3]^T$.
The optimal strategy for the column player is $[1/2, 1/2, 0$.
The value of the game is 2.

3. The optimal strategy for the row player is $[0.2, 0, 0.8]^T$.
An optimal strategy for the column player is $[0, 0.5, 05, 0, 0, 0$.
The value of the game is 1.

5. The optimal strategy for the row player is $[1/3, 2/3, 0]^T$.
The optimal strategy for the column player is $[2/3, 0, 0, 1/3]$.
The value of the game is $-2/3$.

7. The optimal strategy for the row player: $[1/8, 0, 7/8, 0]^T$.
The optimal strategy for the column player: $[0, 1/4, 0, 0, 0, 3/4]$.
The value of the game is -0.25.

§21. Other Methods

1. The first row and column are dominated. The optimal strategy for the column player is $[0, 0.5, 0.5]^T$. The optimal strategy for the row player is $[0, 0.25, 0.75$. The value of the game is 2.5.

3. The optimal strategy for the column player is $[0, 0.4, 0, 0.6]^T$.
The optimal strategy for the row player is $[0, 0.4, 0.6$. The value of the game is 2.8.

5. The optimal strategy for the column player is $[1/3, 1/3, 1/3]^T$.
The optimal strategy for the row player is $[0, 0, 2/7, 3/7, 2/7, 0$. The value of the game is 0.

7. The value of the game is 0 because the game is symmetric.

9. The first two columns and the first row go by domination.
The value of the game is $11/7$.

11. 0 at a saddle point.

13. 0 at a saddle point.

§22. What is Linear Approximation?

1. The mean is $-2/5 = -0.4$. The median is 1. The midrange is $-5/2 = -2.5$.

3. The mean is $5/9$. The median is 0. The midrange is $1/2 = 0.5$.

5(b). 1, 2, 9

5(d). Exercise 1

5(f). Exercise 3

§23. Linear Approximation and Linear Programming
  1. $\min = 0$ at $a = -15, b = 50$ for $w = a + bh$
and
  $\min \approx 19$ at $a \approx 25.23$ for $w = ah^2$
  2. $x + y + 0.3 = 0$
  3. $a = 0.9, b \approx -0.23$

§24. More Examples
  1. The model is $w = ah + b$, or $w - x_2 = a(h - 1988) + b'$ with $b = x_2 + -1988a$ and $x_2 = 37753/45 \approx 838.96$. Predicted production P in 1993 is $x_2 + 5a + b'$.
  For $p = 1, we have a \approx 16.54, b' \approx 31, P \approx 953$.
  For $p = 2, we have a \approx 0, b' \approx 32, P \approx 871$.
  For $p = \infty, we have a \approx 17.59, b' \approx 32, x_5 \approx 959$.
So in this example l^∞-prediction is the best.
  3. $a = \$4875, b = \$1500$

Index